MW00604159

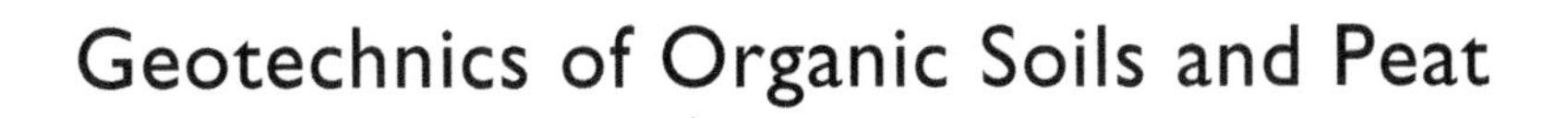

Geotechnics of Organic Soils and Peat

Geotechnics of Organic Soils and Peat

Bujang B.K. Huat

Department of Civil Engineering, Universiti Putra Malaysia, Serdang, Malaysia

Arun Prasad

Department of Civil Engineering, Indian Institute of Technology (BHU), Varanasi, India

Afshin Asadi

HRC, Faculty of Engineering, Universiti Putra Malaysia, Serdang, Malaysia

Sina Kazemian

Department of Civil Engineering, Payame Noor University, Iran

CRC Press is an imprint of the
Taylor & Francis Group, an **informa** business

A BALKEMA BOOK

CRC Press/Balkema is an imprint of the Taylor & Francis Group, an informa business

© 2014 Taylor & Francis Group, London, UK

Typeset by MPS Limited, Chennai, India
Printed and Bound by CPI Group (UK) Ltd, Croydon, CR0 4YY

All rights reserved. No part of this publication or the information contained herein may be reproduced, stored in a retrieval system, or transmitted in any form or by any means, electronic, mechanical, by photocopying, recording or otherwise, without written prior permission from the publisher.

Although all care is taken to ensure integrity and the quality of this publication and the information herein, no responsibility is assumed by the publishers nor the author for any damage to the property or persons as a result of operation or use of this publication and/or the information contained herein.

Library of Congress Cataloging-in-Publication Data

Huat, Bujang B. K.
Geotechnics of organic soils and peat / Bujang B.K. Huat, professor, Department of Civil Engineering, Universiti Putra Malaysia, Serdang, Malaysia, Arun Prasad, associate professor, Department of Civil Engineering, Indian Institute of Technology (BHU), Varanasi, India, Afshin Asadi, HRC, Faculty of Engineering, Universiti Putra Malaysia, Serdang, Malaysia, Sina Kazemian, Department of Civil Engineering, Payame Noor University, I.R. of Iran.
pages cm
Includes bibliographical references and index.
ISBN 978-0-415-65941-3 (hardback : alk. paper)
1. Soil mechanics. 2. Peat soils. I. Prasad, Arun. II. Asadi, Afshin. III. Kazemian, Sina. IV. Title.
TA710.H786 2014
624.1'5136—dc23

2013045959

Published by: CRC Press/Balkema
P.O. Box 11320, 2301 EH Leiden, The Netherlands
e-mail: Pub.NL@taylorandfrancis.com
www.crcpress.com – www.taylorandfrancis.com

ISBN: 978-0-415-65941-3 (Hardback)
ISBN: 978-0-203-38630-9 (e-book PDF)

Table of contents

Foreword

Peat and organic soils commonly occur as extremely soft, wet, unconsolidated surficial deposits that are integral parts of the wetland systems. They may also occur as strata beneath other surficial deposits. These problematic soils are known for their low shear strength and high compressibility characteristics. Access to these surficial deposits is usually very difficult as the water table will be at, near, or above the ground surface. Geotechnical problems that arise with these types of soils can include sampling, stabilization, settlement, stability, *in situ* testing, and construction. Undoubtedly, these result in the tendency to either avoid construction and buildings on these soils, or when this is not possible, to simply remove, replace or displace them, which in some instances may lead to possibly uneconomical design and construction alternatives. However, in many countries of the world, including Malaysia, this material covers a substantial land area. Pressure on land use by industry, housing, infrastructure as well as agriculture is leading to more frequent utilization of such marginal grounds. For a successful design, construction and performance of structures built on such marginal soils, it is highly necessary to predict geotechnical behavior in terms of settlement, shear strength, and stability, with respect to time. It is therefore necessary to expand our knowledge of these soils, which requires reliable characterization of their geotechnical properties and mechanical behavior, and subsequently to devise suitable design parameters and construction techniques for these materials. There is also a need for specialized laboratory and field tests to evaluate geotechnical properties, and some special kind of tools/equipment and/or techniques may be called for.

The mechanical characteristics of peat which distinguish it from most soft mineral soils such as clay are high porosity, extreme compressibility, strong dependence on permeability and porosity, large change in properties under stress, high degree of spatial variability in properties, fibrosity, resulting in strong anisotropy of many geotechnical properties, and high strength due to fiber reinforcement.

The general challenges faced by those responsible for construction on peat and organic soils include: limited accessibility and difficult traffic ability, expectations of very large settlements over an extended time period, and possibility of stability problems.

Although a number of books and collected papers have been published in the past, there has been no up-to-date synthesis on peat and organic soils. We felt that an in-depth analysis of peat and organic soils was needed to help in their development. Finally, we wanted to provide a comprehensive text that would be informative on peat and organic soils throughout the world.

An adequate scientific understanding of the nature and functions of peat and organic soils is critical to their proper and safe use, and this book intends to contribute to this by providing comprehensive knowledge to students, researchers, engineers, and academics involved with these types of soils.

This book draws on our experience with tropical peatlands, and we hope that this book will be useful not only to geotechnical engineers, but also to soil scientists and agriculturalists, who are involved in the development of peatlands.

Bujang B.K. Huat
Arun Prasad
Afshin Asadi
Sina Kazemian

About the authors

Bujang B.K. Huat graduated from the Polytechnic of Central London, UK in 1983, and obtained his MSc and PhD at the Imperial College London and the Victoria University Manchester, UK in 1986 and 1991 respectively. He has spent his professional career as a Professor in Geotechnical Engineering, at the Department of Civil Engineering, Universiti Putra Malaysia, one of Malaysia's five research universities. Currently he serves as the Dean of School of Graduate Studies of the same university. His special area of interest is in the field of geotechnical and geological engineering, especially peat, and slope engineering; he has authored and co-authored 18 books, edited ten conference proceedings, and published more than 100 journal and conference proceedings papers in the field of soil mechanics and foundation engineering.

Arun Prasad is Associate Professor of Geotechnical Engineering at the Indian Institute of Technology (Banaras Hindu University), India. He graduated with a BSc in Civil Engineering in 1986 from Utkal University, India; he obtained his MSc and PhD from Sambalpur University and Devi Ahilya University, India in 1989 and 2000 respectively. He worked as Post-Doctoral Researcher at Universiti Putra Malaysia during 2009–10. His special area of research is the soil stabilization of soft and contaminated soils. He has co-authored three books and co-edited a book in the field of Geotechnical Engineering, and has published more than 60 papers in journals and conference proceedings.

Dr. Afshin Asadi received his BSc in Civil Engineering from IAU, his MSc in Civil Engineering-Environmental Engineering from the Iran University of Science and Technology, and his PhD in Geotechnical Engineering from Universiti Putra Malaysia in 2010. He received an Australia Endeavour Research Fellowship Award in 2011 and completed his postdoctoral studies at the University of Wollongong in 2012. His research areas are mostly ground improvement, electrokinetics, and environmental geotechnics. He is a member of the *Environmental Geotechnics* editorial board published by ICE Publishing, UK. Presently, he is a Research Fellow at the Housing Research Centre (HRC), Universiti Putra Malaysia.

Sina Kazemian is Assistant Professor at the Civil Engineering department of Payame Noor University (PNUM), I.R. of Iran He obtained his PhD (with distinction) in Geotechnical and Geological Engineering from Universiti Putra Malaysia (UPM) and achieved recognition of excellence during his PhD viva; his name was inscribed in the "Hall of Fame" at UPM. He has worked as a lecturer/researcher at Azad University of Bojnourd, Iran and also has more than 10 years of working experience in the industry as Senior Geotechnical Engineer at Sepehr Andishan Sanabad (SAS) Co., Iran and Structure Civil Geotechnics (SCG) Co., Malaysia. Currently, he is also the Principal Geotechnical Engineer and technical associate of Kavosh Pay Co. in Iran. To date he has published more than 100 papers in reputed journals and conference proceedings.

Chapter 1

Introduction

1.1 SOIL ENGINEERING

Soil engineering or ground engineering, more commonly called geotechnical engineering, is a branch of civil engineering that deals with soil, rocks and underground water, and their relation to the design, construction and operation of engineering projects. Nearly all civil engineering projects must be supported by the ground, and thus require at least some form of ground/soil or geotechnical engineering.

1.2 TYPES AND FORMATION OF SOILS

To begin with, let us look at the various types of soils, their composition and how they are formed. In term of composition, soils in engineering are generally considered to comprise four main constituents: gravel, sand, silt and clay. Soils are therefore considered as particulate materials that consist of individual particles, and not as a solid mass. Soils are also considered as three-phase materials, whereby they can simultaneously contain solid, liquid and gas phases. The liquid and gas phases are contained in the voids or pores between the solid particles.

In their natural environment, soils rarely consist of one constituent only, but are often a mixture of two or more constituents. For example, we have gravelly sand, which means sand with some gravel, or sandy-silty clay, which means clay with some sand and silt. Soils may also contain organic matter, which is essentially the remains of plants. These soils are called 'organic soil' if their organic content exceeds 20% of their dry mass. Organic soils in turn are termed 'peat' once their organic content exceeds 75%. The main reason for this definition is that when the organic content of the soil exceeds 20%, its mechanical behaviour will start to depart from that of mineral soils (i.e. gravel, sand, silt and clay). This will be explained in more detail in later chapters.

Soils are also described based on their formation, transportation and deposition. The categories include residual soils, glacial soils, alluvial soils, lacustrine and marine soils, aeolian soils and colluvial soils, as well as organic soils and peat.

1.2.1 Residual soils

If the rate of rock weathering is faster than the water, gravity and wind transport processes, much of the resulting soil will remain in place. Such soil is known as residual

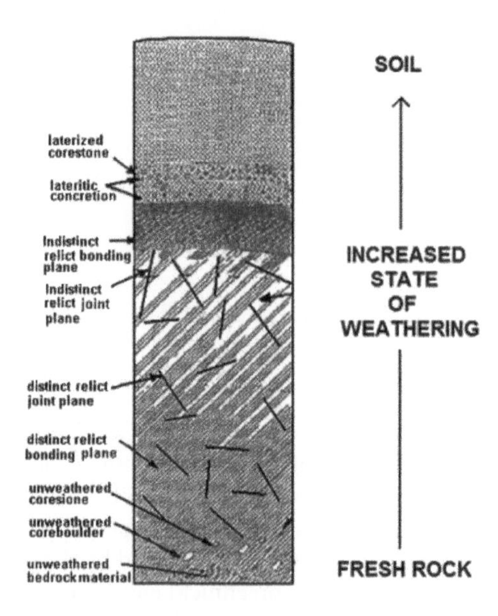

Figure 1.1 Soil profile of a weathering in a metamorphic rock (amphibole schist) (*modified after* Raj, 1993).

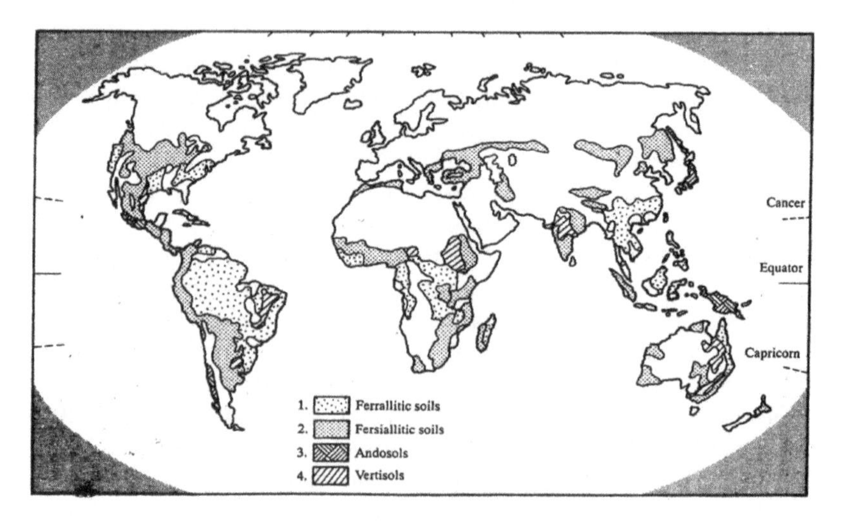

Figure 1.2 Distribution of tropical residual soils (*after* F.A.O. World Soil Map).

soil. This soil typically retains many of the characteristics of the parent rock. For example, decomposed granite will result in sandy residual soil.

In a tropical region, residual soil layers can be very thick, sometimes extending for hundreds of metres before reaching unweathered rock (Figure 1.1). Cooler and more arid regions normally have much thinner layers, and often no residual soil at all. The worldwide distribution of residual soil is shown in Figure 1.2.

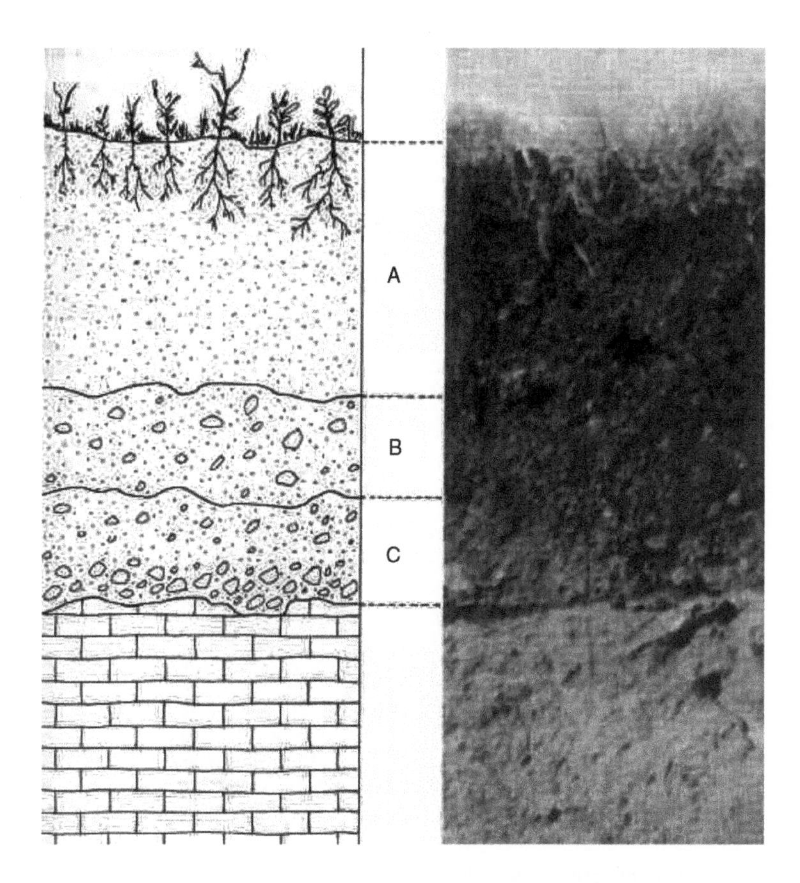

Figure 1.3 Laterite and saprolite profile: A represents soil; B represents laterite, a regolith; C represents saprolite, a less-weathered regolith; and D represents bedrock (*Source*: http://en.wikipedia.org/wiki/Laterite).

Saprolite, sometimes called 'rotten rocks', is a general term for residual soils that are not extensively weathered and still retain much of the structure of the parent rock.

Laterite is a residual soil found in tropical regions. This type of soil is cemented with iron oxides, which gives it a high dry strength (Figure 1.3).

The engineering properties of residual soils range from poor to good, and generally improve with depth.

1.2.2 Glacial soils

Much of the Earth's land area was once covered with huge masses of ice in the form of glaciers. These had a dramatic effect on the landscape and created a category of soil called glacial soils. Glacier ice is not stationary but moves along the ground, often grinding down some areas and filling in others. Glaciers grind down soils and rocks and transport these materials over long distances, up to hundreds of kilometres. The resulting deposit often contains materials from various sources. These deposits can

Figure 1.4 Glacial till (*Source:* http://en.wikipedia.org/wiki/Glacial_till).

Figure 1.5 Alluvial soil deposit in Red Rock Canyon State Park, California (*Source:* http://en.wikipedia.org/wiki/Alluvium).

have a wide range of hardness and particle sizes, resulting in soils which are often very complex and heterogeneous (Figure 1.4).

The term *drift* encompasses all glacial soils, which can then be divided into three categories: till, glaciofluvial and glaciolacustrine.

1.2.3 Alluvial soils

Alluvial soils, also known as fluvial soils or alluvium, are those soils transported to their present position by rivers and streams. These soils are very common all over the world, and many large engineering structures are built on them. The main reason is that many of the world's major cities and civilizations are built close to or along rivers, which used to be an important means of transportation in the past and remain so in many parts of the world. Alluvium also often contains extensive ground water aquifers, so it is important in the development of water supply wells and in geoenvironmental engineering.

When a river or stream is flowing rapidly, the silts and clays remain in suspension and are carried downstream (Figure 1.5). Only sand, gravels and boulders will

Figure 1.6 Lacustrine soil deposit (*Source*: http://soilweb.landfood.ubc.ca/landscape/parent-material/water-environment/lacustrine-environment).

be deposited. However, when the water flows more slowly, more of the finer soils (silt and clay) are also deposited. Rivers flow rapidly after a period of heavy rainfall or snowmelt, and slowly during drought. Alluvial soils often contain alternating horizontal layers of different soil types.

Rivers in relatively flat terrain move much more slowly and often change course, thus creating complex alluvial deposits, such as meander belt deposits.

1.2.4 Lacustrine soils

Lacustrine soils are soils deposited beneath lakes. These deposits may be still under water, or may now be exposed due to lowering of the lake water level (Figure 1.6). Most lacustrine soils are primarily silt and clay. Their suitability for foundation ranges from poor to average.

1.2.5 Marine soils

Marine soils are also deposited under water, except that they are formed in the ocean. Deltas result in a special type of marine soil deposit formed when rivers meet large bodies of water, gradually building up to the surface. This mode of deposition creates a very flat terrain, so water flows very slowly. The resulting soils are primarily silt and

Figure 1.7 Marine soil deposit (*Source*: Parent Materials Mode of Deposition in Atlantic Canada: Marine, sis.agr.gc.ca, Agriculture and Agri-Food Canada ©. Reproduced with the permission of the Minister of Agriculture and Agri-Food Canada, 2014).

Figure 1.8 Aeolian soil deposit (*Source*: http://ec.europa.eu/research/research-for-europe/international-calter_en.html).

clays, and are very soft. Because of their mode of deposition, most marine soils are very uniform and consistent (Figure 1.7). Thus although their engineering properties are often very poor, they may be more predictable than other more erratic soils.

Some sands may also accumulate as marine deposits, especially in areas where rivers discharge into the sea at a steeper gradient. This sand is moved and sorted by waves and currents, and some of it is deposited back on shore as beach sands. These sands are typically very poor graded and very loose.

Deeper marine deposits are more uniform and often content organic materials from marine organisms.

1.2.6 Aeolian soils

Aeolian soils, also known as eolian soils, are those deposited by wind. This mode of transportation generally produces very poorly graded soils. These soils are also usually very loose, and thus have only fair engineering properties (Figure 1.8).

Figure 1.9 Colluvial soil deposit (*Source*: Parent Materials Mode of Deposition in the Yukon Territory: Colluvial, sis.agr.gc.ca, Agriculture and Agri-Food Canada ©. Reproduced with the permission of the Minister of Agriculture and Agri-Food Canada, 2014).

1.2.7 Colluvial soils

Colluvial soils or colluviums are soils transported down slopes by the action of gravity. There are two types of down slope movement: slow and rapid. Both occur only on or near sloping ground (Figure 1.9).

Slow movement, typically of the order of millimetres per year, is called creep. It occurs because of gravity-induced shear stresses, frost action and other processes. Creep typically extends to a depth of 0.3–3 m, with the greatest displacement occurring at the surface. Such slow movement may at first appear to be inconsequential. But with time they can produce significant distortions in structures founded on such soils.

Rapid movements are those created by dramatic events such as landsides or mudflow. Although this rapid movement can occur in any type of soil, the product is always considered to be colluvial soil.

1.2.8 Organic soils and peat

Technically any material that contains carbon is called 'organic'. However, engineers and geologists use a narrower definition when applying the term to soils. An organic soil is one that contains a significant amount of organic material recently derived from plant remains. This implies that it needs to be 'fresh' and still in the process of decomposition, thus retaining a distinctive texture, colour and odour (Figure 1.10). The most chemically active colloidal fraction of organic materials is humus.

Some soils contain carbon, but not from recently derived plant material. Thus they are not considered organic in this context. For example, some sand contains calcium carbonate (calcite), which arrived as a chemical precipitate.

Figure 1.10 Organic soil deposit (*Source*: http://www.teara.govt.nz/en/soils/7/3/1).

The identification of organic soils is very important because they are much weaker and more compressible than inorganic (mineral) soils. As such they do not provide suitable support for most engineering projects. Traditionally, the normal practice is to avoid such soils, or excavate them or drive a pile though them. However, due to the scarcity of more suitable land for infrastructure development and agriculture, avoidance is now no longer an option, as explained later in this book.

The term *peat* refers to highly organic soils derived primarily from plant remains. In other words, peat is unconsolidated superficial deposits with high non-crystalline colloid (humus) content, constituting the subsurface of wetland systems. It normally has a dark brown to black colour, a spongy consistency and an organic odour. Plant fibres are sometimes visible, but in the advanced stages of decomposition they may not be evident.

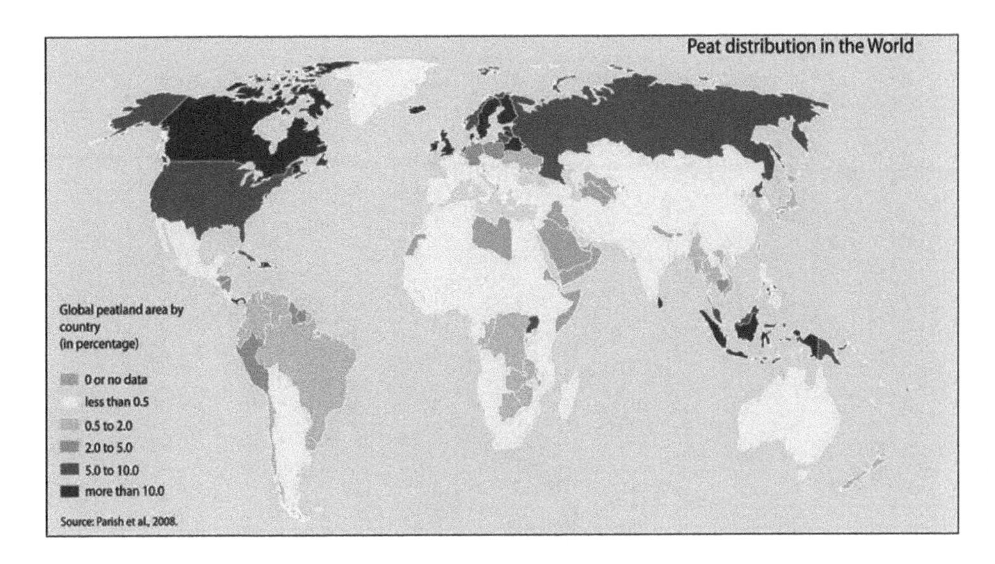

Figure 1.11 Worldwide distribution of peat (*after* Parish *et al.*, 2008).

As mentioned above, soils with an organic content greater than 20% are generally termed organic soils. The organic content is essentially the remains of plants whose rate of accumulation is faster that the rate of decay. The precise definition of peat, however, varies between the disciplines of soil science and engineering. Soil scientists define peat as soil with an organic content greater than 35%. To a geotechnical engineer, all soils with organic content greater than 20% are known as organic soil and peat is an organic soil with an organic content of more than 75% (see further definition in Chapter 2). The engineering definition is essentially based on the mechanical properties of the soil. It is generally recognized that when a soil possesses greater than 20% organic content the mechanical criteria of conventional mineral soil (silt and clay) can no longer be generally applied. To avoid further confusion, we will generally term all organic soils as peat in this book – in other words, the terms *peat* and *organic soil* will be used interchangeably. One main reason for this is that many reference sources about soil distribution are based on agricultural science, whose definition of peat is different from that of engineering. The worldwide distribution of peat is shown in Figure 1.11.

In temperate region, such as in Canada, Europe and the USA, peat deposits are termed bogs and fens. These are pits or basins filled with organic material. Bogs are typically covered with live moss. Thoroughly decomposed peat is sometimes called muck. Swamps are larger than bogs and may contain a wide variety of materials. Slow streams or lakes typically feed them. A noteworthy example is the Everglades in Florida, USA.

In tropical countries like Malaysia and Indonesia, peat is generally termed either basin or valley peat. Basin peat is usually found on the inward edge of mangrove swamps along a coastal plain. Individual peat bodies may range from a few to 100,000 hectares and they generally have a dome-shaped surface. The peat is generally classified as ombrogenous or rain-fed peat, and is poor in nutrients. Due to coastal and

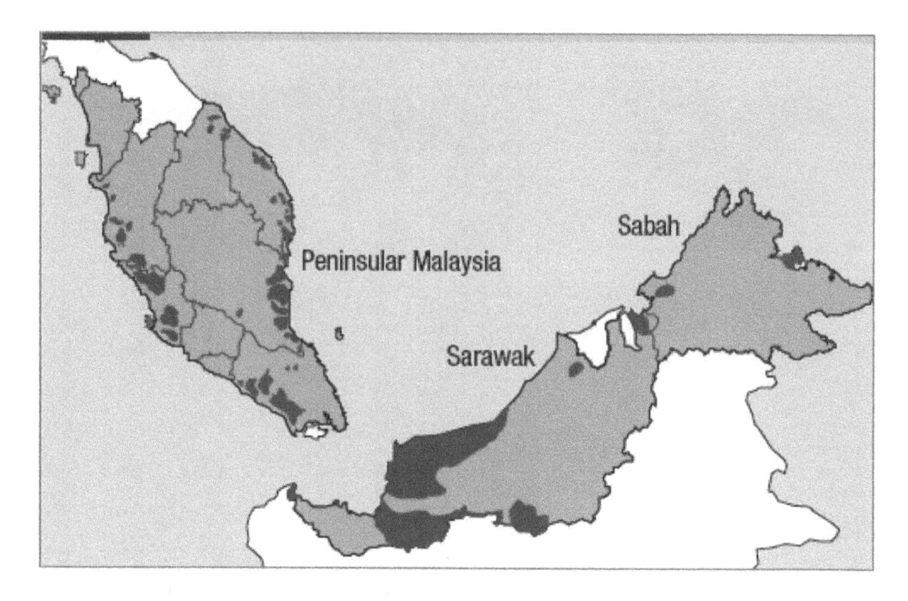

Figure 1.12 Distribution of peat land in Malaysia (*after* Leete, 2006).

alluvial geomorphology they are often elongated and irregular, rather than having the ideal round bog shape. The depth of the peat is generally shallower near the coast and increases inland, locally exceeding more than 20 m. Water plays a fundamental role in the development and maintenance of tropical peat. A balance of rainfall and evapotranspiration is critical to their sustainability. Rainfall and surface topography regulate the overall hydrological characteristics of peat land. Peat land is also generally known as wetland or peat swamp because of its water table, which is close to or above the peat surface throughout the year and fluctuates with the intensity and frequency of rainfall. Chapter 2 describes in further detail the various types of peat land and their mode of deposition. The distribution of peat land in Malaysia is shown in Figure 1.12.

Peat or organic soils may also occur as deposits buried underground and covered with inorganic alluvial soils. These are often difficult to detect, and can be a source of large differential settlements. Organic deposits may also be mixed with inorganic soils, such as silt and clay, producing soils not as bad as peat but worse than inorganic deposits.

1.3 ENGINEERING IN PEAT LAND

Peat has certain characteristics that sets it apart from most mineral soils and requires special considerations for construction over them. These special characteristics include:

1. High natural moisture content (up to 1500%).
2. High compressibility, including significant secondary and tertiary compression.
3. Low shear strength (typically S_u= 5–20 kPa).

4. High degree of spatial variability.
5. Potential for further decomposition as a result of changing environmental conditions.
6. High permeability compared with clay.
7. High charge and high specific surface area of the colloidal fraction of peat.

In terms of development, the vast area of peat land and its occurrence close to or within population centres and existing cropped areas means that some form of infrastructure development has to be carried out in these areas. This would include road crossings and, in some instances, housing developments that encroach on the peat land areas as available land becomes more and more scare. To stimulate agricultural development, for instance, basic civil engineering structures are required. These would include irrigation and drainage, water supply, roads, farm buildings etc.

However, just as soil scientists and agriculturalists have, engineers have recognised that peat land or peat is a very problematic soil that is best avoided as far as possible. But of course this is no longer an option. To an engineer, peat or organic soils represent the extreme form of soft soil. They are subject to instabilities, such as localised sinking and slip failure, and to massive primary and long-term secondary and even tertiary settlements when subjected to even moderate load increases. In addition, there is discomfort and difficulty of access to the sites, tremendous variability in material properties and difficulty in sampling.

As stated above, these materials may also change chemically and biologically with time. For example, further humification of the organic constituents would alter the soil's mechanical properties, such as compressibility, shear strength and hydraulic conductivity. Lowering of ground water may cause shrinking and oxidation of peat, leading to humification and a consequent increase in permeability and compressibility.

It is obvious that the mechanical properties of peat are very different from those of the mineral soils (silt and clay) that are familiar to engineering graduates throughout the world. Thus it is critical to understand fully the characteristics of peat – high water tables, lack of topographic relief and dynamics in their soil properties – that set them apart from mineral soils. Criteria based on mineral soils cannot be generally applied to peat conditions.

However, with the valuable experience gained in peat land development, and research into the mechanical properties of the material, we are now in better position to understand peat and hence develop safer and cheaper engineering design and construction techniques. In the past, we use to hear tales of 'disappearing roads' or roads that disappeared 'overnight' in peat swamps. Fortunately, there are now a number of methods that can be used for infrastructure construction in areas where peat predominates. These are: excavation and replacement methods; surface reinforcement and preloading; vertical drains; injection and deep mixing stabilization; cement/stone columns; and geomaterials or lightweight fill. The preferred method will depend on working out the best solution considering economic and technical factors, the available construction time and the target performance standards. Again, this calls for better and deeper understanding of peat's characteristics to make the construction more manageable.

The excavation and replacement method is suitable for peat with a depth of less than about six metres. In this process the peat is excavated and replaced with stable fill, such as sand.

The surface reinforcement and preloading methods are essentially for road construction. They use geotextile, geogrids, timber or bamboo mattresses that separate the granular fill from the soft peat on which they are placed. The geotextile also serves to prevent the fill from becoming contaminated by the weak subsoil material, thus preserving the better load-bearing properties of the fill. As the fill becomes thicker, the need for separation decreases and the geotextile then becomes more of a reinforcement.

The vertical drain technique uses geosynthetic vertical drains or sand drains, and even stone columns, in order to speed up the dissipation of pore water pressures and hence speed up the settlement process.

Piled supports are a fundamental means of construction in all soft soils. By carrying the structural forces to a competent layer, the problem of settlement of the structure can largely be avoided. But this method is generally very costly. A newer alternative is geomaterials or lightweight fill, an example of the latter being EPS (expanded polystyrene). This technique has been used for a number of road constructions in countries like Japan and Norway, and has been successfully applied for high bridge approach embankments in Malaysia. Because they are lightweight (about 20 times lighter than conventional fill materials) very little pressure is actually exerted on the existing ground, hence minimising the stability and settlement problem.

There are also several other alternatives, some of which have been applied while others are still very much at an experimental stage. Examples of these are geocells, geo mattress rafts, vibrated concrete columns, deep stabilization techniques with cement and lime or chemical additives using injection or vacuum preloading, thermal compression, electrokinetics and biogrout.

Having said all the above, whatever techniques that we may employ, unless the structure's load is exclusively transferred to a hard stratum, the stability of any structure built on peat will inevitably depend on the overall change of the peat with time. For example, over-drainage of a peat land will result in general land subsidence, and this in turn may result in the instability of structures such as roads and buildings built there. As another illustration, there is a conflict between the need to drain peat for traffic, such as farm machinery, to gain access and the moisture supply required for crop growth. The water table has to be lowered sufficiently to attain the required bearing capacity for mechanization, which takes water from the crops. Thus an integrated approach is vital for peat land development. It cannot be done in isolation.

Chapter 2

Development of peat land and types of peat

2.1 INTRODUCTION

Peat represents an accumulation of disintegrated plant remains which have been preserved under conditions of incomplete aeration and high water content. It accumulates wherever the conditions are suitable, that is, in areas with excess rainfall and poorly drained ground, irrespective of latitude or altitude. It forms when plant material, usually in marshy areas, is inhibited from decaying fully by the acidic conditions and an absence of microbial activity. For example, peat formation can occur along the inland edge of mangroves, where fine sediments and organic material become trapped in the mangrove roots. Nonetheless, peat deposits tend to be most common in those regions with a comparatively cool wet climate. Physico-chemical and biochemical process cause this organic material to remain in a state of preservation over a long period of time. In other words, waterlogged poorly drained conditions not only favour the growth of a particular type of vegetation but also help preserve the plant remains.

Peat is found in many countries throughout the world. It is found in 42 states of the USA, with a total acreage of 30 million hectares (Mesri and Ajlouni, 2007). Canada and Russia are the countries with the largest areas of peat: 150 and 150 million hectares respectively. Table 2.1 shows the distribution of peat deposits throughout the world.

The total area of tropical peat swamp forests or tropical peat lands in the world amounts to about 30 million hectares, two thirds of which are in Southeast Asia.

In Malaysia, some 3 million hectares (about 8%) of the country's land area, is covered with peat (Figure 2.1). It is found in several states in Malaysia, including Perak, Selangor, Johor, Terengganu, Penang, Sarawak and Sabah. Sarawak has the largest area of peat in the country, covering about 1.66 million hectares and constituting 13% of the state. This is followed by Peninsular Malaysia which has about 984,500 hectares, comprising 7% of its total area, while Sabah has 86,000 hectares, representing 1% of the state. Of the 1.66 million hectares in Sarawak, 1.5 million are deep peat. In peninsular Malaysia, approximately 0.8 million hectares (or 89%) are deep peat. Table 2.2 shows the peat soil distributions in Malaysia.

In Indonesia, peat covers about 26 million hectares of the country's land area, with almost half of the total in Kalimantan (Figure 2.2). Peat is also found in many other parts of Asia, such as Japan, Bangladesh and China.

In Japan, peat is widely distributed throughout Hokkaido, which is the northernmost of Japan's four main islands, with an area of approximately 2,000 km^2 (Figure 2.3), approximately 6% of the flat area on the island. The peat in Japan,

Table 2.1 The area of peat land in some countries of the world (*after* Mesri and Ajlouni, 2007).

Country	*Peat land (km^2)*	*Percentage of land area*
Canada	1,500,000	18
USSR (former)	1,500,000	
USA	600,000	10
Indonesia	170,000	14
Finland	100,000	34
Sweden	70,000	20
China	42,000	
Norway	30,000	10
Malaysia	25,000	
Germany	16,000	
Brazil	15,000	
Ireland	14,000	17
Uganda	14,000	
Poland	13,000	
Falklands	12,000	
Chile	11,000	
Zambia	11,000	
26 other countries	220 to 10,000	
Scotland		10
15 other countries		1 to 9

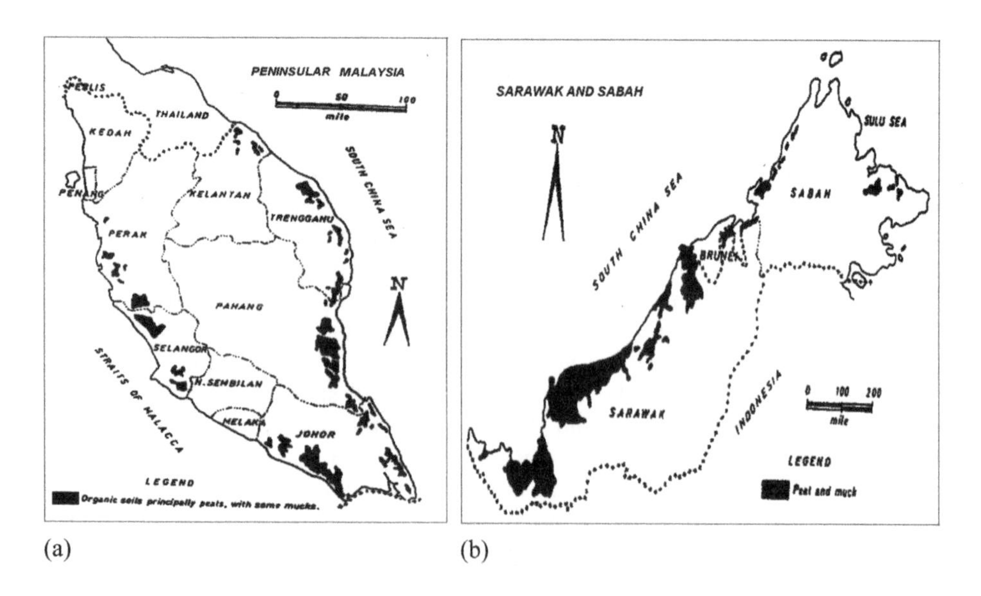

(a) (b)

Figure 2.1 Distribution of peat land in Malaysia: (a) West Malaysia; (b) East Malaysia (*after* Huat, 2004).

in many cases, is basin peat, formed when lakes and marshes become filled with dead plants growing around them, which then turns into land. This type of peat is characterized by the spongy formation of plant fibre.

All present-day surface deposits of peat in Europe, Asia, Canada and the USA are believed to have accumulated since the last ice age and have therefore been formed

Table 2.2 Peat distribution in Malaysia (*adapted* from Leete, 2006).

State	*Total peat (ha)*
Johor	228,960
Negeri Sembilan	6,300
Selangor	194,300
Perak	107,500
Pahang	219,561
Terengganu	81,245
Kelantan	7,400
Sarawak	1,657,600
Sabah	86,000

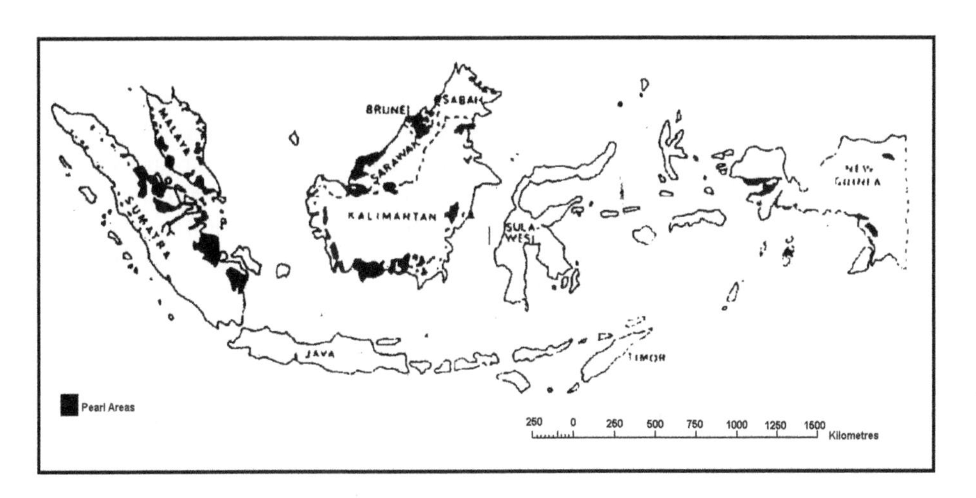

Figure 2.2 Location of peat land in Indonesia and vicinity (*after* Rieley, 1991).

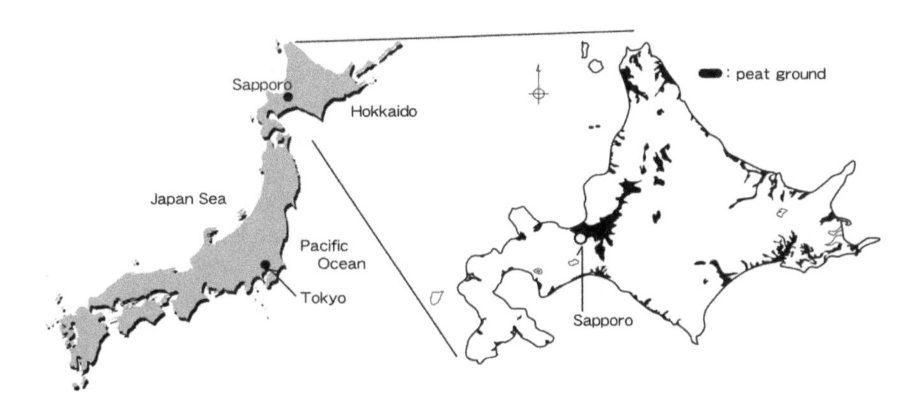

Figure 2.3 Distribution of peat land in Hokkaido, Japan (*after* Huat, 2004).

during the last 20,000 years. On the other hand, some buried peat may have been developed during the inter-glacial period. Peat also has accumulated in postglacial lakes and marshes, where they are interbedded with silts and muds.

2.2 DEFINITION OF PEAT AND ORGANIC SOILS

Peat commonly occurs as extremely soft, wet, unconsolidated superficial deposits, normally as an integral part of wetland systems. It may also occur as strata beneath other superficial deposits. The term *peat* describes a naturally occurring highly organic substance derived primarily from plant materials. It is formed when organic (usually plant) matter accumulates more quickly than it humifies (decays). This usually occurs when organic matter is preserved below a high water table, as in swamps or wetlands.

Peat is predominantly made up of plant remains such as leaves and stems. It is produced by the incomplete decomposition and disintegration of sedges, trees, mosses and other plants growing in wet places and marshes where there is a lack of oxygen. Therefore the colour of peat is usually dark brown or black, and it has a distinctive odour (Craig, 1992). Since the main component is organic matter, peat is very spongy, highly compressible and combustible. These characteristics also lend peat its own distinctive geotechnical properties compared with inorganic soils, such as clay and sandy soils, which are made up only of soil particles (Deboucha *et al.*, 2008).

Commonly, peat is classified based on its fibre, organic and ash content. Decomposition is the breakdown process of plant remains by the soil microflora, bacteria and fungi via aerobic decay. In this procedure, as mentioned earlier, the peat's structure breaks down and its primary chemical composition changes. The end products of the decomposition process are carbon dioxide and water. The degree of decomposition varies throughout peat, since some plants (or parts of plants) are more resistant than others. Also, the degree of decomposition depends on a combination of conditions, such as the chemistry of the water supply, the temperature of the region, aeration and the biochemical stability of the peat-forming plant (Lishtvan *et al.*, 1985).

However, the cut-off value of the percentage of organic matter necessary to classify a superficial deposit or soil as peat varies throughout the world, usually depending on the purpose of classification. This cut-off value also serves to differentiate peat from superficial deposits or soils with lesser amounts of organic content. The terms *peat* and *organic soils*, used to describe soils with an organic content, were once synonymous, but the latter is now used for superficial deposits or soils that contain organic matter.

As described in Chapter 1, soils with an organic content greater than 20% are generally termed organic soils. The precise definition of peat, however, varies between of the disciplines of soil science and engineering, as well as between countries. Soil scientists define peat as soil with an organic content greater than 35%. To a geotechnical engineer, all soils with an organic content greater than 20% are known as organic soil. 'Peat' is an organic soil with organic content of more than 75%. The engineering definition is essentially based on the mechanical properties of the soil. It is generally recognized that when a soil possess organic content greater than 20%, the mechanical criteria of conventional mineral soil (silt and clay) can no longer be generally applied. Table 2.3 shows the ASTM (D4427) classification of peat samples by laboratory testing.

Table 2.3 Organic content ranges (ASTM D4427).

Basic soil type	*Description*	*Organic content (%)*
Clay or silt or sand	Slightly organic	2–20
Organic soil	–	25–75
Peat	–	>75

Table 2.4 Organic soil classification based on organic content ranges (*after* Jarrett, 1995).

Basic soil type	*Description*	*Symbol*	*Organic content (%)*
Clay or silt or sand	Slightly organic	O	2–20
Organic soil	–	O	25–75
Peat	–	Pt	>75

Slightly organic silts or clays will most probably appear as inorganic fined-grained soils, probably black to dark brown in colour, with an organic odour and possibly some visible organic remains. Their plasticity limits should be evaluated as for other fine-grained inorganic soils. These soils would then be classified as silts or clays of low, medium or high plasticity.

Peat, on the other hand, may well appear to be completely organic, contain recognizable plant remains, have a low density and also be black to dark brown in colour. Organic soils, however, are more difficult to sub-divide. Under the Unified Soil Classification System (UCS), organic soils are recognized as a separate soil entity and have a major division called Highly Organic Soils (Pt), which refers to peat, muck and highly organic soils.

Jarrett (1995) gives a classification for organic soils which purportedly can be integrated with the UCS to bridge the gap between peat, as defined above, and purely inorganic clays, and silts, and is shown in Table 2.4.

Hobbs (1986) illustrates the various classification of peat between countries as follows:

1. Russian geotechnical engineers assume that peat is a soil containing more than 50% of particle weight of vegetable origin (organic matters), while peaty soil contains 10–50% of particles of vegetable origin.
2. ASTM, D2607-69 assumes peat as a soil having organic matter of more than 75%, as stated above.
3. Hobbs suggests that peat is a soil having organic matter more than 27.5%.

LPC (France), on the other hand, describes organic soils as those having greater than 10% organic content (Figure 2.4).

Figure 2.5 is a comparison of various classification systems made by Andrejko *et al.* (1983).

To avoid further confusion, we will generally term all organic soils (soils having an organic content greater than 20%) as peat in this book, meaning that the terms *peat* and *organic soils* will be used interchangeably.

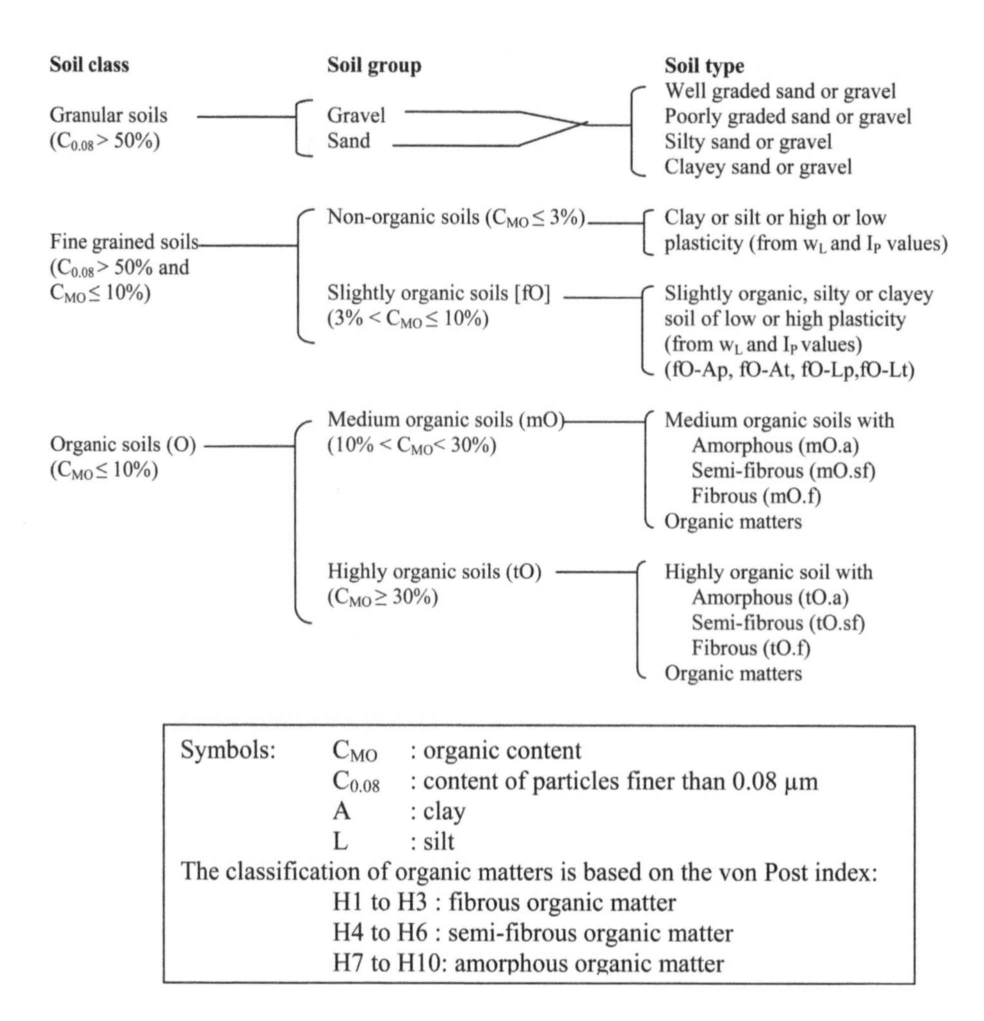

Figure 2.4 LPC classification of organic soils (*after* Magnan, 1994).

According to the American Society for Testing and Materials (ASTM) standard (ASTM, 1990), peat classification has been narrowed to only three classes, based on fibre content, ash content and the acidity of the soil. In the fibre content classification, peat is divided into three groups: (i) fibric (fibrous; least decomposed, with more than 67% fibre content), (ii) hemic (semi-fibrous; intermediate decomposition) and (iii) sapric (amorphous; most decomposed, with less than 33% fibre content).

Fibrous peat has a high organic and fibre content, with a low degree of humification. It consists of undecomposed fibrous organic materials, is easily identifiable and extremely acidic. Sapric peat contains highly decomposed materials. The original plant fibres have mostly disappeared and the water-holding capacity is generally less than that of either fibrous or hemic peat. It is generally very dark grey to black in colour and is quite stable in its physical properties. Compared with fibrous peat deposits, sapric peat deposits are likely to exist at lower void ratios and display lower permeability,

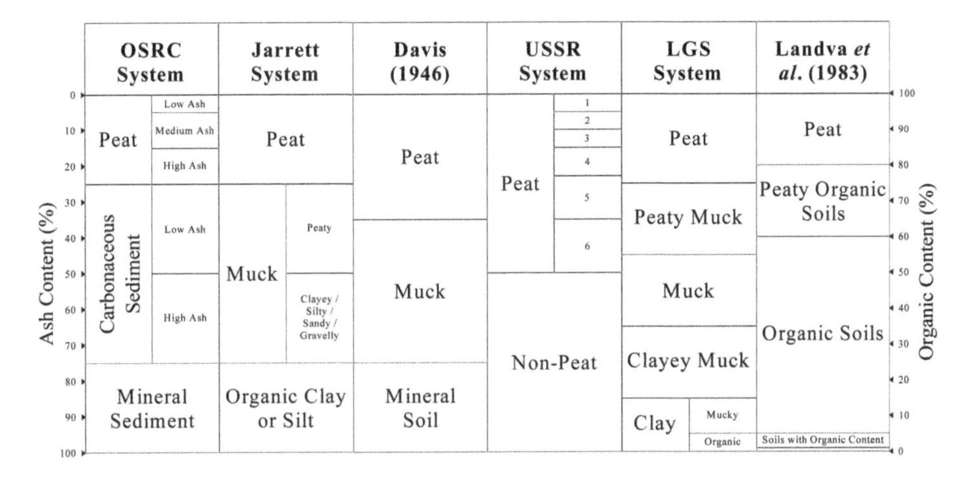

Figure 2.5 Comparison of classification systems used for peat and organic soils (*after* Andrejko *et al.*, 1983).

Figure 2.6 Scanning electron micrographs of peats: (a) fibrous, (b) sapric and (c) hemic (*after* Kazemian *et al.*, 2011).

lower compressibility, a lower friction angle, and a higher coefficient of earth pressure at rest. Hemic peat has properties intermediate between fibrous and sapric peats. Figure 2.6 shows scanning electron micrographs (SEM) of fibrous, hemic, and sapric peats.

2.3 CLASSIFICATION BASED ON FIBRE CONTENT AND DEGREE OF HUMIFICATION

Another useful means to classify peat or organic soils as opposed to mineral soils (silts and clay) is to base it on their fibre content and degree of humification or decomposition (also known as the von Post scale) (see Table 2.5). The US Department of Agriculture's (USDA) three-point scale classification based on fibre content resulting from decomposition is described in Table 2.6.

Table 2.5 Qualifying terms and symbols for peat soils.

Organic	*von Post scale*	*Qualifying terms*	*Symbol component*
Peat			Pt
	H1–H3	Fibric/fibrous	f
	H4–H6	Hemic/moderately decomposed	h
	H7–H10	Sapric/amorphous	a

Table 2.6 USDA classification of peat.

Type of peat	*Fibre content*	*von Post scale*
Fibric peat	Over 66%	H4 or less
Hemic peat	33–66%	H5 or H6
Sapric peat	Less than 33%	H7

Fibre content is determined typically from dry weight of fibres retained on a #100 sieve (>0.15 mm opening size) as a percentage of oven-dried mass (ASTM Standard D 1997). Fibres may be fine (woody or non-woody) or coarse (woody).

The organic fibre content is also referred to as the fabric of organic soil. A fibre is defined as >0.15 mm in diameter. An appreciation of the constituent matter and its attributes, such as orientation, aids in the constitutive modelling of this soil type for basic understanding of its mechanical behaviour (Molenkamp, 1994).

Fibric peat is mostly undecomposed and typically tan to light reddish brown in colour. Hemic peat is intermediate between fibric and sapric peat in degree of decomposition or humification, organic content and bulk density, and typically dark reddish brown in colour. Sapric peat, on the other hand, is generally of darker colour than the above two types of peat, and the most decomposed/humified. It generally has the highest organic content and bulk density of the three types of peat.

Decomposition or humification involves the loss of organic matter either as gas or in solution, the disappearance of physical structure and change in chemical state. As previously described, the breakdown of plant remains is brought about by soil microflora, bacteria and fungi, which are responsible for aerobic decay. Therefore the end products of humification are carbon dioxide and water, the process being essentially biochemical oxidation. Immersion in water reduces the oxygen supply enormously, which in turn reduces aerobic microbial activity and encourages anaerobic decay, which is much less rapid. This results in the accumulation of partially decayed plant material as peat. Peat degree of decomposition or humification is usually assessed by means of the von Post scale. This will be further described in details in Chapter 3.

Another three-way division of peat based on the von Post scale is described as follows (see also Table 2.7):

- ***Fibrous peat*** is low humified and has a distinct plant structure. It is brown to brownish yellow in colour. If a sample is squeezed in the hand, it gives brown to colourless, cloudy to clear water, but without any peat matter. The material

Table 2.7 Classification of peat on the basis of decomposition on the von Post scale (*after* Karlsson and Hansbo, 1981).

Designation	*Group*	*Description*
Fibrous peat	H1–H4	Low degree of decomposition. Fibrous structure. Easily recognized plant structure, primarily of white masses.
Pseudo-fibrous peat	H5–H7	Intermediate degree of decomposition. Recognizable plant structure.
Amorphous peat	H8–H10	High degree of decomposition. No visible plant structure. Mushy consistency.

remaining in the hand has a fibrous structure. (Degree of decomposition on the von Post scale: H1–H4.)

- ***Pseudofibrous peat*** is moderately humified and has an indistinct to relatively distinct plant structure. It is usually brown. If a sample is squeezed in the hand, less than half of the peat mass passes between the fingers. The material remaining in the hand has a more or less mushy consistency, but with a distinct plant structure (H5–H7).
- ***Amorphous peat*** is highly humified and the plant structure is very indistinct or invisible. It is brown to brown-black in colour. If a sample is squeezed in the hand, more than half of the peat mass passes between the fingers without any free water running out. When squeezing, only a few more solid components, such as root fibres and wood remnants, can be felt. These constitute any material remaining in the hand (H8–H10).

A Malaysian Soil Classification for organic soils includes the two factors mentioned above, i.e. organic content and degree of humification, as shown in Table 2.8.

Peat is also referred to as topogenous or clayey peat and ombrogenous peat. The former, as its alternative name suggests, is composed of slightly to moderately decomposed plant matters and fine clastic sediments (mineral matter). The latter is described as essentially a pile-up of loose trunks, branches and roots (Yogeswaran, 1995). Ombrogenous peat is typically acidic, characterized by low pH.

Table 2.9 shows the other classification of peat mentioned in ASTM according to fibre content (ASTM D1997), ash content (ASTM D2974), and acidity (ASTM D2976).

In addition to the organic and fibre content and degree of humification, other index parameters, such as water content, liquid limits, specific gravity and unit weights are also useful for peat and organic soils. Hobbs (1986) and Edil (1997) suggested that the following characteristics should be included in a full description of peat.

1. Colour, which indicates the state of the peat.
2. Degree of humification, which (as described above) represents the degree to which the organic content has decayed (fibric, hemic, sapric).
3. Water content determined by oven drying method at 105°C.

Table 2.8 Organic soils and peat section of Malaysian Soil Classification Systems for engineering purposes (*after* Jarrett, 1995, *based on* BS 5930:1981).

Soils group (see note 1)		*Sub-group and laboratory identification*						*Field Identification*
		Description	*Group Symbol*	*Sub-group symbol*	*Liquid Limit %*	*Degree of Humification*	*Sub-group name*	
ORGANIC SOILS and PEATS	SLIGHTLY ORGANIC SOILS Organic Content 3–20%	Slightly Organic SILT Slightly Organic Clay	Mo Fo Co	Mo Clo Clo Cho Cvo CEo	<35 35–50 50–70 70–90 >90		Slightly Organic SILT (Sub-divide like Co) Slightly Organic CLAY of low plasticity Slightly Organic CLAY of intermediate plasticity Slightly Organic CLAY of high plasticity Slightly Organic CLAY of very high plasticity Slightly Organic CLAY of extremely high plasticity	Usually very dark to black in colour, small amount of organic matter may be visible. Often has distinctive organic smell.
	ORGANIC SOILS Organic Content 20–75%	ORGANICS SOILS	O				Subdivision of Organic Soil is difficult, as neither the plasticity tests nor the humification tests are reliable for them. As such a 'best attempt' is the probable outcome of subdivision leading to descriptions such as 'Fibrous ORGANIC SOILS' or 'Amorphous ORGANIC SOIL of Intermediate Plasticity'.	
	PEATS Organic Content More than 75%	PEAT	Pt	Ptf Pth Pta		H1–H3 H4–H6 H7–H10	Fibric or Fibrous Peat Hemic or Moderately Decomposed Peat Sapric or Amorphous Peat	Dark brown to black in colour. Material has low density so seems light. Majority of mass is organic so if fibrous the whole mess will be recognizable plant remains. More likely to smell strongly if highly humified.

Table 2.9 Classification of peat in ASTM according to fibre content, ash content and acidity (ASTM, 1990).

Fibre content (ASTM D1997)	Fibric: Peat with greater than 67% fibres Hemic: Peat with between 33% and 67% fibres Sapric: Peat with less than 33% fibres
Ash content (ASTM D2974)	Low ash: Peat with less than 5% ash Medium ash: Peat with between 5% and 15% ash High ash: Peat with more than 15% ash
Acidity (ASTM D2976)	Highly acidic: Peat with a pH less than 4.5 Moderate acidic: Peat with a pH between 4.5 and 5.5 Slightly acidic: Peat with a pH greater than 5.5 and less than 7 Basic: Peat with a pH equal or greater than 7

4. Principal plant components, namely coarse fibre, fine fibre, amorphous
5. Organic content as percentage of dry weight, determined from loss of ignition at 450–550°C as percentage of oven-dried mass at 105°C.
6. Liquid limit and plastic limit.
7. Fibre content determined from dry weight of fibre retained on #100 sieve (>0.15 mm) as percentage of oven-dried mass.

We will discuss this in detail in Chapter 3.

2.4 DEVELOPMENT OF PEAT LAND

The simple definition of a peat land is an area where peat is found. On the Irish Environment and Heritage Service website (www.peatlandsni.gov.uk), peat lands, mires, bogs and fens are defined as follows:

- **Peat land**
  An area with a naturally accumulated peat layer at the surface
- **Mire**
  A peat land where peat is currently forming and accumulating
- **Bog**
  A peat land which receives water solely from rain and/or snow falling on its surface. Raised bogs are found in lowland areas, generally below 150 m, such as river valleys, lake basins, and between drumlins. They are known as raised bogs because the bog surface is raised in the middle, like a dome. The surface of a raised bog is a mixture of pools, raised mossy hummocks and flatter lawns, and is colonized by plants and animals adapted to the acidic conditions and low levels of nutrients found there.Blanket bogs usually form in upland areas above 200 m with heavy rainfall and low temperatures. Because of the undulating nature of this ground, the thickness of peat can vary between 1 m and 6 m. Like raised bogs, blanket bogs receive their nutrients from rainwater and the atmosphere, and are also acidic.

- Fen
  A peat land which receives water and nutrients from the soil, rock and groundwater as well as rain and/or snow. Fens generally form in natural basins that have been flooded and developed into lakes. Plants start to grow around the edges of these lakes and eventually extend over most of the surface, often with the only area of open water at the centre or deepest part. The fen peat forms as plants, such as sedges, reeds and herbs, die and accumulate at the bottom of the lake.

In temperate regions, peat often occurs as fen, transition and bogs, which are lakes or pits filled with organic material. Fen peat and bog peat are respectively sometimes known as low-moor and high-moor. Peat lands may pass through a number of morphological stages, each with its own particular plant communities, which characterize the type of peat that develops. Figure 2.7 illustrates the various morphological stages and the associated properties of British peat (after Hobbs, 1986).

In the above example, sediments gradually filled lakes or basins, but surface run-off continues to bring both nutrients and sediments into the area allowing colonization of vegetation. The vegetation contributes plant remains in greater and greater abundance to the sedimentary deposit. The peat land eventually adopts a marsh-like landscape, which has been referred to as fen. As can be inferred from above, fen peat is frequently underlain by very soft organic mud.

Next follows a transition stage when, because of upward growth, the water to the peat lands is supplied more and more by direct precipitation. The peat is generally mixed and woody, forming what is termed as basin bog, or transition peat. Lastly, the ombrotropic stage is reached, when the peat land grows beyond the maximum physical limits of its groundwater supply and therefore relies entirely for its water supply on direct precipitation. The peat itself acts as a reservoir holding water above ground water level. The water associated with such peat land is typically acidic. The peat deposit formed is termed *raised bog* or *bog*.

The above process of peat land development is also called a lake-filled process (also referred to terrestrialization; see Figure 2.8). The differences between fen and bog peat are attributable to the types of plant remains in the peat and their mode of origin. The differences involve the degree of humification, structure, fabric and proportion of mineral material contained in the peat, and this in turn affects their engineering behaviour. Raised bogs or bogs are typically fibric (fibrous). The above successive stages of peat land development are also known as wetland succession.

According to Hobbs (1986), some fen peats in Britain, because they occur in areas of carbonate rocks such as chalk or limestone, were supplied with water which was slightly alkaline. As such, the plant communities are more diverse, giving rise to what is called rich fen peat. This develops a much higher degree of humification than acid peat. Because the strength and permeability of peat declines significantly as humification increases, rich fen peat presents more problems to engineers than acid peat.

Valley bogs are formed along the flatter parts of valley bottoms and generally occur as a result of water draining from relatively acidic rocks. These bogs have complex lateral zonation due to differences in the vegetation that developed: for example, it is richer along the border of the bog and along streams flowing in the valley.

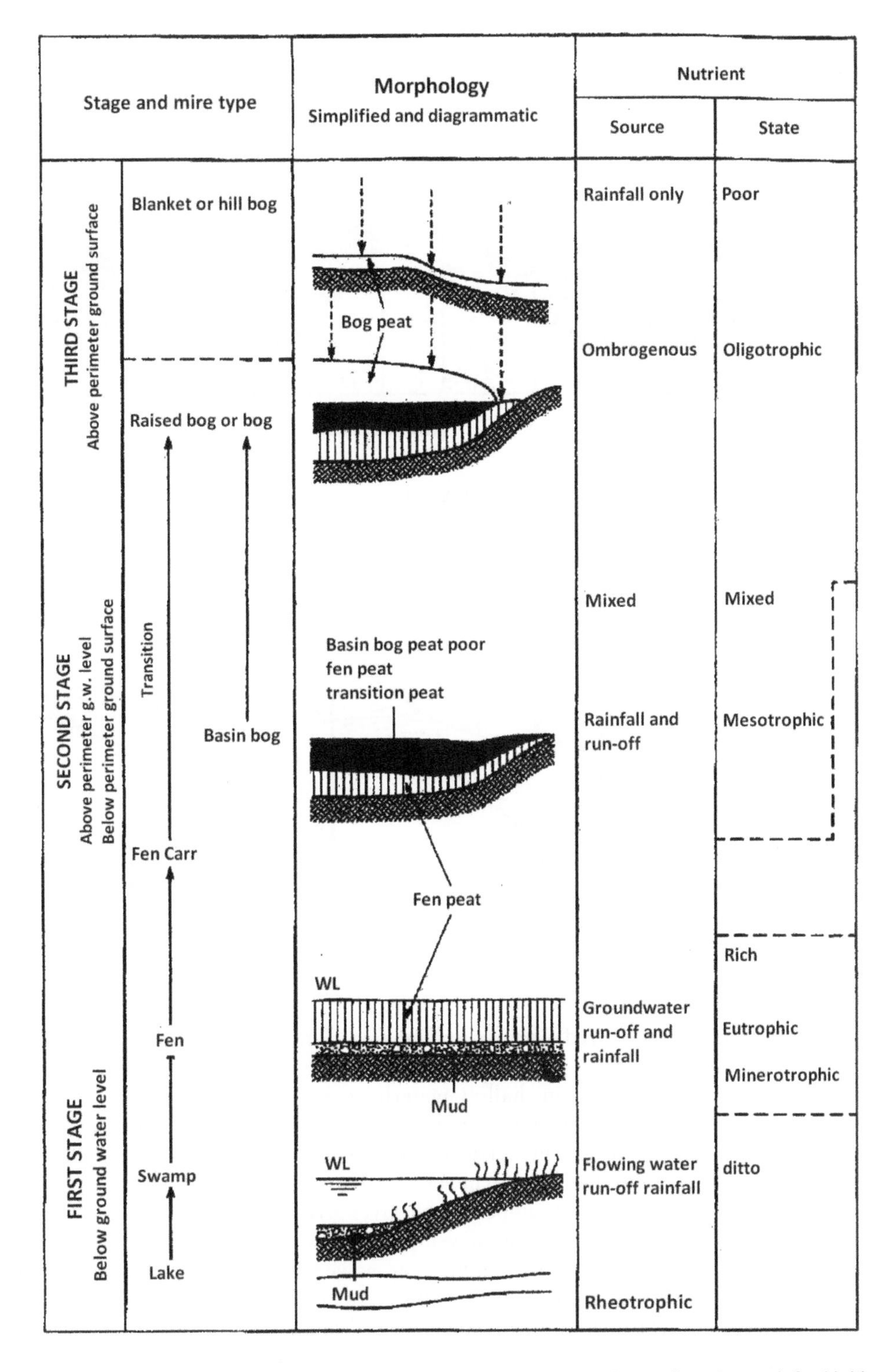

Figure 2.7 Mire stages, morphology, flora and associated properties of some British peat (*after* Hobbs, 1986).

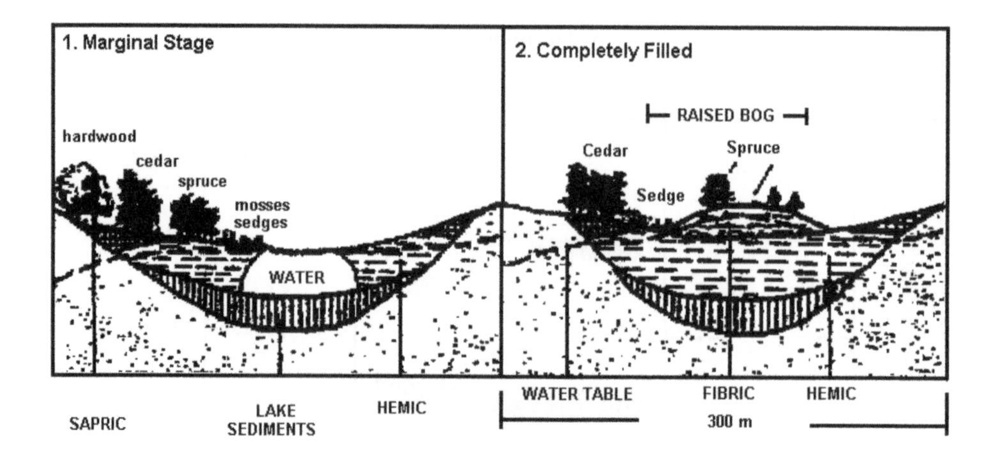

Figure 2.8 Lake fill process.

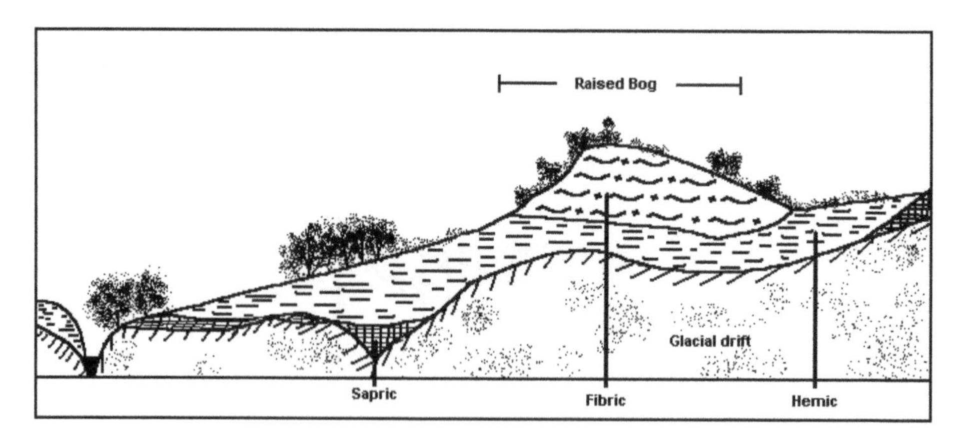

Figure 2.9 Paludification process.

Blanket bogs are associated with wet upland areas. The peat develops where slopes are not excessive and drainage is impeded. The process is often one of paludification, or swamping, and it may start in shallow waterlogged depressions. The bog extends down-slope if poorly drained surface water gives rise to waterlogging. In fact, high rainfall in such cool upland areas gives rise to leaching, which leads to the accumulation of impervious humus colloid and iron pan at small distances below the surface, usually between 0.3 m and 1.0 m. Such an impermeable layer gives rise to waterlogging, which represents the ideal condition for the development of ombrotropic peat. Soft sand and clays generally do not underlie blanket bog, as vegetation grows directly on the ground beneath. Generally these peats are thinner at higher altitude and on steeper slopes. Significant thickness is attained only in large deep depressions in the flat surface topography. A noteworthy example of this type of peat land is the Red Lake peat land in northern Minnesota (Figure 2.9). It began with the onset of a cooler and wetter climate about 3,500 year ago. Because of poor drainage on flat or gently sloping land, such as old glacial lakebeds, peat begins to accumulate.

In tropical areas, such as Malaysia and Indonesia, peat deposits also occur in both highlands and lowland areas. They are generally termed basin and valley peat respectively. However, lowland or valley peat is more extensive and occurs in low-lying, poorly drained depressions or basins in coastal areas. Basin peat is usually found on the inward edge of the mangrove swamps along a coast. The individual peat bodies may range from a few hectares to 100,000 hectares, and they generally have a dome-shaped surface. The peat is generally classified as ombrogenous or rain-fed peat, and is poor in nutrients (oligotrophic). Due to coastal and alluvial geomorphology they are often elongated and irregular, rather than having the ideal round bog shape. The depth of the peat is generally shallower near the coast and increases inland, locally exceeding more than 20 m. The coastal peat land is generally elevated well above adjacent river courses. Steep gradients are found at the periphery, while the central peat plain is almost flat. Water plays a fundamental role in the development and maintenance of tropical peat. A balance of rainfall and evapotranspiration is critical to their sustainability. Rainfall and surface topography regulate the overall hydrological characteristics of the peat land. Peat land is also generally known as wetland or peat swamp because of its water table, which is close to, or above, the peat surface throughout the year and fluctuates with the intensity and frequency of rainfall. Peat swamps are an important component of the world's wetlands - the dynamic link between land and water, a transition zone where the flow of water, the cycling of nutrients and the energy of the sun combine to produce a unique ecosystem of hydrology, soils and vegetation. The build-up of layers of peat and degree of decomposition depends principally on the local composition of the peat and the degree of waterlogging (Figure 2.10).

Peat formed in very wet conditions accumulates considerably faster and is less decomposed than peat accumulating in drier places. The peat acts as a natural sponge, retaining moisture at times of low rainfall but, because it is normally waterlogged already, with a very limited capacity to absorb additional heavy rainfall during periods such as a tropical monsoon. Peat swamp forests develop on these sites where dead vegetation has become waterlogged and is accumulating as peat. Water in peat swamps is generally high in humic substances (humus and humic acids) that give a typically dark brown to black colour to the water. Peat swamps are characterized by diverse features that relate to the nature of the water supply, such as flooding by surface or groundwater, or solely from rainfall; the type of landscape in which the peat swamp occurs, such as shallow depressions close to rivers; and the type of landscape that the swamp creates, such as accumulation of peat above groundwater level so that vegetation, often with prominent aerial roots, becomes wholly dependent on rainfall.

Basin peat forms domes, which according to Mutalib *et al.* (1991) are up to 15 m high whilst valley peats are flat or interlayered with river deposits. Normally, sandy ridges bound basin peat at their seaward side or they gradually merge into muddy coastal flats. Low-lying levees flank these domes along the rivers. The complexity of the domes becomes more pronounced as the distance from the sea increases, as shown in Figure 2.11.

Tropical (basin) peat domes were found to have typically well-developed internal stratification. An example is shown in Figure 2.12. The peat deposit is shown to be lenticular and dome surfaced, with a typical concave base. The centre of the dome,

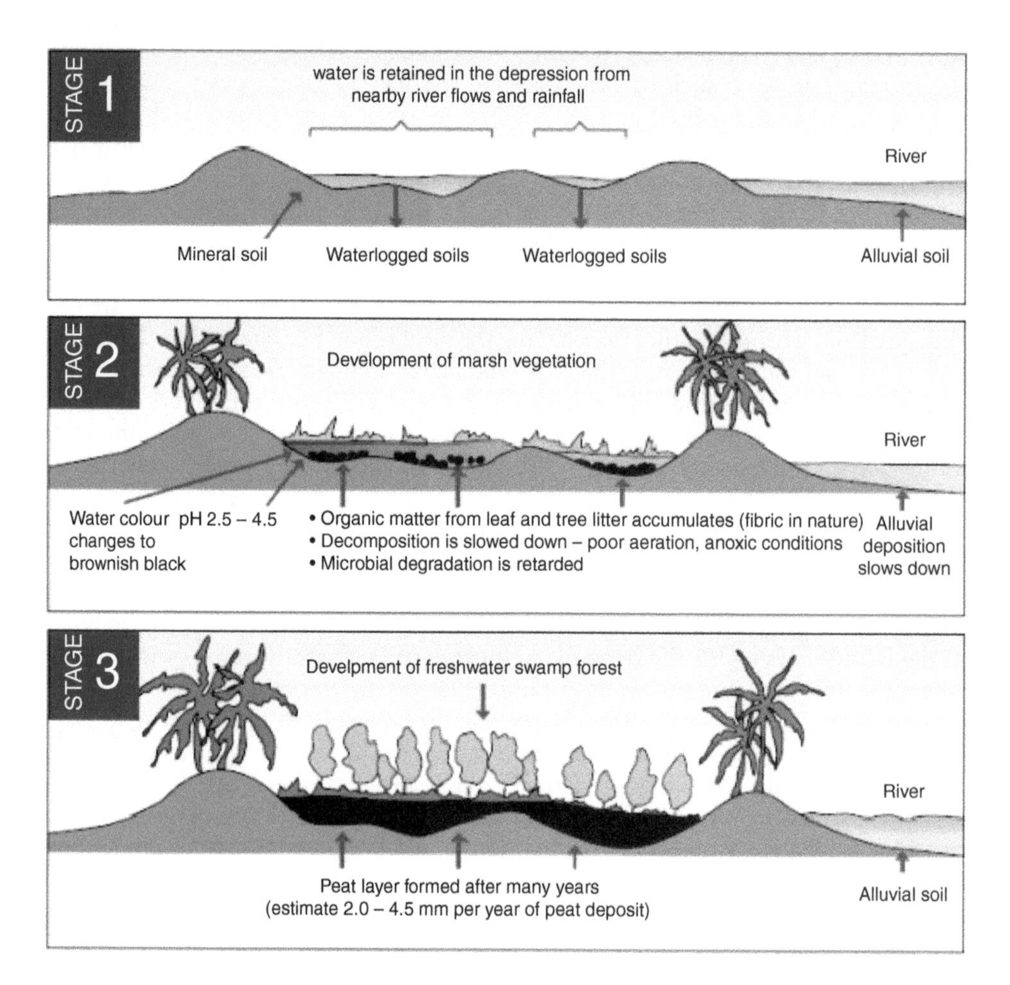

Figure 2.10 Peat swamps formation (*after* Leete, 2006).

however, is usually flat. The internal stratification is typically threefold, with a fine-grained hemic/sapric layer overlying a thick zone of fine- to medium-grained woody hemic-fibric, over-fine grained hemic, peat. The base of the peat dome is typically dark gray clay and sand with a thin layer of clayey peat or peaty clay.

Lam (1989) postulates the possible events leading to the development of peat deposits as a result of sea level changes. The last global glaciations resulted in rapid denudation and deep incision of the parent rock formations. After the last maximum glaciations (some 20,000 years BP), sea level rose rapidly and reached a maximum level 5,500 years BP. This resulted in the transportation and deposition of a large amount of sediment, which formed deltas and flood plains. Peat swamps were initiated in the depressions and basins between isolated hills and levees, and in the deltas. During the initial stage, plants developed in mineral soils. The areas were still under influence of rivers, with an influx of clastic (mineral) sediments during flood. The

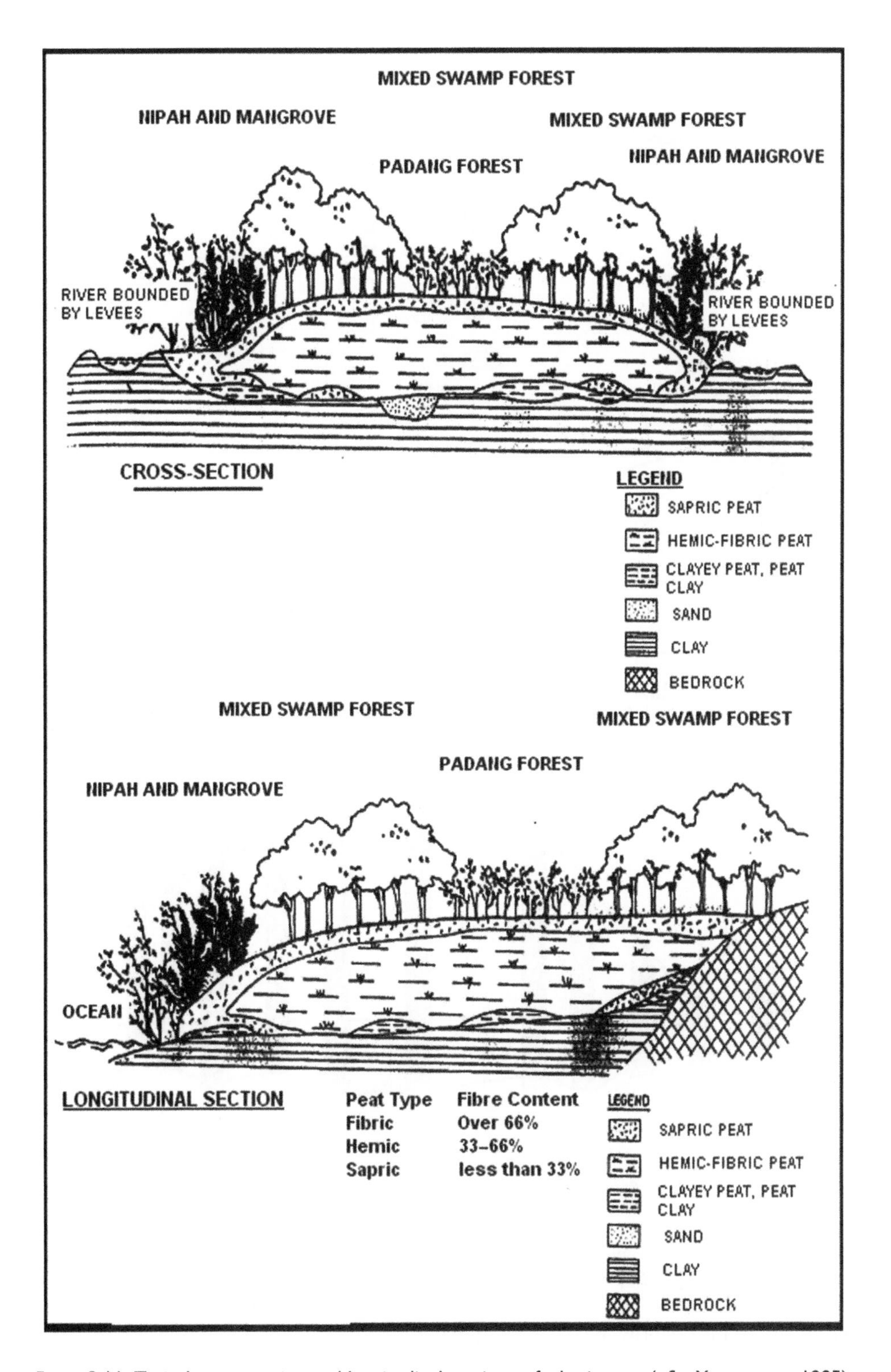

Figure 2.11 Typical cross-section and longitudinal sections of a basin peat (*after* Yogeswaran, 1995).

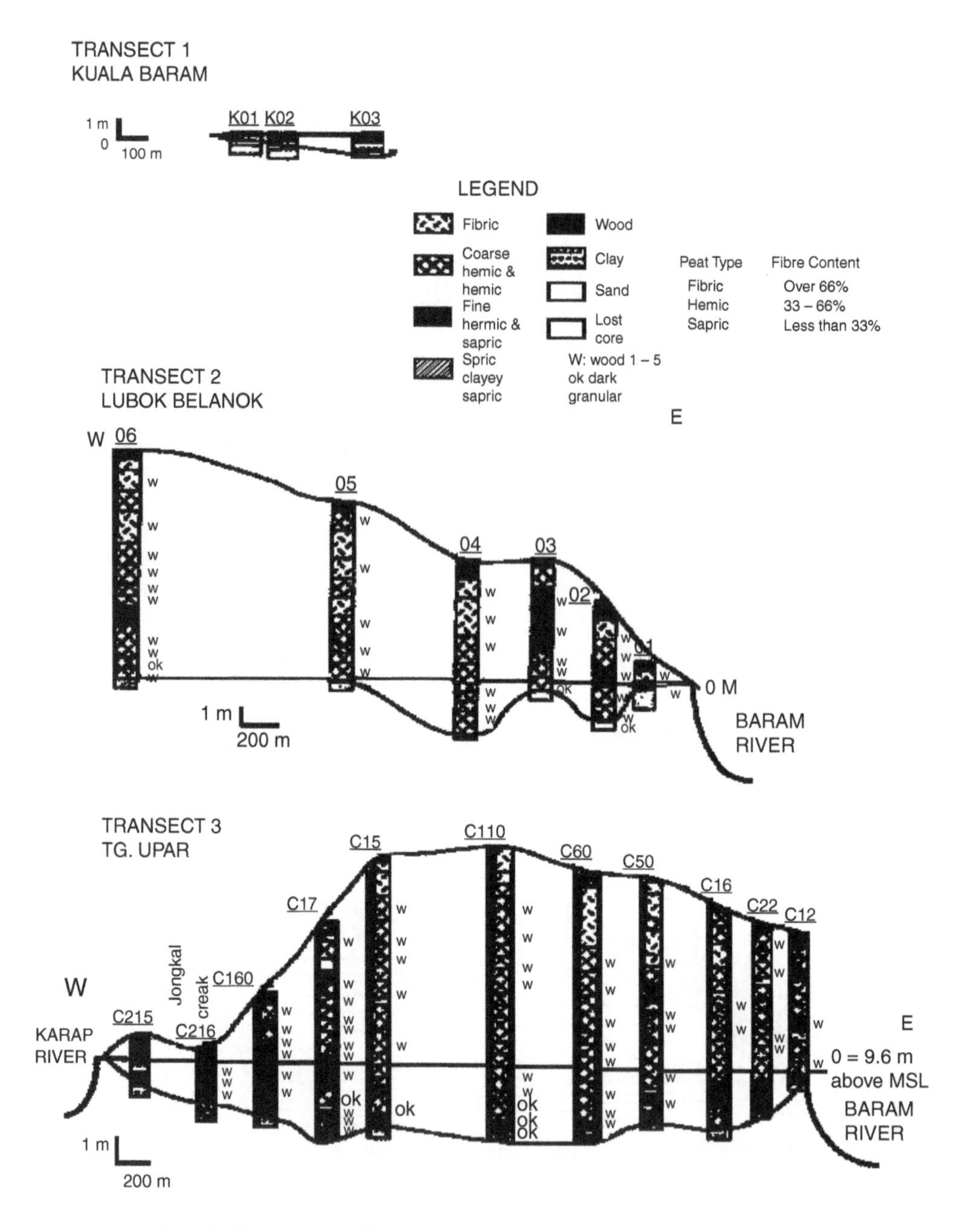

Figure 2.12 Vertical profile of a basin peat dome (*after* Esterle *et al.*, 1991).

accumulation of clastic sediments and plant remains resulted in the formation of clayey peat (topogenous peat). As plant remains accumulated, the ground surface levels were elevated. This led to the formation of peat, which was free of (or low in) clastic sediments (ombrogenous peat) and highly acidic.

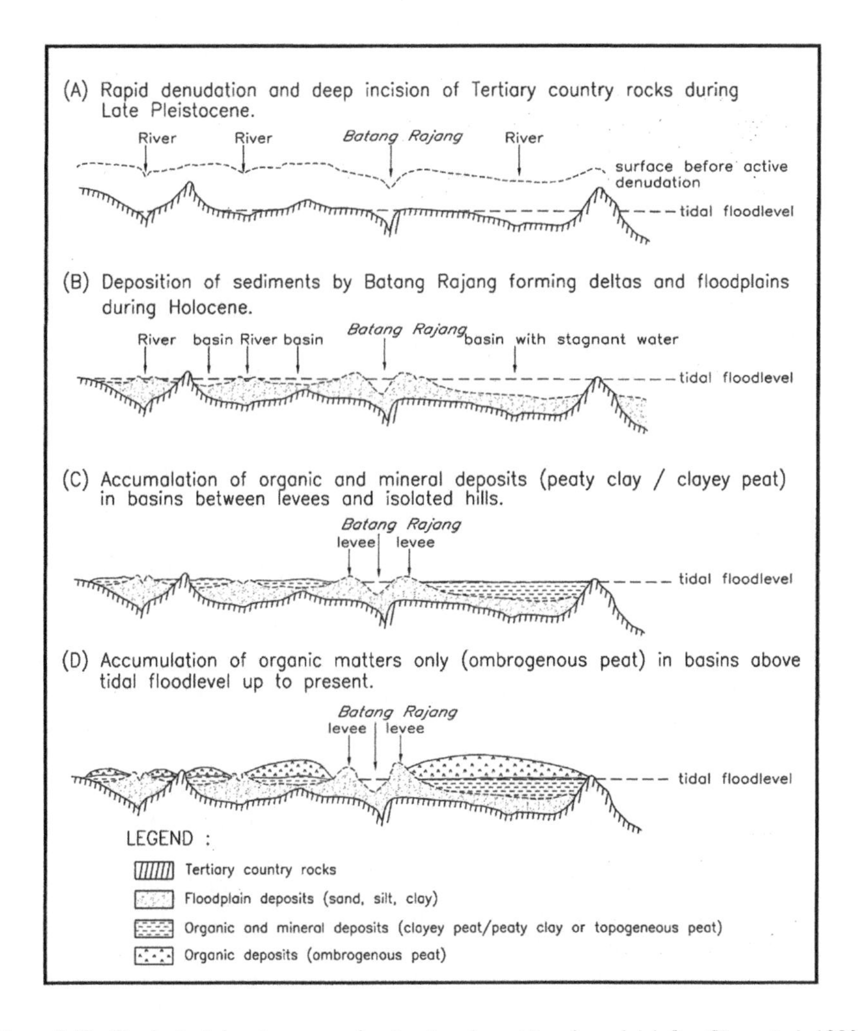

Figure 2.13 Geological development of a riverine depositional model (*after* Chen *et al.*, 1989).

The peat-forming vegetation consisted mainly of large trees, resulting in a high lignin content, which according to Anderson (1983) is twice that of bog peat. Figure 2.13 illustrates the geological development of a riverine depositional model leading to the deposition of the basin peat.

2.5 SITE INVESTIGATIONS AND SAMPLING OF PEAT

Before embarking on a project and locating a structure or embankment, it is important to have an overview of the distribution of soils, and in particular the location of peat land. Use of archival materials, earlier investigations, aerial photos and maps is important in delineating areas of rock outcrops, coarse or fine-grained soils and peat land. Peat land can be identified on aerial photos if it is not covered by mineral soils. It typically has a flat topography (except for raised bogs) that coincides with low areas, plants requiring a great amount of water, and wetlands, coastal lowlands and isolated river marshes. Mapping peat land is particularly important in the selection of highway routes.

A soil investigation aims at delineating not only the aerial distribution but also the subsurface stratigraphy. In peat areas, determination of the depth of soft organic substrata is important. This is best accomplished by sounding. Several methods of sounding are available, from simple to more sophisticated. The practice of sounding varies regionally. In Scandinavian countries, the Light Dynamic Probing Method (LDP) and the Weight Sounding Test (WST) are two of the simpler sounding methods. A variety of dynamic probing methods are also used in Germany and the Netherlands, especially for sounding deposits containing hard or gravelly layers. In Japan, a single-tube cone penetrometer was developed for peat deposits in 1952 and a double-tube cone penetrometer in 1958 (Noto, 1991). However, currently the Dutch cone is the most common method of sounding. More sophisticated methods of sounding include the cone penetration test (CPT), which is widely used in Europe and Japan and to a limited extent in the USA, and the standard penetration test (SPT), widely used in the USA. In addition to sounding, CPT provides strength information and SPT soil samples.

Ground-penetrating radar can be used to determine the thickness of strata with highly contrasting electromagnetic properties (i.e. dielectric constant) from the surface down to a depth of 10 m. During investigation, the antenna is moved over the ground surface, while the reflection of the electromagnetic waves is recorded. The method is quick and effective in delineating bottom layers. This method is gaining greater interest and much development is under way. Down-hole radar provides a three-dimensional distribution of soil strata.

Sampling, i.e. extraction of soil samples, is a necessary complement to other investigation methods, both for the identification of soils and for laboratory investigation. For the case of peat, collection of an undisturbed organic soil core is certainly a more difficult task compared with soft mineral soils such clays and silts. Conventional thin-wall sampling tubes (typically 76.2 mm in diameter) used for taking 'undisturbed' samples of soft clays may not be suitable for sampling peat. Sample size is important with respect to both sampling disturbance and representative sample volume. Peat samplers 100 to 250 mm in diameter have been developed, including a block sampler (Landva *et al.*, 1983).

There are a number of different tools for taking soil samples in different types of soil. However, the quality of the samples will vary between the samplers and between the types of soil. When choosing sampling quality and tool, it is necessary to consider the subsequent laboratory investigation. For soil identification only, disturbed or remoulded samples can be used. If the deformation and strength characteristics of the soil are to be investigated in the laboratory, it is necessary to obtain an undisturbed sample. In 1981, the ISSMFE subcommittee on soil sampling presented an international manual on soil sampling of soft cohesive soils that describes the definition of undisturbed sample as 'undisturbed samples, the soil retains the same fabric, type and proportion of constituents and physical and mechanical properties as in the field'.

The sample collection process must fulfil numerous regional requirements as well. These requirements are determined by a variety of factors that affect how samples should be collected for an individual sampling event. These factors include:

1. The type of samples being collected (organic/inorganic, aqueous or soil/sediment)
2. How the samples will be analyzed
3. Acceptance or performance criteria

The processes by which a sample can be disturbed during sampling and laboratory testing are (Hvorslev, 1949):

1. Change in stress conditions
2. Change in water content and void ratio
3. Disturbance of the soil structure
4. Mixing and segregation of the soil constituents

This disturbance can be related to sampler design, sampling, handling and storage. Many authors have discussed the factors that influence sample quality, in particular Hvorslev (1949) and Kallstenius (1963).

Peat often contains more than 90% water but has fibrous/spongy layers that are difficult to cut without causing compression, particularly if using a tube sampler. This has to be relatively wide, sharp, possibly toothed and inserted gradually, preferably by rotation or rapid vibration. There are two primary potential problems: improper sample collection and selection of the sampler for specific soil conditions. This causes disturbance of the matrix, resulting in compaction of the sample or inadequate homogenization of the samples where required, resulting in variable, non-representative results.

Soil samples may be collected using a variety of methods and equipment depending on the depth of the desired sample, the type of sample required and the soil type. Near-surface soils may be easily sampled, while sampling at greater depths may be performed using a hand auger, continuous flight auger, a trier or a split-spoon.

2.5.1 Disturbed but representative sampling

Samples that are disturbed in a geotechnical sense (changed density and mechanical properties) but representative of the type and proportion of the constituents and water content can be obtained at shallow depths by manually operated samplers such as a Hiller borer, Davis sampler, Finnish piston-type sampler or Scottish Macaulay sampler (*Muskeg Engineering Handbook*, 1969; Peat Testing Manual, 1979). Macaulay sampler has a vane or cover. It is inserted into position by pushing with the vane closed. Rotating the sampler opens the vane and the edge of the vane removes a one-half cylinder of material that is relatively undisturbed by rotation. It is evaluated to be the best available sampler for one-person operation (Peat Testing Manual, 1979). Two types of shallow sampler are reportedly used in Japan: a sampler with a cover (similar to a Macaulay sampler) and a piston sampler (Noto, 1991). Split-spoon samplers can obtain deeper disturbed but representative samples as part of the Standard Penetration Test (SPT). Screw augers also provide disturbed samples where a drill rig can be mobilized.

2.5.2 Undisturbed sampling

As mentioned above, it is virtually impossible to obtain undisturbed samples of any type of soil, including peat. Both physical intrusions of the sampler and the removal of *in situ* stresses cause disturbance. However, using certain sampling techniques, disturbance can be minimized. It can be said that there is a reasonably well-established understanding of the causes of disturbance during the sampling, transport and handling of inorganic clays and the corresponding accepted practices of sampling such soils. There are additional factors that need to be considered in sampling peat. These include compression while forcing the sampler into the ground, tensile resistance of fibres near the sampler edge during extraction of the sampler, and drainage and internal redistribution of water.

The thin-walled piston sampler with a fixed piston is normally the most suitable tool for non-fibrous soils. A number of piston samplers with different sample diameters are available, c.f. ISSMFE Subcommittee on Soil Sampling (1981).

The *piston sampler* consists mainly of sampler head, piston, piston rod, piston extension rod or wire, sampling tube or liner and thrust equipment or machine. During sampling, the closed sampler is pushed into the soil to a level just above the where the sample is cut out (Figure 2.14). After a while, when part of the generated excess pore pressure has disappeared and the sampled soil adheres to the inside of the walls, the sampler can be extracted. To obtain high-quality samples it is, among other things, very important to have a very sharp edge on the sampler and to have the piston carefully fixed to the ground or thrust machine. It is also important to have the sample cut out steadily and slowly (Kallstenius, 1963). Samplers with small inner clearances and small area ratios should be used. One difficulty with the small-diameter sampler is taking good samples in fibrous peat. Because of the force applied to cut off the fibres, the sample may be compressed at sampling.

A *peat sampler*, Ø100 mm, has been developed in Sweden. This sampler consists of a sharp wave-toothed edge mounted on a plastic tube with a driving head at the upper end (Figure 2.15). Samples are taken from the ground surfaces or the bottom of pre-bored holes. After extraction of the sampler, the cutting edge and driving head are removed and the sample in the plastic tube is sealed. Laboratory tests show that samples of fibrous soils taken with this peat sampler are of better quality than samples taken with a small-diameter piston sampler. Practical experience has also shown good correlation between laboratory test data from this kind of sample and measured field behaviour under embankment on fibrous peat.

The NGI 54 mm fixed piston sampler (composite version) was developed and designed by NGI and it is the most common sampler used in Norway. It is a composite piston sampler using plastic inner cylinders to prevent corrosion and avoid practical difficulties with the production and use of steel cylinders. Pre-augering through the dry crust is usually done before starting the sampling. Beneath the dry crust the displacement method is used, wherein the sampler (with the piston in front of the sample tube) is pushed down to the desired sampling depth without preboring. During sampling

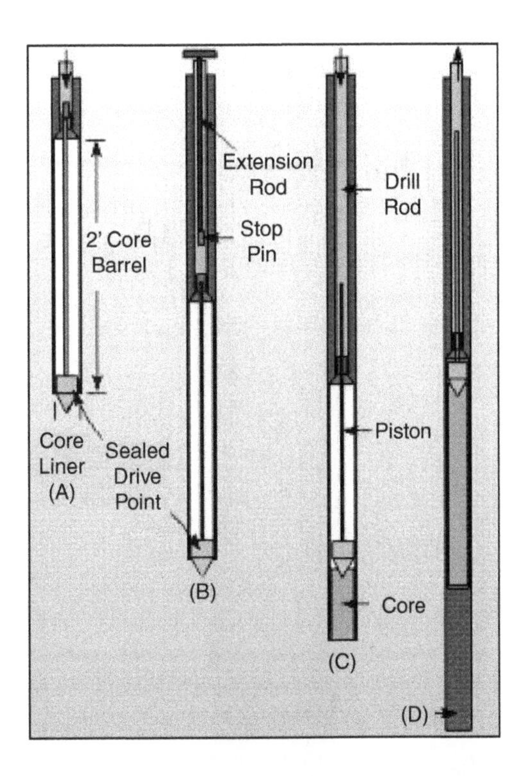

Figure 2.14 Piston sampler: (A) piston sampler entering the hole with a sealed piston drive point locked in place; (B) pin is then released from the surface with an extension rod, releasing the drive point; (C) sampler is further driven into the soil; (D) discrete core sample is collected (*Source*: http://www.esnnw.com/discrete.html).

the inner rods and piston are fixed in a locked position and the outer rods are pushed down at a constant rate. After withdrawal of the sampler, the sample is sealed at the top by not removing the piston when the cylinder is disconnected from the sampler (Andersen, 1981).

The NGI 95 mm piston sampler was developed and designed by NGI for use in both soft and stiff clays. A tower is used during sampling in order to minimize the number of connections and disconnections. The sampling tube can be connected and disconnected to the sampler when it is hanging from the tower. The sampler tube is made of cold-drawn mild steel tubes, with a polished inside surface. As for the 54 mm piston sampler, the displacement method is used for powering the sampler down to the sampling depth. During penetration the piston is kept in its lower position by locking the inner rods and pushing the outer rods at a constant rate. To prevent problems with vacuum at the bottom of the sampling tube during withdrawal, in some cases a thin plastic tube is fixed to the outer wall of the sampling tube, bringing water from the surface down to the cutting edge (Andresen, 1981).

The Japanese Committee of Soil Sampling has proposed the Japanese 75 mm fixed piston sampler as the standard in Japan. The boring rod is connected to the upper end of the sampler head and the sampling tube is connected to the lower end. A locking

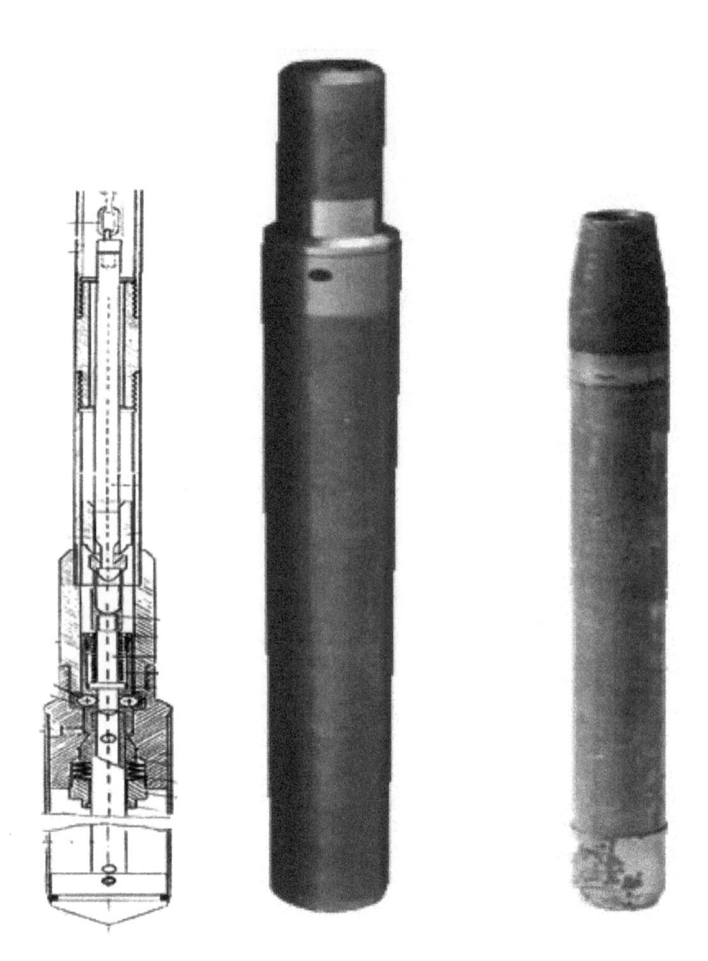

Figure 2.15 Cross-section of piston sampler (*after* Huat, 2004).

mechanism in the head allows only the upward movement of the piston rod, thus preventing loss of the sample when the sampler is withdrawn from a borehole. A ball cone clamp is typically used as a locking mechanism. Before each sample is taken, the standard Japanese method of pre-augering down to the sampling depth is used. Special care is exercised to make sure that the bottom of the borehole is clean and straight. When sampling, the boring rod and the piston extension rod are connected while the sampler is lowered smoothly to the bottom of the borehole. Details of the sampler are presented in the article by Tanaka *et al.* (1996).

Seaby (2001) has presented a sampler designed not for geotechnical investigation but for forestry. Despite this, it can be used to take undisturbed samples from medium depths to obtain peat cores more than 1 m long without compacting them. The sampler is described as comprising two halves inserted separately (Figure 2.16). A length of 3.5 mm thick PVC pipe, 80 or 110 mm in diameter, is halved and the tip of each half is chamfered and provided with a pointed, sharp metal blade attached to inside. On

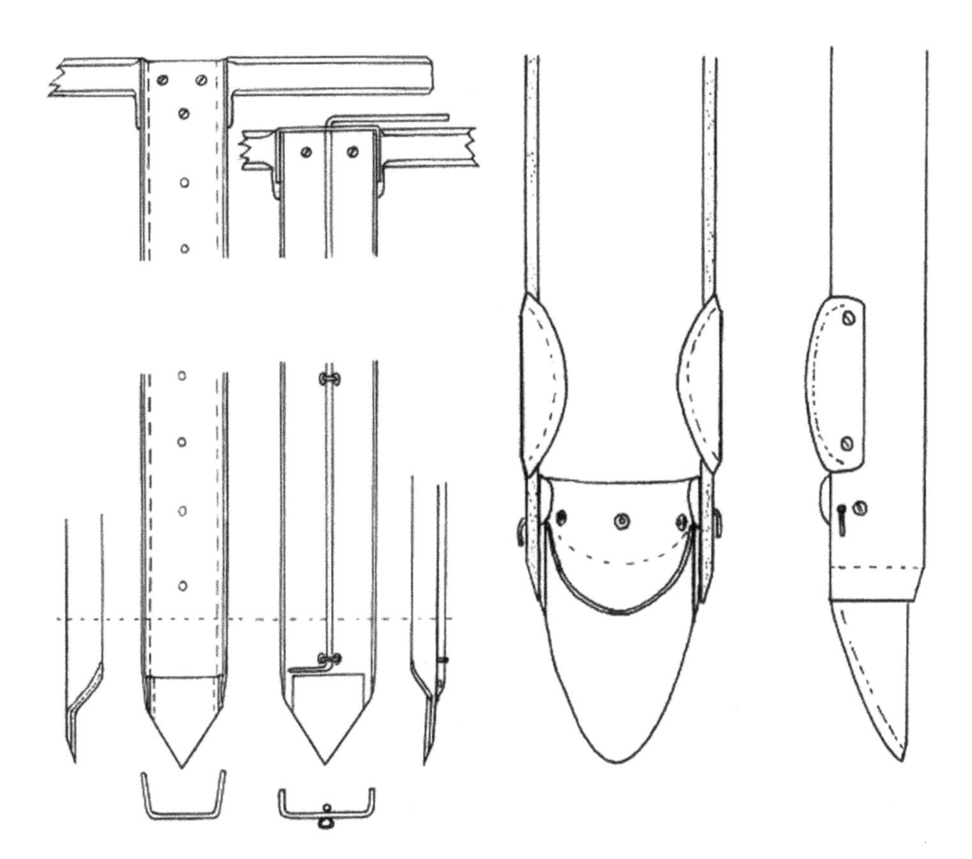

Figure 2.16 Plan and side view showing the end of the second half to be driven; the main blade, attached to the inside circumference of the plastic tube, has two flanges that internally slightly overlap the first half to be driven (*after* Seaby, 2001).

this blade is a loop of spring steel wire, which, due to its hinging through holes in the outer edges of the blade, is pushed against its inside circumference during insertion; during extraction, a slight downward movement, relative to the sampler, swings it out to help grip the core firmly. To align the two halves during insertion the second half has wider cutting blades, one attached to each side near the tip. These closely overlap the first half externally, acting as a guide. Both halves have cross handles near the top, but to aid extraction and reduce the risk of back strain, the halves are bolted together and a series of holes along one length allows a lever with a pointed tip, acting over a fulcrum, to ease the sampler out.

Deformation of soil during sampling causes disturbance of the soil sample. The volume displacement can be minimized by adjusting the area ratio (the ratio of the external to the internal diameter of the sampler). An optimal combination of area ratio and angle of cutting edges is essential for better quality sampling. A good sampler for low- to high-fibrous peat should ideally have good cutting edges and excellent driving technique to collect high-quality samples with different dimensions while keeping the sampling cost low (Al-Raziqi *et al.*, 2003).

For shallow samples, block samplers can be obtained. Typically a pit is excavated and blocks of peat are removed from the pit wall. Large block samples (250 mm square) can be obtained from below the ground and groundwater surface (down to a depth of 7 m) using a block sampler for peat described by Landva *et al.* (1983). It provides somewhat disturbed samples, but retains their general structure. There are other large-size down-hole block samplers, such as the Sherbrooke sampler (250 mm in diameter) and Laval sampler (200 mm in diameter), which have been developed for sampling clays but can also be used for organic soils and probably for peat. Hebib (2001) describes a novel method that was employed to obtain 1 m^3 block samples of Ballydermot peat, Co. Offaly, Ireland. Samples were taken from a vertical face, 2.5 m high, using a custom-made block sampler attached to the lifting arm of a tracked excavator. The sampler was fabricated from a steel box that had three faces removed. Each of the remaining exposed faces had sharpened edges so as to minimize disturbance during sampling. The retrieved samples were wrapped in polythene to prevent loss of moisture and placed in wooden boxes for transportation to the laboratory.

A sampler for collecting undisturbed peat was developed by Duraisamy *et al.* (2009). It consisted of a hollow cylinder of 150 mm internal diameter connected to a 100 cm long stem. The top of the stem was provided with a handle 60 cm wide for pushing the sampler into the peat. The sampler was formed from a hollow cylindrical body with a cover plate at the top, and a thin tube with a valve was provided. The valve was designed to be left open during sampling to release both air and water pressure. Meanwhile, the lower part of the cylindrical tube was kept sharpened to cut roots as the auger slowly rotated into the peat ground during sampling. Once the augur reached the desired depth of sampling, the valve was closed prior to withdrawal of the tube with the peat sample enclosed, thus providing a vacuum effect to help hold the sample in place. Soon after the sampler was withdrawn, the sample was sealed in the cylindrical tube with paraffin wax (Figure 2.17).

Once in the laboratory, the top cover of the cylindrical tube is opened to extract the sample. The auger enables the extraction of peat core samples 150 mm in diameter and 230 mm long. The top and bottom of the sample were trimmed. As fibrous soil, such as peat, is easily disturbed, the trimming process was done very carefully and quickly to minimise any change to the soil sample's water content. All tests reported here were performed inside the lab maintained at a constant temperature of 20 ± 2°C (Kazemian and Huat, 2009; Duraisamy *et al.*, 2009).

2.5.3 In situ tests

There are no special tools available for determining the *in situ* properties of peat. Therefore select methods that have been developed for use in soft clays are used either directly or in a somewhat modified manner to test peat and organic soils. Because of the nature of these soils, certain methods have gained prominence over others. The methods of interpreting the *in situ* test results as applied to peats and organic soils are limited in the literature, and direct use of methods primarily developed on the basis of mineral soil experience should be conducted with great caution. Because of the greater variability and fabric effects in organic deposits, larger numbers of *in situ* tests and complementary sampling and laboratory testing are required to arrive at design properties. Some of the more common approaches to *in situ* testing in such deposits

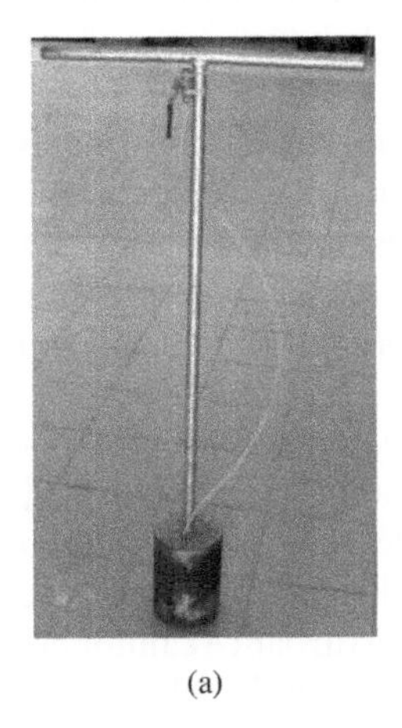

(a)

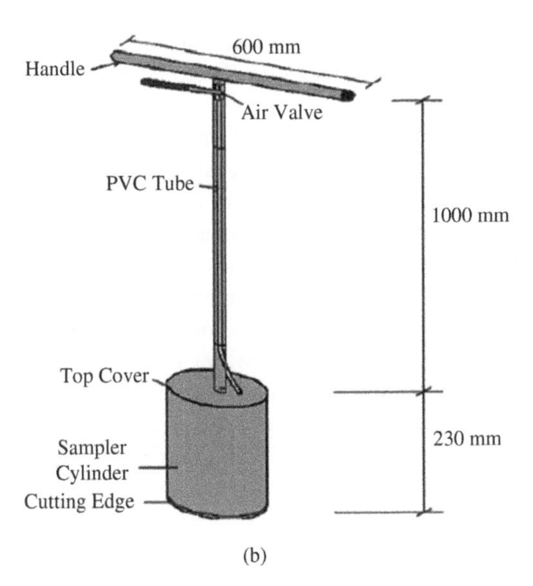

(b)

Figure 2.17 Sampler fabricated to collect undisturbed peat samples. (a) Actual photograph (*after* Duraisamy *et al.*, 2009); (b) line diagram (*after* Kazemian *et al.*, 2012b).

include the following tests, and their use in peat and organic soils was reviewed by Edil (2001).

(a) Vane shear test

The miniature vane shear test may be used to obtain estimates of the undrained shear strength of fine-grained soils. The test provides a rapid determination of the shear strength on undisturbed, remoulded or reconstituted soils. Field vane shear tests provide an indication of the *in situ* undrained shear strength of fine-grained clays and silts or other fine geomaterials, such as mine tailings, organic muck and substances where undrained strength determination is required. The test is applicable to soils with undrained strengths of less than 200 kPa (2 tsf). Very sensitive soils can be remoulded during vane insertion (http://astm.org).

(b) Cone penetration test

This test method determines the end bearing and side friction, the components of penetration resistance that are developed during steady slow penetration of a pointed rod into soil. This test method is sometimes referred to as the Dutch Cone Test or Cone Penetration Test and is often abbreviated as CPT. This test method tests the soil in place and does not obtain soil samples (http://astm.org).

(c) Pressuremeter test

This test method provides a stress-strain response of the soil *in situ*. A pressuremeter modulus and a limit pressure are obtained for use in geotechnical analysis and foundation design. The results of this test method depend on the degree of disturbance during drilling of the borehole and insertion of the pressuremeter probe. Since disturbance cannot be completely eliminated, the interpretation of the test results should include consideration of conditions during drilling. This disturbance is particularly significant in very soft clays and very loose sands. Disturbance may not be eliminated completely, but should be minimized for the prebored pressuremeter design rules to be applicable (http://astm.org).

(d) Dilatometer test

Soundings performed using this test method provide a detailed record of dilatometer results, which are useful for evaluating site stratigraphy, homogeneity, depth to firm layers, voids or cavities, and other discontinuities. The penetration resistance and subsequent membrane expansion are used for soil classification and correlation with the engineering properties of soils. When properly performed at suitable sites, the test provides a rapid means of characterizing subsurface conditions. The DMT test provides measurements of penetration resistance, lateral stress, deformation modulus and pore water pressure (in sands). However, the *in situ* soil properties are affected by the penetration of the blade. Therefore published correlations are used to estimate soil properties for the design and construction of earthworks and foundations for structures, and to predict the behaviour of soils subjected to static or dynamic loads (http://astm.org).

(e) Plate load test and screw plate load tests

The plate load test is used to decide the safe bearing capacity. While this procedure may be adequate for light or less important structures under normal conditions, relevant laboratory tests or field tests are essential in the case of unusual soil types and for all heavy and important structures. The load test included in the standard is also used to find the modulus subgrade reaction, which is useful in the design of raft foundations and in pavements. The plate load test, although useful for obtaining the necessary information about the soil with particular reference to the design of foundations has some limitations. The test results reflect only the character of the soil located within a depth of less than twice the width of the bearing plate. Since the foundations are generally larger than the test plates, the settlement and shear resistance will depend on the properties of a much thicker stratum. Moreover, this method does not give the ultimate settlement, particularly in the case of cohesive soils. Thus the results of the test are likely to be misleading if the character of the soil changes at shallow depths, which is not uncommon.

The screw-plate load test permits a convenient method for measuring the *in situ* vertical compressibility, bearing capacity and dynamic stability of loose to medium dense cohesionless soils. It also provides a practical means for avoiding some of the significant technical difficulties associated with the use of rigid plate bearing load tests for these purposes (http://astm.org).

A review of the literature on *in situ* testing in peat indicates that the state of the art and practice are not as developed as for inorganic soils. It is important to recognize the differences between various organic deposits. Organic soils (with organic content of 25% or less) can be treated similarly to inorganic clays, and many of the same *in situ* testing tools, such as vane shear, cone penetrometer, pressuremeter, dilatometer and regular or screw plate load tests can be applied. While there may be some questions regarding the interpretation of the test data in determining mechanical parameters for design, it appears that the standard approaches can be followed with a greater degree of cross-calibration of the various tests and care for greater material variability and compressibility. At the other extreme, there are serious questions regarding the applicability of the conventional *in situ* tests to fibrous, high-organic content surficial peat (i.e. not buried and compressed). The presence of fibres, inherent anisotropy, tendency for high compressibility and rapid drainage, and the low and highly variable strength of these materials make using conventional field tests and interpretation of mechanical parameters unviable. The use of large test tools (vanes, cones etc.), more sensitive measuring devices, and more rapid loading rates to minimize compression may improve the prospects. However, the irrelevance of the various modes of failure induced in the field as well as laboratory tests relative to fibre interaction, anisotropy and compressibility results in unusual values for mechanical parameters and inconsistencies between various tests. For instance, the shearing surface is vertical in vane shear, ill-defined in cone penetration, inclined in triaxial compression and horizontal in direct, simple or ring shear tests. This situation has led some investigators, such as Landva, to recommend test fills as opposed to solely relying on laboratory or field tests when designing embankments on such deposits. Amorphous peat is somewhat intermediate between organic soils and fibrous peat. There are reports of successful use of *in situ* tests. Such materials must be handled on a case-by-case basis. Interpretation of *in situ* test results requires corrections usually calibrated based on local experience with organic deposits and laboratory strength tests. The combined use of extensive sampling for the definition of site variability, *in situ* tests and laboratory mechanical property tests, and where possible the use of test fills, provides a reasonable approach to dealing with these difficult organic deposits (Edil, 2001).

Chapter 3

Engineering properties of peat and organic soils

3.1 INTRODUCTION

As mentioned in Chapter 2, degree of humification, surface charge characteristics, resistivity and other index parameters, such as water content, liquid limits, specific gravity, unit weights, zeta potential and pH, are also useful parameters for peat and organic soils. We will look at how these parameters are useful for describing peat soil, how they are assessed and measured, and their relation with each other, as well as with other engineering parameters such as void ratios, deformation parameters and shear strength that are dealt with in later chapters.

3.2 PHASES OF PEAT

In conventional soil mechanics, soils are considered as particulate materials, as opposed to rocks, and can simultaneously contain three phases: solid, liquid and gas (Figure 3.1). The liquid and gas phases are contained in the voids or pores between the solid particles. From these phases, the weight-volume relationships of soils are derived. These are the moisture content, degree of saturation, unit weight, density, void ratio and specific gravity of soils.

For the case of peat and organic soil, the solid phase consists of two components: organic matter and inorganic earth minerals. The relative proportion of

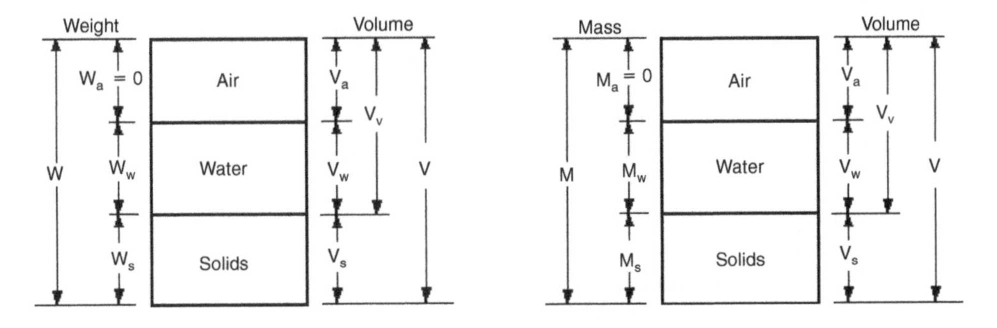

Figure 3.1 Soil phase diagram.

these components and their specific nature determine the physical and geotechnical properties of these soils (Edil, 2003).

The mineral component is similar to that of inorganic soils consisting mainly of clay minerals, but non-clay minerals are also encountered. Clay minerals consist of small particles, typically less than 0.002 mm in diameter. The mineral constituent is generally incombustible and ash-forming, while the organic matter is generally combustible carbonaceous matter. Soil organic matter has been defined (Russell, 1952) as '...a whole series of products which range from undecayed plant and animal tissues through ephemeral products of decomposition to fairly stable amorphous brown to black material bearing no trace of the anatomical structure of the material from which it was derived; and it is this latter material that is normally defined as the soil humus'. Additionally, soil organic matter also contains products of microbial synthesis. In summary, soil organic matter includes: (1) fresh plant and animal residues (decomposable), (2) humus (resistant) and (3) inert forms of nearly elemental carbon (charcoal, coal or graphite). Carbon is the chief element of soil organic matter that is readily measured quantitatively by combustion (C is determined as CO_2 emitted). The organic carbon content may be used to estimate total organic matter by multiplying the figure for organic carbon by a factor of 1.724 or somewhat higher figures up to 2.

Alternatively, ignition of the soil at high temperature (450 °C to 550 °C) to achieve destruction of organic matter can be used to determine organic content directly. The loss-on-ignition method gives quantitative oxidation of organic matter, but because it may decompose some of the inorganic constituents of the soil it may give a number in excess of the actual organic content (Edil, 2003). Procedures for obtaining the organic content of peat by the loss on ignition method will be described later in this chapter.

The organic matter, when extracted, can be fractionated into components, primarily those characteristic of plant tissues and those based on humus. The first group (nonhumic matter) includes fats, waxes, oils, resins, water-soluble polysaccharides, hemicellulose, cellulose and protein. The second group includes the humus fraction consisting of basically humic and fulvic acids and humin and exists in both solid and liquid phases (Huttunen *et al.*, 1996).

3.3 BOTANICAL ORIGIN AND FIBRE CONTENT

Botanical terms are used in describing peat especially for horticultural purposes, but such terms also have relevance to engineering because of the texture they imply. *Sphagnum peat* designates a material with predominantly sphagnum moss (>60–75%). *Sedge peat* has one or more species of sedge (plants that are grass-like in appearance). *Woody peat* is indicates dominance of woody pieces. Other terms include *taxodium peat, reed peat* and combinations of botanical terms. Of course, for highly decomposed peat with low fibre content (<33%), it would be difficult to identify the botanical origin.

The structure of peat and organic soils is an arrangement of primary and secondary elements that make up the soil. On a qualitative basis, however, organic ground can be described in terms of its fibre content and type. Fibre content is determined typically from dry weight of fibres retained on a #100 sieve (>0.15 mm opening size) as a percentage of oven-dried mass (ASTM Standard D 1997). Fibres may be fine (woody

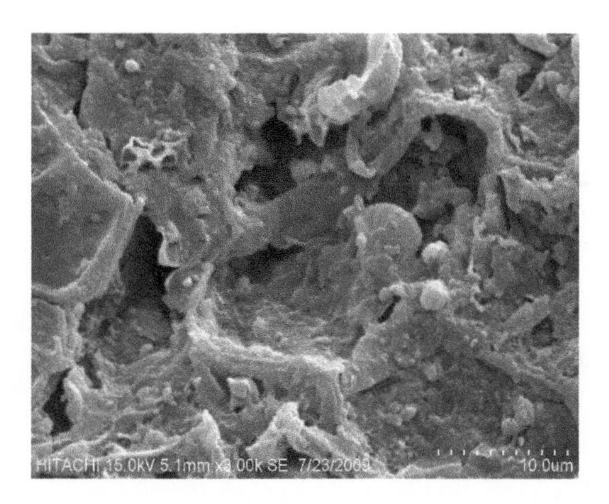

Figure 3.2 SEM images of fibrous peat (H2–H3) from Malaysia (*after* Kazemian *et al.*, 2011).

or non-woody) or coarse (woody). Organic matter that does not have an identifiable fibre shape is finer in size than 0.15 mm and designated as amorphous-granular matter. It is also referred as *peat humus*.

3.4 FABRIC OR STRUCTURE

Fabric or microstructure refers to the morphology and spatial arrangement of the constituent soil elements. Macro fabric or macrostructure refers to those features visible to the naked eye, whereas micro fabric or microstructure involves much smaller features at the particle or fibre level. Radforth, in *Muskeg Engineering Handbook* (1969) provides a classification of peat structure based on two structural elements: fibres and granules. In this system, peat structure is classified into three predominant characteristics: (1) amorphous granular (dominated by highly disintegrated formless botanical tissues), (2) fine fibrous (dominated by an open framework of highly preserved fossilized plant remnants), and (3) coarse fibrous with 17 categories further subdividing these characteristics. The fibres of peat consist of cellular structures giving rise to a two-level structure involving macro and micro pores, i.e. those between and within fibres, respectively (Dhowian and Edil, 1980).

It is the structure of peat, in its various aspects, that affects the engineering behaviour (*Muskeg Engineering Handbook*, 1969; Dhowian and Edil, 1980). There has been virtually no research to correlate different structural types and their physical and engineering properties. However, MacFarlane and Radford (*Muskeg Engineering Handbook*, 1969) suggest that the engineering behaviour of peat can be grouped broadly into fibrous and amorphous granular. This suggestion is supported by other investigators (Dhowian and Edil, 1980). The two-level structure described by Adams (1965) may be used to describe the unusual compression behaviour of fibrous peat.

Scanning Electron Microscopy (SEM) images can be used for a visual appreciation of the soil microstructure. Figure 3.2 shows typical SEM image of fibrous peat from Malaysia. Images such as those shown in Figure 3.2 indicate that the peat is extremely

Figure 3.3 SEM image of hemic peat from Malaysia (H6–H9) (*after* Kazemian *et al.*, 2011).

Figure 3.4 SEM image of sapric peat from Malaysia (H10) (*after* Kazemian *et al.*, 2011).

fibrous, with relatively large pore spaces at macroscopic level between stems and leaves and at microscopic level within the open and perforated plant structures.

Figure 3.3 shows the same fibrous peat but with a higher degree of humification. Figure 3.4 shows SEM image of an amorphous peat.

Unlike the SEM images of fibrous peat, the state of the organic matter for the above peat suggests that it is amorphous-granular. Evidence of plant-derived organic matter is more difficult to identify in this image, despite the relatively high organic content (80%). This may be at least partially attributed to the high degree of humification (H10) and to the fact that well humified organic matter is generally difficult to distinguish (Landva and Pheeney, 1980).

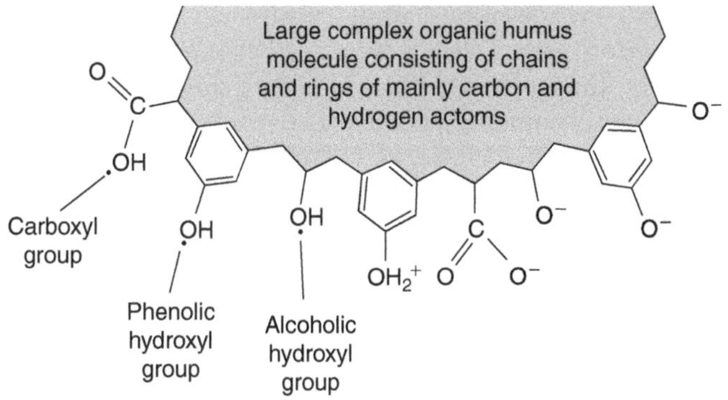

Figure 3.5 Organic humus structure ((*after* Horowitz, 1991).

3.5 SOIL ORGANIC COLLOIDS

Colloids are the most chemically active fractions of soils. They are very small: less than 2 μm in diameter. They are mineral (clays) or organic (humus). They can be crystalline (definite structure) or amorphous. Clay and humus are dynamic and very active in charge by comparison with sand and silt that are static. Colloids can impart chemical properties (the source of ions, source of electro-negativity and buffering capacity) and physical properties (the large surface area per unit of mass and the plasticity) to soils. The types of soil colloid are: (i) crystalline silicate clays, (ii) non-crystalline silicate clays, (iii) iron or aluminium oxide and (iv) organic material (humus).

Electrical charges are carried by the surface of soil colloids, and these surface charges are the main cause of a series of surface properties. The quantity of ions that are adsorbed on the surface of soil colloids can be determined by the quantity of surface charge. In addition, the surface charge properties of the soil can affect the migration of ions in a soil, dispersion, flocculation and swelling. Therefore the surface charge properties have a key role in soil structure. The sources of charges on colloids are: (i) permanent or constant charges due to isomorphous substitutions (montmorillonite, illite, zeolite etc.), and (ii) variable or pH-dependent charges due to broken edges and OH and COOH groups (kaolinite, humus and Al or Fe oxides) (Yu, 1997; Brady and Weil, 2007).

For soils with high organic content, humus is the most important source of variable charge. For soils with variable surface charge, oxides, hydrated oxides of iron and aluminium, and the edge surfaces of clay minerals of soil are responsible.

There is a large quantity of acid groups in humus. The origin of charge is the dissociation of the acid groups. A large quantity of charge is carried by humus, in the range of 200 to 500 cmol kg^{-1}. These charges come mainly from carboxyl groups. Hydroxyl groups, including phenolic hydroxyl, quinonic hydroxyl and enolic hydroxyl groups can also produce charge (Figure 3.5). The role of phenolic hydroxyl groups under alkaline conditions is considerable. In soil humus, carboxyl and hydroxyl groups account for around 50% and 30% of the total functional groups, respectively (Yu, 1976). Furthermore, at high alkaline conditions, amino groups can also produce negative charge on the soil surface (Horowitz, 1991; Stevenson, 1994; Yu, 1997). Humic substances are the main body of humus (Dai, 2004). There are three kinds of humus, including loosely, stably and tightly combined humus. The main components of organic matter in soil are humic acid, fulvic acid and humin. Humin is the main composition of tightly combined humus. Humic acid and fulvic acid are stably combined humus (Lu, 2000; Chen and Wang, 2006).

The high charge ranges (200 to 500 cmol kg^{-1}) carried by humus may play an important role in the surface charge of organic soil colloids (Schnitzer and Khan, 1989; Yu, 1997; Stevenson, 1994; Tan, 2008).

In an organo-mineral complex, the amount of surface charge is smaller than the sum of charges carried by the two components separately because of various bonding forces. The mechanisms of this phenomenon are not known. Charged organic colloids can join with positively charged iron and aluminium oxides or with positive sites at the edges of clay minerals. Organic matter as an important charge component can also affect the zero point of charge of the mineral soil (Yu, 1997; Stevenson, 1994; Majzik and Tombacz, 2007).

It is noteworthy that some humic nitrogen compounds can carry positive charges, which can change the behaviour of humus colloids (Fuchsman, 1986a).

The concentrations of electrolytes, types of electrolyte, valences of ions and pH are important factors that can influence surface charge. Among them, pH is the most important one (Alkan *et al.*, 2005; Li and Xu, 2008; Yukselen and Erzin, 2008). The dissociation of H^+ ions from the hydroxylated surface and the adsorption of H^+ ions by the hydroxylated surface are pH dependent. As a result of this condition, variable charge minerals, such as iron and aluminium oxides, carry a negative charge when the pH is higher than their zero point of charge.

3.6 HUMIFICATION OF PEAT

As described above, decomposition or humification involves the loss of organic matter either as gas or in solution, with the disappearance of the physical structure and changes in chemical state. The breakdown of plant remains is brought about by soil microflora, bacteria and fungi, which are responsible for aerobic decay. Therefore the end products of humification are carbon dioxide and water, the process being essentially biochemical oxidation. Immersion in water reduces the oxygen supply enormously, which in turn reduces aerobic microbial activity and encourages anaerobic decay, which is much less rapid. This results in the accumulation of partially decayed plant material as peat (Fuchsman, 1986b)

The influences on metabolic activity, apart from the supply of oxygen, are temperature, acidity and the availability of nitrogen. Normally, higher temperatures and pH values enhance the decomposition activity. Decomposition tends to be most active in neutral to slightly alkaline conditions. The more acid the peat, the better the plant remains are preserved. The acidity of the peat depends on the rock types in the area draining into the peat land, the types of plant growing there, the supply of oxygen and the concentration of humic acid. In temperate regions, bog peat (blanket and raised bogs) is generally acidic with pH values in the range 3 to 4. Fen peat, on the other hand, is generally neutral or slightly alkaline. Bog peat is generally more fibrous compared with fen peat. In the tropics, peat is generally acidic, with pH values in the range 3 to 4.5 (Mutalib *et al.*, 1991).

The loss of organic matter and the accompanying changes in chemical state result from the breakdown of cellulose within plant tissues, so that the detritus gradually becomes increasingly fine until all trace of fibrous structure disappear. The peat then has an amorphous granular appearance, the material consisting principally of organic acids, which have a sponge-like fabric. The degree of humification varies throughout the peat since some plants are more resistant than others, and certain parts of the plants are more resistant than others. The change undergone as a result of increasing humification is not uniform, since the fibres are reduced in size and strength in an irregular manner as the quantity of totally humified peat increases. Generally, the fresher the peat, the more fibrous material it contains, and as far as engineering is concerned, the more fibrous the peat, the higher are the shear strength, void ratio and water content.

Under normal field conditions, total degradation of the organic fraction under water is limited due to the volatile acid toxicity and nutrient imbalance. The introduction of nutrients by groundwater seepage may initiate or sustain decomposition over long periods. The process is finally complete when only humus (nondegradable residue) and microbial cells are left. Decomposition causes a decrease in solid volume, i.e. compression.

From a physical point of view, the humification processes cause some changes, including reduction of the total water content, increase in specific gravity, increase in compaction, decrease in the pore space, and changes in colour towards dark brown and black (Gunther, 1983).

As described above, the degree of decomposition or humification is usually assessed by means of the von Post scale. von Post (1922) proposed a classification system based on a number of factors, such as degree of humification, botanical composition, water content, content of fine and coarse fibres and woody remnants. There are 10 degrees of humification (H1 to H10) in the von Post system, which are determined based on the appearance of soil water that is extruded when the soil is squeezed in hand. For geotechnical purposes, these 10 degrees of humification are often reduced to three classes: fibric or fibrous (least decomposed), hemic or semi-fibrous (intermediate) and sapric or amorphous (most decomposed), respectively (Magnan, 1980; ASTM Standard D 5715).

Table 3.1 shows the von Post scale for assessing the degree of humification. To perform the test, a sample of peat is squeezed in the hand. The colour and form of fluid that is extruded between the fingers is observed together with the pressed residue remaining in the hand after squeezing, with reference to the 10 point scale mentioned above.

However, the von Post scale is adapted to pure peat containing little or no mineral matter. Its use in organic soils with more than 20–25% mineral mater is difficult. As a result, various coarser scales have been devised with 3–5 degrees of humification (see also Chapter 2).

Figure 3.6 illustrates an example of an internal stratification of a tropical basin peat dome based on the degree of humification.

Table 3.1 von Post degree of humification.

Symbol	*Description*
H1	Completely undecomposed peat which, when squeezed, releases almost clear water. Plant remains easily identifiable. No amorphous material present.
H2	Almost entirely undecomposed peat which, when squeezed, releases, clear or yellowish water. Plant remains still easily identifiable. No amorphous material present.
H3	Very slightly decomposed peat which, when squeezed, releases muddy brown water, but from which no peat passes between the fingers. Plant remains still identifiable, and no amorphous material present.
H4	Slightly decomposed peat which, when squeezed, releases very muddy dark water. No peat is passed between the fingers but the plant remains are slightly pasty and have lost some of their identifiable features.
H5	Moderately decomposed peat, which, when squeezed, releases very 'muddy' water with a very small amount of amorphous granular peat escaping between the fingers. The structure of the plant remains is quite indistinct, although it is possible to recognize certain features. The residue is very pasty.
H6	Moderately highly decomposed peat with a very indistinct plant structure. When squeezed, about one-third of the peat escapes between the fingers. The residue is very pasty, but shows the plant structure more distinctly than before squeezing.
H7	Highly decomposed peat. Contains a lot of amorphous material with very faintly recognizable plant structure. When squeezed, about half of the peat escapes between the fingers. The water, if any is released, is very dark and almost pasty.
H8	Very highly decomposed peat with a large quantity of amorphous material and very indistinct plant structure. When squeezed, about two-thirds of the peat escapes between the fingers. A small quantity of pasty water may be released. The plant material remaining in the hand consists of residues such as roots and fibres that resist decomposition.
H9	Practically fully decomposed peat in which there is hardly any recognizable plant structure. When squeezed it is fairly uniform paste.
H10	Completely decomposed peat with no discernible plant structure. When squeezed, all the wet peat escapes between the fingers.

3.7 OXIDATION

Lowering of the groundwater table for sustained periods of time allows the organic fraction come in contact with air, which leads to shrinkage and oxidation. Consequently, the fibre structure is destroyed due to accelerated decomposition and it becomes more amorphous-granular (Vonk, 1994). This activity may result in significant changes in engineering properties.

Pore water

Usually peat has an acidic reaction caused by the presence of carbon dioxide and humic acid resulting from decay. Peaty waters are practically free of salts and generally show pH values of 4 to 7 (Lea, 1956).

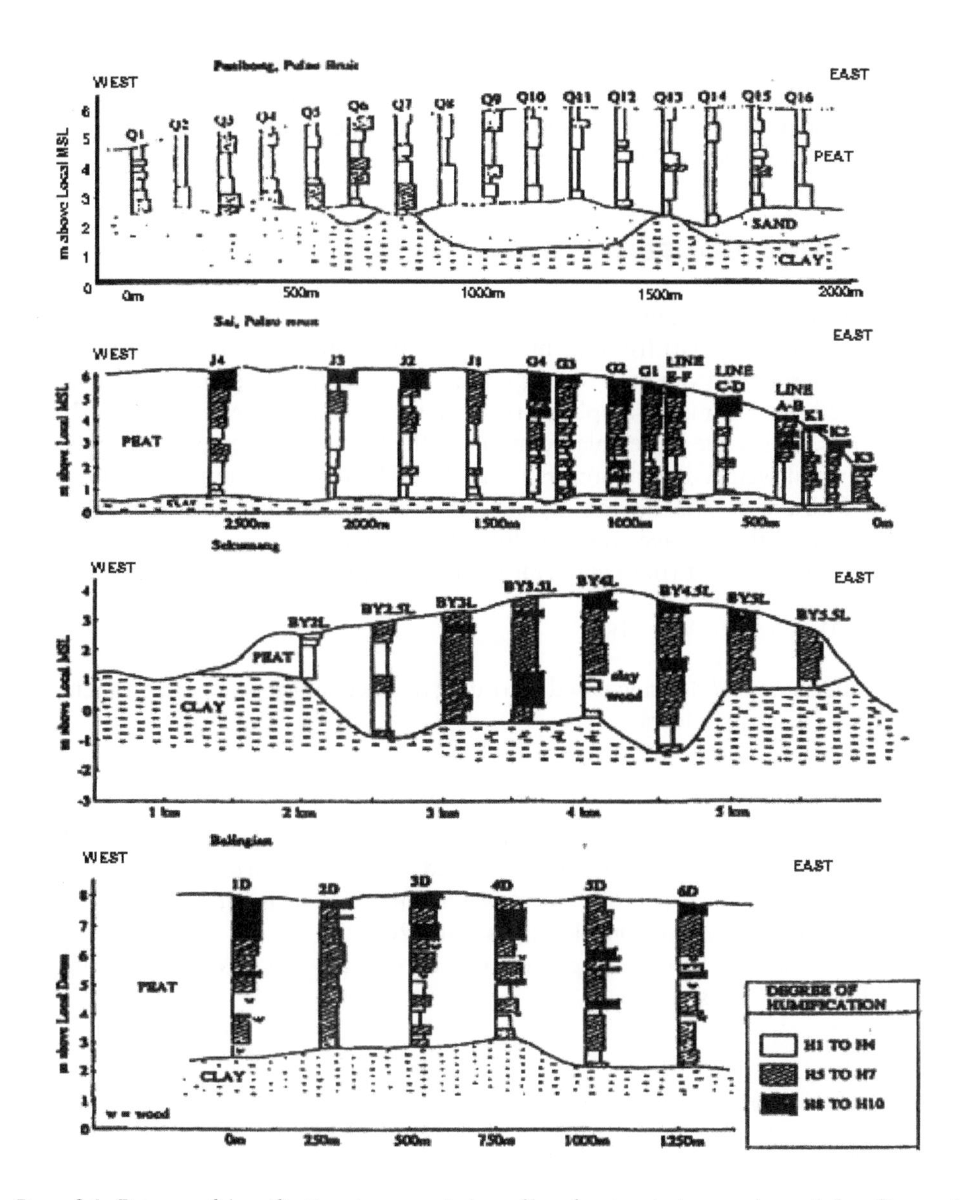

Figure 3.6 Degree of humification in a vertical profile of a tropical peat dome (*after* Ong and Yogeswaran, 1991).

Pore gas

The submerged organic component of peat is not entirely inert but undergoes very slow decomposition, accompanied by the production of marsh gas (methane) with lesser amounts of nitrogen and carbon dioxide (*Muskeg Engineering Handbook*, 1969). Hydrogen sulfide is another gas encountered in deposits containing sulfur. Gas content is of considerable practical importance since it affects all the physical properties measured and field performance that relates to compression and water flow. Consolidation test results are particularly impacted by gas. The gas content of peat is difficult to determine and no widely recognized method is yet available. A gas content of 5–10% of the total volume of the soil is reported for peat and organic soils (*Muskeg Engineering Handbook*, 1969).

3.8 ORGANIC CONTENT

As mentioned earlier, organic content is an important parameter for peat and organic soils, which set them apart from the mineral soils (silts and clay). The organic content is usually determined from the loss of ignition test as a percentage of oven dried mass (ASTM D 2974). The moist sample is first dried in the oven at a temperature of 105°C for 24 hours. A crucible is then placed in the muffle furnace at a temperature of 450°C for 1 hour and weighed to obtain its mass, M_1. The dry soil from the oven is then weighed with the crucible giving M_2. The soil sample together with the crucible is finally heated in the furnace at 450°C for 5 hours, and the mass is then recorded as M_3 after it has cooled at room temperature. Loss of ignition, N, is calculated as

$$N = (M_2 - M_3)/(M_2 - M_1) \tag{3.1}$$

The organic content (H) is calculated according to an equation proposed by Skempton and Petley (1970) as follows:

$$H\% = 100 - C(100 - N) \tag{3.2}$$

where C is the correction factor. For a temperature of 450°C, $C = 1.0$ (Arman, 1971). In Europe, a higher temperature of 550°C is used for combustion of peat, and $C = 1.04$ is then applied as the correction. However, the difference is usually small and hence not significant for practical considerations (Edil, 2003).

Table 3.2 shows the organic content of various peat deposits found throughout the world for comparison.

Table 3.2 Organic content of various peat deposits.

Peat deposit	*Organic content (%)*
Antoniny fibrous peat, Poland	65–85
Co. Offaly fibrous peat, Ireland	98–99
Cork amorphous peat, Ireland	80
Cranberry bog peat, Massachusetts	60–77
Italy peat	70–80
Japan peat	20–98
Canada peat	17–80
Hokkaido peat	20–98
West Malaysia peat	65–97
East Malaysia peat	50–98
Central Kalimantan peat	41–99

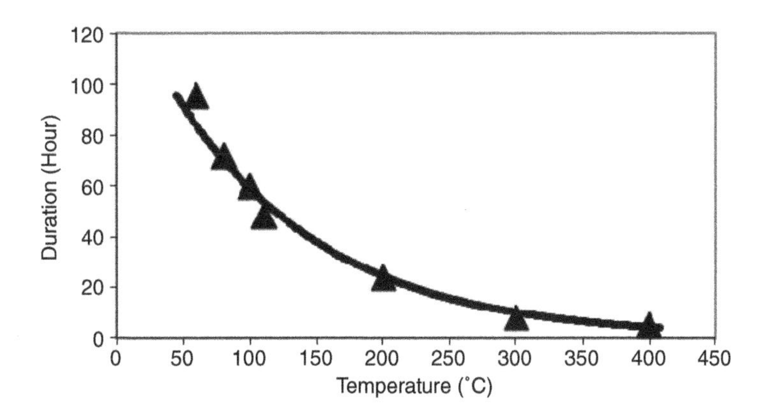

Figure 3.7 Duration versus drying temperature (°C) (*after* Zainorabidin and Bakar, 2003).

3.9 WATER CONTENT

Water content (also commonly referred to as natural moisture content) is one of the most common soil parameters, expressed in term of mass or weight.

$$w = w_w / w_s \times 100\% \quad (3.3)$$

where w = water content
w_w = weight of water
w_s = weight of dry soil

A small value of w indicates dry soil, while a large value indicates a wet one.

The water content can easily measured in the laboratory by conducting a moisture content test (ASTM D2216 or BS 1377: 1990), which involves drying a soil sample in an oven at 105°C for 24 hours.

For peat soils, there is a general fear that standard drying of the soil at 105°C during 24 hours will lead to charring of the organic component in peat, thus producing too large a figure for water content. Some therefore advocate a lower temperature, between 50 and 95°C. Skempton and Petley (1970) and Kabai and Farkas (1988) investigated the effect, and concluded that the loss of organic matter at 105°C is insignificant, while drying at lower temperatures retains a small amount of free water. Zainorabidin and Bakar (2003) investigated the drying temperature effect for hemic peat soil and suggested that for temperatures between 100 and 200°C the drying duration required is between 24 and 60 hours (Figure 3.7).

Standard practice is to dry at a temperature of 105°C. After a 24 hour drying period, the water content is then calculated using the formula

$$w = \frac{w_2 - w_3}{w_3 - w_1} \times 100\% \quad (3.4)$$

where w_1 = weight (or mass) of container + lid
w_2 = weight (or mass) of container + lid + wet soil
w_3 = weight (or mass) of container + lid + dry soil

Table 3.3 Natural water content of various soil deposits.

Soil deposits	*Natural water content (%)*
Malaysia west coast clay	70–140
Malaysia east coast clay	36–73
Quebec fibrous peat	370–450
Antoniny fibrous peat, Poland	310–450
Co. Offaly fibrous peat, Ireland	865–1400
Cork amorphous peat, Ireland	450
Cranberry bog peat, Massachusetts	759–946
Austria peat	200–800
Japan peat	334–1320
Italy peat	200–300
America peat	178–600
Canada peat	223–1040
Hokkaido peat	115–1150
West Malaysia peat	200–700
East Malaysia peat	150–2207
Central Kalimantan peat	467–1224

For peat, the water is held in the organic matter and cells of the plant remains. Generally the water content will decline with increase in mineral content. On the other hand, fibrous peat also tends to have a higher water content than humified peat. Peat soils generally have a very high natural water content, which can be in excess of 1500%, compared with mineral soils (sand, silt and clay), whose values in the field may range between 3 and 70%, but values greater than 100% are sometimes found in soft soils below the groundwater table. Table 3.3 shows typical values for the natural water content of various soil deposits.

3.10 ATTERBERG LIMITS

In 1911, the Swedish soil scientist Albert Atterberg developed a series of tests to evaluate the relationship between water content and soil consistency. Then in the 1930s, Terzaghi and Casagrande adapted these tests for civil engineering purposes, and they soon become a routine part of civil engineering. These tests include three separate tests: the liquid limit test (LL), the plastic limit test (PL) and the shrinkage limit test (SL). Together they are known as Atterberg limit tests (ASTM D 427, D 4318, BS1377: 1990) (Figure 3.8). The liquid limit and plastic limit tests are performed in many soil mechanics laboratories, especially for cohesive mineral soils (silts and clays). However, the shrinkage limit test is less useful and is rarely performed by civil engineers.

Two methods normally used to determine the liquid limits of soils are the Casagrande method and the cone penetrometer method. Figure 3.9 illustrates the equipment used for both of these methods. In the Casagrande method, the liquid limit is the water content of the soil when it exactly closes a groove over a distance of half an inch for 25 drops of the Casagrande cup or bowl. In the cone penetrometer method, the liquid limit is the water content of the soil when the standard cone penetrates the soil sample by exactly 20 mm deep.

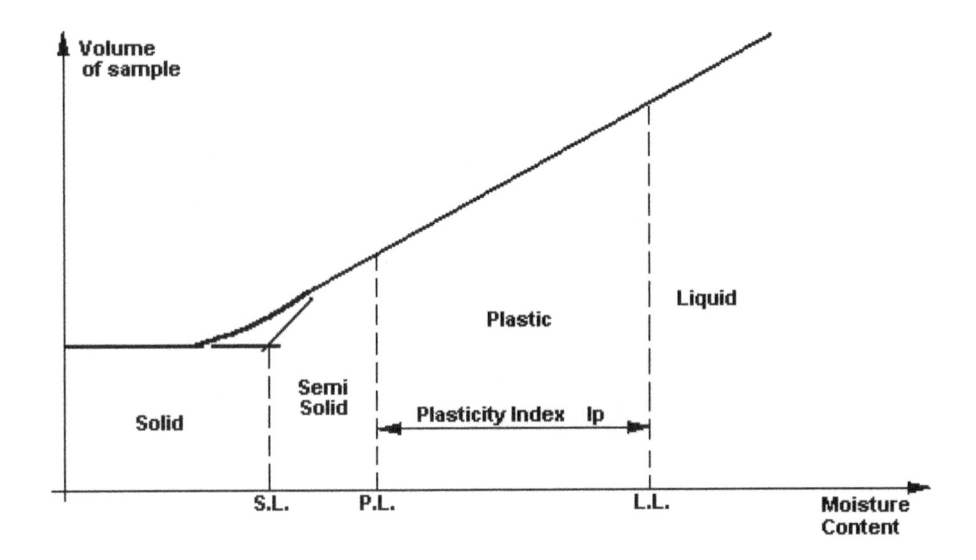

Figure 3.8 Atterberg limit.

The plastic limit test procedures involve carefully rolling the soil samples into threads. As this rolling process continues, the thread becomes thinner and eventually breaks. If the soil is dry, it breaks at a large diameter. If it is wet, it breaks at a much smaller diameter. By definition the soil is at its plastic limit when it breaks at a diameter of 1/8 of an inch (3 mm) (Figure 3.9 (c)).

For peat, the liquid limit depends on type of plant detritus contained, degree of humification and the proportion of clay soil present. For temperate peat, the liquid limit of fen peat ranges from 200–600%, and for bog peat from 800–1500% (Hobbs, 1986). Skempton and Petley (1970) put the boundary at approximately H3 of the von Post scale for the liquid limit and H5 for the plastic limit. In any case, according to Hobbs (1986), there is little point in performing plastic limit tests on peat soils since the plasticity gives little indication of their character. Table 3.4 shows the Atterberg limits of various soil deposits.

3.11 DENSITY AND SPECIFIC GRAVITY

Density or unit weight is another useful soil parameter for a geotechnical engineer. Density (ρ) is simply defined as total mass (M) over total volume (V):

$$\rho = M/V \tag{3.5}$$

Unit weight (γ) is total weight (W) over total volume (V). Weight (often expressed as kN) is equal to mass (kg) times acceleration due to gravity, g ($g = 9.81\ \mathrm{m\,s^{-2}}$).

Peat's unit weight is both low and variable compared with mineral soils, being related to the organic content, mineral content, water content and degree of saturation. The average unit weight of peat is typically slightly higher than water. Amorphous peat

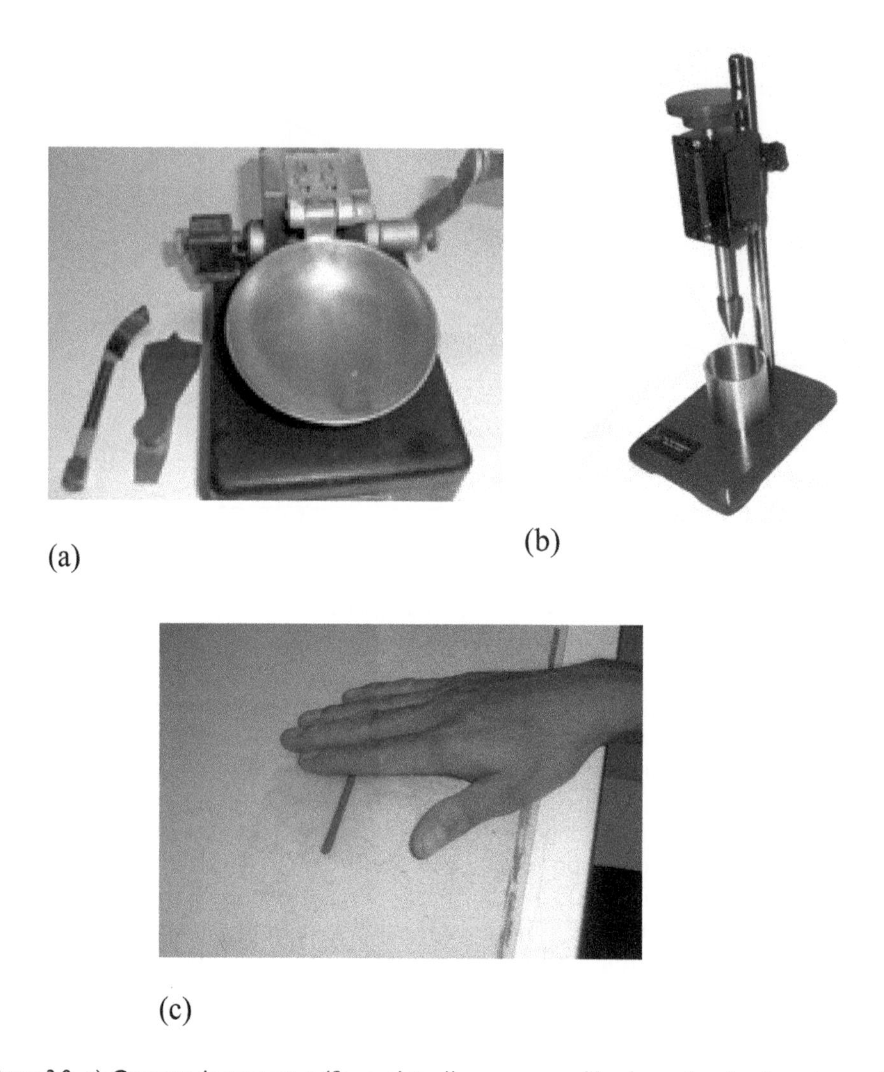

Figure 3.9 a) Casagrande apparatus (*Source*: http:// www.env.auckland.ac.nz/uoa/engineering-geology-equipment), (b) cone penetrometer (*Source*: http://www.ramaas.in/products.php?id=3) and (c) plastic limit (*Source*: http://www.denichsoiltest.com/Plastic-Limit-Test.html).

has a higher bulk density than fibrous peat. For instance, in the former it can range up to 10 kN m^{-3}, whilst in woody peat it may be half this figure. Table 3.5 shows the typical unit weight for various soils.

Dry density, ρ_d (dry unit weight, γ_d) is mass, M_s (or weight of solid particle), over total volume (V):

$$\rho_d = M_s/V \tag{3.6}$$

Dry density is a more important engineering property of peat, influencing its behaviour under load. The dry density itself is influenced by the effective load to which a deposit

Table 3.4 Atterberg limits of various soil deposits.

Soil deposits	*Liquid limit (%)*	*Plastic limit (%)*
Malaysia west coast clay	56–90	35–30
Fen peat	200–600	–
Bog peat	800–1500	–
Antoniny fibrous peat, Poland	305–310	–
Cork amorphous peat, Ireland	690	561
Cranberry bog peat, Massachusetts	580–600	375–400
West Malaysia peat	190–360	100–200
Samarahan (East Malaysia) (hemic) peat	210–550	125–297

Table 3.5 Typical unit weight of various soils.

Soil Type	*γ(kN m^{-3}) (above water table)*	*γ(kN m^{-3}) (below water table)*
Mineral soils		
Gravel	17.5–20.5	19.5–22.0
Sand	15.0–21.0	19.0–23.0
Silty sand	12.5–21.0	17.5–22.0
Clayey sand	13.5–20.5	17.5–21.0
Silt	11.5–17.5	11.5–20.5
Clay	12.5–17.5	11.0–19.5
Organic soil		
Bog peat		9.5–10.5
Fen peat		8.5–11.0
Peaty clay		10.0–13.0
Quebec fibrous peat		8.7–10.4
Antoniny fibrous peat, Poland		10.5–11.1
Co. Offaly fibrous peat, Ireland		10.2–11.3
Cork amorphous peat, Ireland		10.2
Cranberry bog peat, Massachusetts		10.1–10.4
Austria peat		9.8–13.0
Italy peat		10.2–14.3
Hokkaido peat		9.5–11.2
West Malaysia peat		8.3–11.5
East Malaysia peat		8.0–12.0
Kalimantan peat		8.0–14.0
Sumatra peat		4.0–9.0

of peat has been subjected. The dry density of peat is usually about 1.3 kN m^{-3}, but it can be as low as 0.7 kN m^{-3}. For West Malaysia hemic peat the value ranges from 1.2 to 1.5 kN m^{-3}.

The specific gravity of any material is the ratio of its density to that of water. For soil it is computed for the solid phase only:

$$G_s = M_s/V_s * \rho_w \tag{3.7}$$

where ρ_w = density of water

Figure 3.10 Density bottle used for measuring soil G_s in the laboratory (*courtesy* www.jaytecglass.co.uk).

The specific gravity of soil solids can be tested in the laboratory using the specific gravity bottle method or the gas jar method (ASTM D854, BS1377: 1990). For most mineral soils (sand, silt and clay), the specific gravity ranges from 2.60–2.80. For organic soils, however, it is affected by the organic constituents, and cannot simply be set to somewhere near that of mineral soils. Cellulose has a specific gravity of approximately 1.58, while for lignin it is approximately 1.40. These low values reduce the compounded specific gravity of organic soils. Figure 3.10 shows the density bottle used for measuring soil G_s in the laboratory. Table 3.6 shows the typical specific gravity of various soils.

Table 3.7 summarizes the characteristics of some British peat.

3.12 SURFACE CHARGE PROPERTIES OF ORGANIC SOILS AND PEAT

3.12.1 Cation exchange capacity

The cation exchange capacity (CEC) is the total amount of cations that a soil can hold on its absorption complex and exchange under conditions of pH and shows the ability of the soils to supply cations. For a soil, the CEC is due to electronegative colloidal substances, such as clay minerals, organic matter and colloidal silica. In other words, the quantity of exchangeable cations in the zone adjacent to the charge surface that can be exchangeable for other cations is termed the CEC. The CEC of the soil is expressed

Table 3.6 Typical specific gravity of various soils and minerals.

Soil type/mineral	*Specific gravity*
Non-clay	
Quartz	2.65
Mica	2.76–3.20
Gypsum	2.32
Clay minerals	
Kaolinite	2.62–2.66
Montmorillonite	2.75–2.78
Illite	2.60–2.96
Peat	
Bog peat	1.40–1.60
Fen peat	1.80
West Malaysia peat	1.38–1.70
Samarahan peat	1.07–1.63
Central Kalimantan peat	1.50–1.77

in units of specific adsorbed charge (Chapman, 1965; Sposito, 2008). Exchangeable cations must be removed from the soil to be detected and measured (Conklin, 2005). The unit of CEC is either cmol kg^{-1} or meq per 100 g soil. The determination of CEC gives an indication of the total amount of negative charges of the soil.

The CEC value depends on the pH and the technique which is used (Conklin, 2005). A realistic CEC value is that determined at the field pH value. This is difficult because the soil pH varies greatly.

The ease of cation replacement depends on the valence, ion size and ion relative amount. The range of soil colloid CEC is from a mean minimum about 4 for Al or Fe oxides to a mean maximum about 200 cmol kg^{-1} of colloid for humus (Fuchsman, 1986). Thus, humus has the highest CEC in comparison with minerals like kaolinite, montmorillonite, smectite and even vermiculite. Soil organic matter can contribute a significant fraction of CEC even in clay soils with low amount of organic matter (Magdoff and Weil, 2005).

Unlike clay minerals, organic matter does not have a fixed capacity for binding exchangeable cations. It is now well accepted that to measure the actual CEC of a soil, the pH must not be changed during the procedure. The main problem with measurement of CEC at pH 7 is that it buffers the soil at pH 7, causing large overestimates of CEC for acidic soils (Gillman and Sumpter, 1986; Fang and Daniels, 2006).

The cation exchange capacity (CEC) of organic soils was measured at pH 7 with ammonium acetate method (Chapman, 1965) and by the $BaCl_2$ compulsive exchange method (Gillman and Sumpter, 1986).

To measure the CEC at pH 7, about 125 ml of 1 M NH_4OAc is added to 25.0 g of soil samples in a 500 mL Erlenmeyer flask shaken and allowed to stand overnight. The soil samples are washed gently with NH_4OAc using a Buchner funnel filtration, followed by washing with 95% ethanol. The $NH4^+$ is extracted by leaching the soil with eight separate 25 mL additions of 1 M KCl. The concentrations of NH^{4+}-N can be determined by an auto analyzer.

Table 3.7 Characteristics of some British peat (*after* Hobbs, 1986).

Common plant communities		CEC$	pH of peat	Specific gravity	Organic content	Water content (%)	Liquid limit (%)	Plastic limit test	Permeabilit y	Remarks
Cloudberry Ling Heather Purple moon grass Cottage sedge* Deer srdge* Sphagna Bog asphodel	The Sphagnum cover plant community	High in bog plant communities	<4	1.4	>98% ↑	2000 to 1000	1500 to 900			Bogs invaded by pine and birch under drier climatic conditions.
Willow Aldec Sallow	Fen Carr invaded by Sphagna and	↑		1.4 to 1.6	>80%	1000 to 500	900 to 600		↑	Transition peats can be very variable. Properties intermediate between fen and bog.
Fen mosses Spearwort Meadow rue Purple loosestrife Saw sedge Sedges Common reed Reedmaces Rushes Water lily Submerged plants		Low in fen and transition plant communities	<5	1.6	Increasing with height above the substrate ↑	500 to 200	600 to 200	Not possible on pure bog peat Generally possible on fen peat	Permeability is higher in more fibrous less humified peats	Very rich in species particularly under calcareous conditions pH>6 Acid fens also occur poorer in species pH ≤ 5

$Cation exchange capacity
** Mire water tends to have somewhat high values

To measure the CEC at soil pH, 2.0 g of soil sample and 20 mL of 0.1 M $BaCl_2$–$2H_2O$ are mixed in a centrifuge tube, shaken for 2 hours in a shaker and centrifuged at 1,000 rpm. 20 mL of 2 mM $BaCl_2$-$2H_2O$ is then added and shaken for 1 hour. The samples are centrifuged and the supernatant discarded again. The addition of

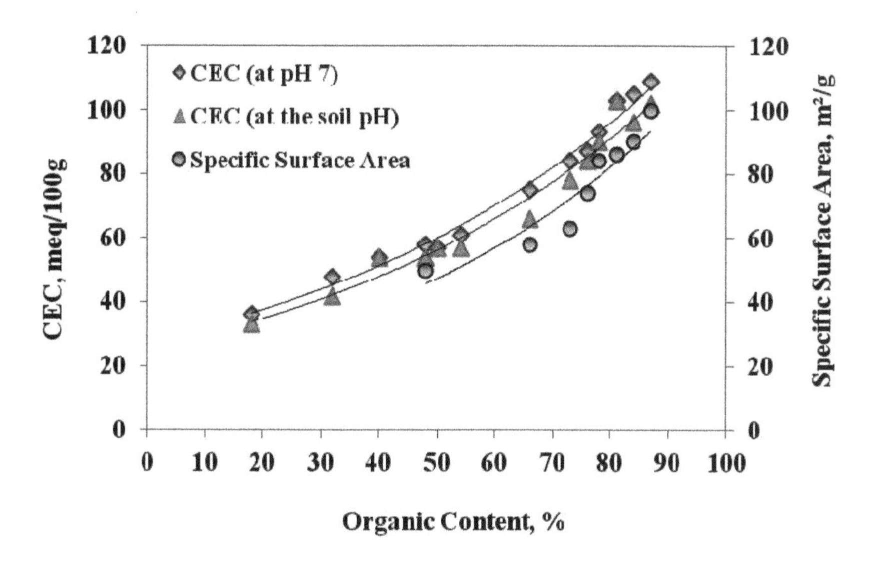

Figure 3.11 Organic content – CEC (*after* Asadi *et al.*, 2009d).

2 mM $BaCl_2$-$2H_2O$ and centrifuging is repeated twice. Before the third centrifuging the pH of the slurry is obtained. After decanting 2 mM $BaCl_2$-$2H_2O$, 10 mL of 5 mM $MgSO_4$ is added and shaken gently for one hour. The conductivity of the 1.5 mM $MgSO_4$ is determined. If the conductivity of the sample is not 1.5 times this value, 0.1 mL increments of 0.1 M $MgSO_4$ are added until it is. The pH of the solution is determined. If it is not within 0.1 units of the previous measurement, 0.05 M H_2SO_4 is added dropwise until the pH is in the appropriate range. Finally, distilled water is added until the solution's conductivity is that of 1.5 mM $MgSO_4$ and the tube is dried and weighed. The CEC is calculated based on the total solution, Mg in solution and the total Mg added, i.e. CEC (meq/100g) is equal to the total Mg (meq) added minus Mg (meq) in the final solution.

Figure 3.11 shows the cation exchange capacity of various organic soils.

The CEC of organic and peaty soils generally increases with increasing organic content. The water adsorption potential of organic soils generally increases with increasing CEC (Asadi *et al.*, 2009d).

Figure 3.12 shows the cation exchange capacity of various organic soils according to the degree of humification. The CEC of organic and peaty soils increases with increasing degree of humification (Asadi *et al.*, 2009d).

Both clay and organic materials can contribute to the CEC of organic soils. Because cations are positively charged, they are attracted to surfaces that are negatively charged. Although there is a qualitative difference between the clay and organic matter fractions of organic soil, the role of the clay fraction is also very important for providing cation exchange sites.

Since the charge in organic soils is strongly pH dependent, the resulting soils develop a greater CEC at near-neutral pH than under acidic conditions. Therefore the main problem with measurement of CEC at pH 7 is that it buffers the soil at

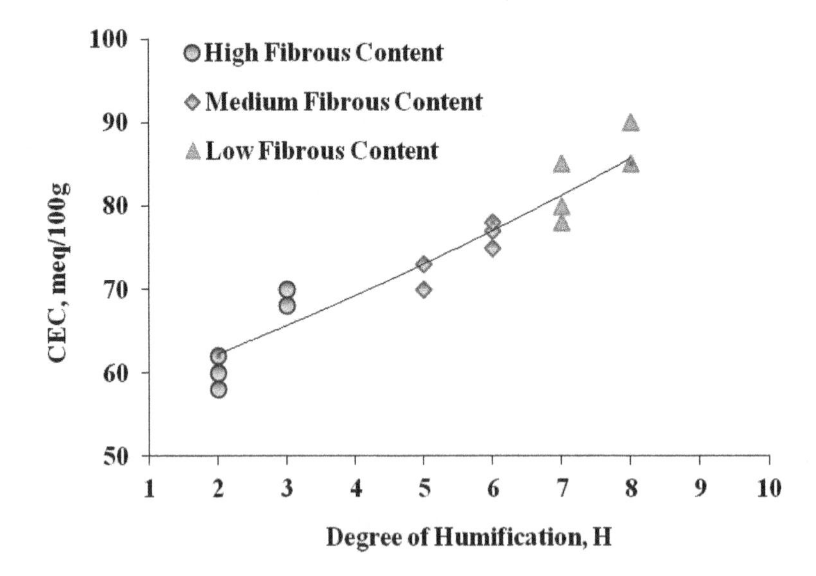

Figure 3.12 CEC vs. von Post degree of peat humification (*after* Asadi *et al.*, 2009d).

pH 7, causing large overestimates of CEC for acidic organic soils. Fibrous peat is low-humified and undecomposed peat is more acidic. Therefore fibrous peat can show a lower CEC at the soil pH (Asadi, 2009b).

The CEC ranges of soils with high fibre content are usually less than the CEC ranges of soils with low fibrous content. Fibric peat is mostly undecomposed, and low-humified, and has a distinct plant structure. A greater degree of humification results in a higher CEC (Asadi, 2010).

3.12.2 Zeta potential of organic soils and peat

The ζ (zeta) potential is the electric potential developed at the solid-liquid interface in response to the movement of colloidal particles; i.e. ζ is the electrical potential at junction between the fixed and mobile parts of the electrical double layer. ζ is less than the surface potential of the particles and shows the value at the slip plane, which is located at a small unknown distance from the colloidal surface (Hunter, 1981; Hunter, 1993). Figure 3.13 shows the zeta potential.

The magnitude and sign of ζ depends on the interfacial chemistry of both liquid and solid phase (Eykholt and Daniel, 1994). This potential is also influenced by ion exchange capacity, size of ion radius, and the thickness of the double layer (Fang and Daniels, 2006).

The concentration of electrolyte, type of electrolyte, valence of ions and pH are important factors and can affect ζ values.

Since the electroosmotic flow correlates with ζ and pH (Bowen *et al.*, 1986; Beddiar *et al.*, 2005), it is crucial to know the microscopic electrical properties of the soil samples. The zeta potential can be determined by electrophoresis (Hunter, 1981; Hunter, 1993). Soil samples with different organic content can be prepared by sieving

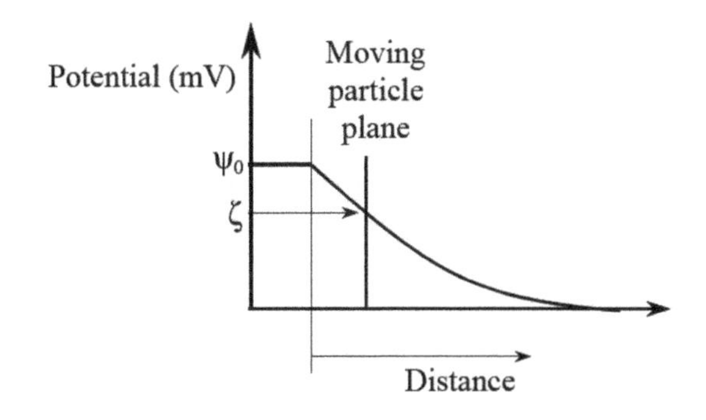

Figure 3.13 Potential distribution showing the slipping plane (zeta) potential (*after* Mitchell and Soga, 2005).

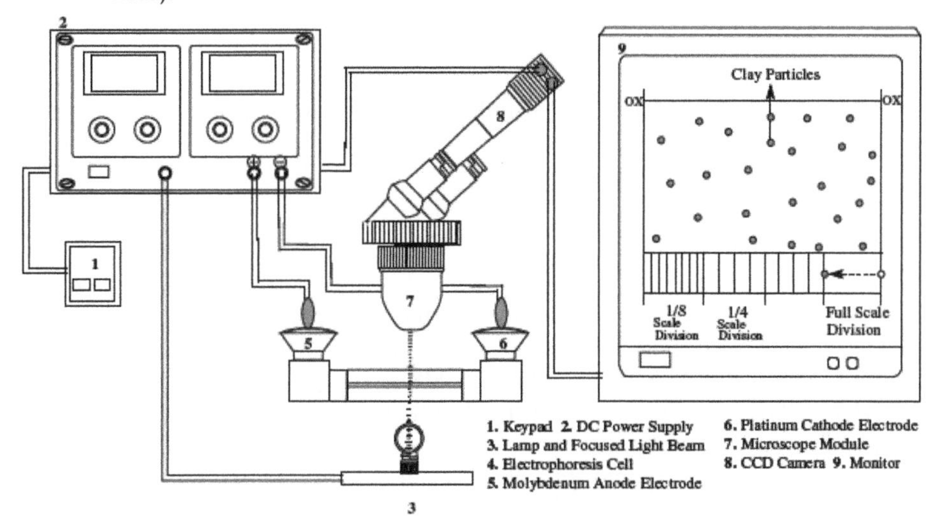

Figure 3.14 Arrangement of Zeta-Meter 3.0+ Unit (*after* Zeta-Meter System 3.0 Operating Instructions).

through a #100 (150 μm) sieve. For each sample, a solution of 0.15 to 0.2 g/L of the pretreated soil in 0.0001 M NaCl can be prepared.

ζ is measured with a zeta-meter as a function of pH values. A zeta-meter is a microprocessor-based instrument (shown schematically in Figure 3.14). The unit automatically calculates the electrophoretic mobility of the particles and converts it to the zeta potential using the Smoluchowski equation, which provides a direct relation between the zeta potential and electrophoretic mobility.

The measurement of Zeta Potential is important in understanding the electrical charge characteristics of sample particles. The electrical charge properties control the interactions between particles and therefore determine the behaviour of the soil particles.

All measurements are usually made in 0.0001 M NaCl solutions and pH adjustments are made using dilute HCl or NaOH solutions (West and Stewart, 1995).

The ζ can be measured by electrophoresis, the measurement is very direct. The sample is placed in a chamber which is called an electrophoresis cell. Electric field is then activated. Then the colloids move with a velocity that is proportional to their zeta potential. The direction of the movement can show whether their charge is positive or negative.

The Smoluchowski equation, the most elementary expression for ζ, gives a direct relation between the zeta potential and electrophoretic mobility:

$$\zeta = 4\pi V_t \frac{EM}{D_t} \tag{3.8}$$

where
ζ = Zeta potential
EM = Electrophoretic mobility at actual temperature
V_t = Viscosity of the suspending liquid
D_t = dielectric constant

It is preferable to calculate ζ in millivolts instead of in electrostatic units. The formula then becomes:

$$\zeta = 11300 V_t \frac{EM}{D_t} \tag{3.9}$$

ζ for a colloid is usually negative, but the magnitude and sign depend on the interfacial chemistry of both the liquid and the solid phase (Eykholt and Daniel, 1994; Vane and Zang, 1997; Yukselen and Erzin, 2008). ζ is a good indicator of the thickness of the double layer. As ζ increases, the thickness of the double layer increases (Hunter, 1981).

The ζ potential of organic soils varies from +41 mV at pH 1.91 to −43 mV at pH 11.5 (Asadi *et al.*, 2009d). It is almost zero at pH 2.5 to 3.5 (Figure 3.15). The variations in ζ with pH are probably related to the nature of the electrical energy field in organic soils. The sign of the natural ζ in organic soils is negative. The charge is affected by pH. As the pH goes up, the net negative charge is produced. As the pH drops, there is less and less negative charge. The ζ behaviour of organic soils in the presence of the NaOH is possibly related to dissociation of H^+. The negative charge of humus is generally believed to be due to the dissociation of $H^+$from carboxylic and phenolic functional groups (Stevenson, 1994; Yu, 1997). All charge on humus is strongly pH-dependent, with humic substances behaving like polyprotonated weak acids. Humic substances are a series of relatively high molecular weight, brown- to black-coloured substances formed by secondary synthesis reactions (Madaeni *et al.*, 2006).

The natural ζ of peats increase with increase in the natural organic content. In the peats, organic contents has more influence over increase of the natural ζ in comparison with organic soils having less organic content, which are under influence of mineral portions. Thus, the relationship between organic contents and zeta potential are not only under influence of organic contents, but also under influence of minerals portions and degree of humification (Asadi *et al.*, 2009d).

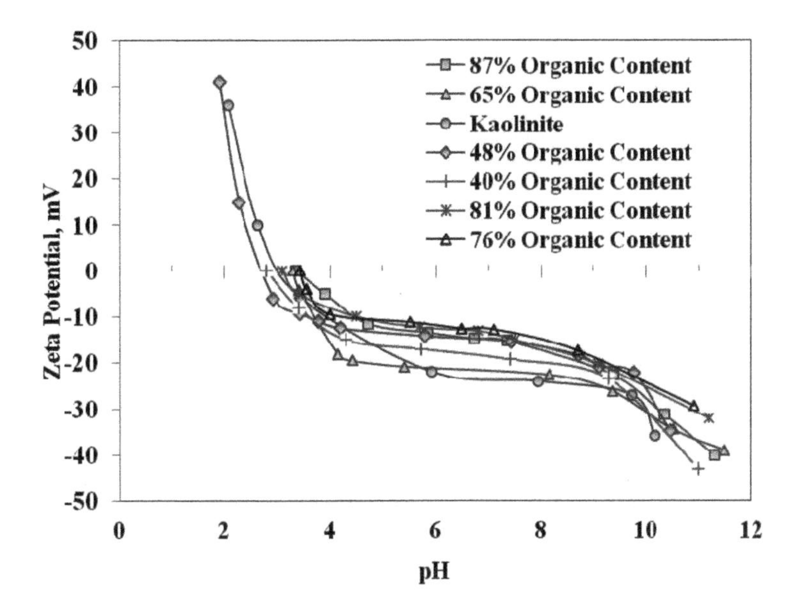

Figure 3.15 Zeta potential vs. pH (*after* Asadi *et al.*, 2009d).

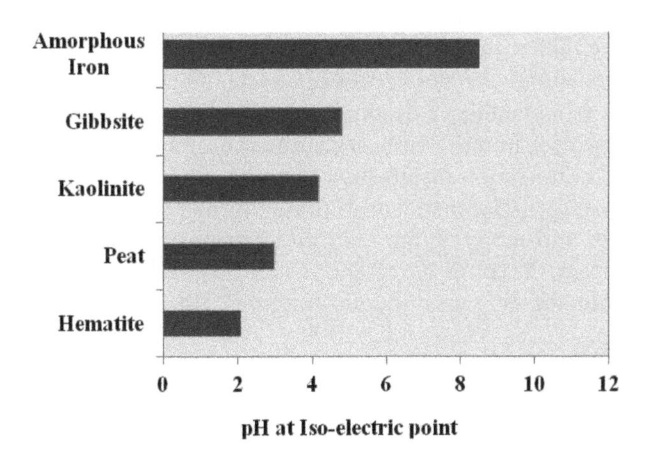

Figure 3.16 pH at iso-electric point of peat in comparison with some minerals (*after* Asadi *et al.*, 2009d).

At a certain pH, the soil surface charge could drop to zero, rendering a zero ζ, or what is called the iso-electric point (Lorenz, 1969). The peat surface charge drops to zero ζ at pH 2.5 to 3.5 (Figure 3.15).

Figure 3.16 shows the values of the iso-electric point of some minerals in comparison with peats from this study (Mohamed and Anita, 1998). Since all charge in humus is strongly pH-dependent, the sensitivity of organic soils to pH changes is greater than that of mineral soils. Despite this high sensitivity, the iso-electric point of organic soils is less than that of amorphous iron, gibbsite and kaolinite.

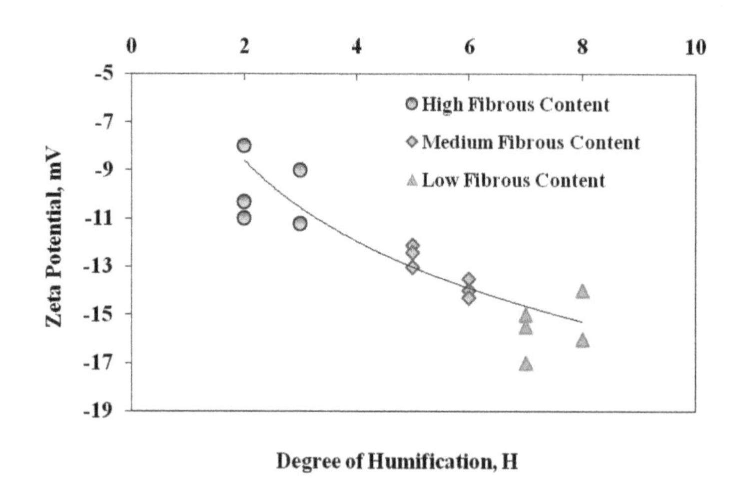

Figure 3.17 Zeta potential – von Post degree of peat humification (*after* Asadi *et al.*, 2009d).

The ζ potential of organic and peaty soils increases with increasing humification (Figure 3.17). The natural ζ ranges of soils with high fibrous content are usually less than the natural ζ ranges of soils with low fibrous content. Decomposition involves the loss of organic matter in either gas or solution form through the disappearance of the physical structure and changes in chemical state. Normally, the higher temperature of a tropical area would enhance decomposition activity.

The degree of peat humification can influence ζ, as depicted in Figure 3.17. A higher degree of humification means lower fibrous content and higher pH. Since the charge peat is strongly pH-dependent, humified peat has a higher ζ with a negative mathematical sign and more charge than undecomposed peat (Asadi *et al.*, 2009a). However, not only are the composition and structure of soil humus complex and incompletely known, but the clay and organic fractions also strongly affect the electrical properties of organic soils Asadi *et al.*, 2009e).

Hunter (1981) provides a detailed description of experiments to show the significance of pH for the ζ of soils. Vane and Zang (1997) showed that pH and ζ vary depending upon the types of mineral for mineral soils. In peat, ζ is dependent mainly on organic content, degree of humification and mineral fraction (Asadi *et al.*, 2009d).

It is noteworthy that with increasing degree of decomposition, the pH at the iso-electric point decreases (Figure 3.18). The pH ranges at the iso-electric points of soils with high fibrous content are higher than those for soils with low fibrous content.

The zeta potentials of organic soils and peat are affected by the type of cation, valence of cations, concentration of cations and pH (Asadi *et al.*, 2010; Asadi, 2010). The zeta potential of the peat decrease as the pH or the concentration of cations increases (Tables 3.8 and 3.9).

The effects of pH on the zeta potential of the peat are in good agreement with soils that have variable surface charges (Hunter, 1981; Forsberg and Alden, 1988; Hamed *et al.*, 1991; West and Stewart 1995; Vane and Zang, 1997; Yu, 1997; Alkan *et al.*, 2005; Fang and Daniels, 2006; Yukselen and Erzin 2008). However, in peat

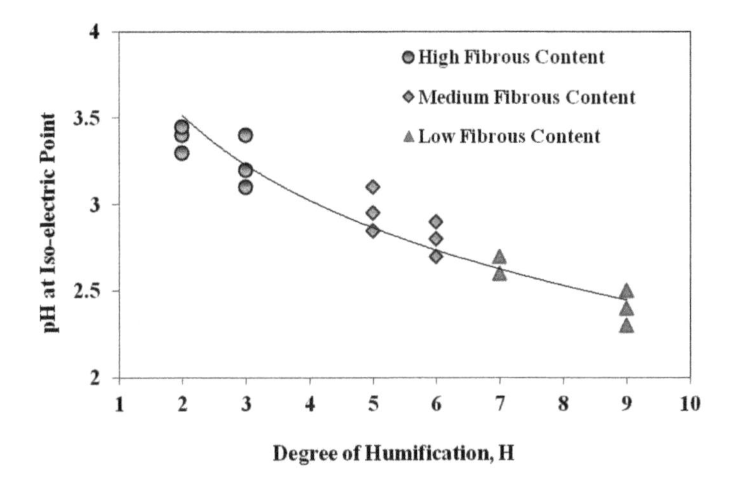

Figure 3.18 pH at iso-electric point vs. von Post degree of peat humification (*after* Asadi *et al.*, 2009d).

Table 3.8 Zeta potential of very slightly decomposed peat (*after* Asadi, 2010).

Cation	*Hydrated radius (nm)*	*Cation valence*	*Concentration (mol L^{-1})*	*pH*	*Zeta potential (mV)*
Na	0.36	1	1.00E-02	3.04	−7.1
			1.00E-02	9.1	−19.9
			1.00E-02	10.46	−25.23
			1.00E-04	3.04	−6.7
			1.00E-04	10.2	−24.2
Ca	0.41	2	1.00E-02	3.42	−2
			1.00E-02	7.42	−9.3
			1.00E-03	3.5	−3.7
			1.00E-04	9.65	−11.3
			1.00E-04	11.2	−14.4
Al	0.48	3	1.00E-03	7.54	−3.1
			1.00E-03	11.63	−17.2
			1.00E-04	11.77	−22.8

the pH effects can possibly be ascribed to dissociation of H^+ from the functional groups. Many carboxylic groups are sufficiently acid to dissociate below pH 6 leaving a negative charge on the functional group:

$$R - COOH = R - COO^- + H^+ \tag{3.10}$$

R represents organic species whose differing electronegativities change the tendency for H^+ to dissociate. Thus the various R–COOH units dissociate at different pH values. As the pH of the system increases, still weaker carboxylic groups and other very weak acids dissociate (Stevenson, 1994).

Table 3.9 Zeta Potential of highly decomposed peat (*after* Asadi, 2010).

Cation	*Hydrated radius (nm)*	*Cation valence*	*Concentration mol L^{-1}*	*pH*	*Zeta potential (mV)*
Na	0.36	1	1.00E-02	3.1	−10.1
			1.00E-03	3.08	−11
			1.00E-03	6.9	−21.2
			1.00E-04	3.08	−12.3
			1.00E-04	11.2	−32.37
Ca	0.41	2	1.00E-02	3.32	−3
			1.00E-02	4.4	−7.1
			1.00E-02	10.89	−19.1
			1.00E-03	3.16	−4.1
			1.00E-04	9.78	−19.6
Al	0.48	3	1.00E-03	5.04	−1.5
			1.00E-04	7.4	−9.3
			1.00E-04	9.63	−21.2

The zeta potential of the highly decomposed peat is higher than that of very slightly decomposed peat, indicating that the higher degree of peat humification results a higher zeta potential. The same conclusions are found in the presence of K^+, Mg^{2+}, Ca^{2+} and Al^{3+}, indicating that the higher concentration of ions results in a zeta potential, and that higher pH results in a higher zeta potential.

The zeta potential of peat in the presence of Al^{3+} at the same concentration and pH is lower than the zeta potential of peat in the presence of Na^+, indicating that the higher the valence of the cation the lower the zeta potential. However, since variations in peat arise from the variety of plants whose residues contribute to peat formation and from the environmental conditions in which humification takes place, the degree of humification can affect the zeta potential. As a result of those variations, a higher degree of humification results in a higher zeta potential (Asadi, 2010; Asadi and Huat, 2010).

The simple alkali metal ions, such as Na^+ and K^+, are known as indifferent ions (Hunter, 1993). These ions are attracted to a charged surface by simple electrostatic forces. The Na^+ and K^+ions can accumulate as counter ions in the electrical double layer. Consequently, they compress the electrical double layer and change the magnitude of the zeta potential. However, they could also lower the surface potential by charge neutralization, reducing the zeta potential.

The effect of ion concentration in decreasing the thickness of the diffuse layer could be higher than the valence effect according to the formula of Usui (1984):

$$\frac{1}{\kappa} = \frac{3}{ZC^{1/2}} \tag{3.11}$$

where
$\frac{1}{\kappa}$ = Thickness of the diffuse layer (Å)
Z = Valence of ion
C = Concentration of ion (mol L^{-1})

The formula indicates that for the same valence the concentration contributes to the thickness of the electrical double layer and consequently could reduce the zeta potential.

The effects of divalent and trivalent electrolytes are higher than those of Na^+ and K^+because of their valence. The effect of Na^+ in decreasing the zeta potential is lower than the effect of K^+ and the effect of Mg^{2+} in is lower than that of Ca^{2+}, indicating the possibility that cations with a higher hydrated radius have a lower effect on decreasing the zeta potential. The hydrated radius of Na^+ (0.79 nm) is higher than the hydrated radius of K^+ (0.53 nm), and the hydrated radius of Mg^{2+} (1.08 nm) is also higher than the hydrated radius of Ca^{2+} (0.96). Thus, the higher hydrated radius ions show a lower capacity to decrease the zeta potential (Asadi, 2010).

3.12.3 Resistivity of organic soils and peat

The resistivity of a soil depends on the surface conductivity of the colloids (i.e. clay or/and humus), presence of ions, porosity, moisture content, and temperature; and is determined according to Ohm's law. Resistance is that property of a conductor which opposes electrical current when a voltage is applied across the two ends and is given by the ratio of the applied voltage to the resulting current flow. The resistance of a conductor depends on the atomic structure of the material or its resistivity, which is that property of a material that measures its ability to conduct electricity. The resistivity is measured in ohm metres, and can be derived from the resistance, length and cross-sectional area of the conductor. The mathematical equation that describes this relationship is:

$$\rho = \frac{E}{I}\frac{A}{L} \tag{3.12}$$

where
ρ = resistivity of soil (Ω m)
E = applied voltage across the sample (V)
A = cross-sectional area (m^2)
I = current (amp)
L = Length of the sample (m)

A resistivity cell can be used to measure the resistivity of organic and peaty soils. In order to increase the degree of accuracy, different constant electrical potentials of 40, 70 and 90 V can be applied across the specimen.

The resistivity of organic soils is affected by the water content and temperature as depicted in Tables 3.10 and 3.11. The resistivity decreases as the water content or the temperature increase. A higher degree of peat humification results in a lower peat resistivity (Asadi *et al*, 2009c, Asadi, 2010).

The resistivity of both very slightly decomposed and highly decomposed peat increases as the organic content increases. The porosity of peat is also an important factor in its resistivity. Peat is a high-porosity material. Since it tends to have a high water content due to the high level of organic matter and plant remains, most of the void could be peat water. Therefore, as the porosity of the peat increases, the potential

Table 3.10 Resistivity of a sample of very slightly decomposed peat (*after* Asadi, 2010).

Organic content (%)	*Porosity (%)*	*Water content (%)*	*Temperature (°C)*	*Resistivity (Ωm)*
94	90	806	23	44.2373
			33.5	35.5126
	91	827	22	40.18
			27.5	33.2156
62	90	542	24.5	46.9426
			28	44.9266
	91	605	23	43.63
			27.5	41.3303
31	83	241	23.5	18.1896
			31	14.857

Table 3.11 Resistivity of a sample of highly decomposed peat (*after* Asadi, 2010).

Organic content (%)	*Porosity (%)*	*Water content (%)*	*Temperature (°C)*	*Resistivity (Ωm)*
85	80	285	21	25.053
			33	20.9616
	83	343	22.5	23.638
			33	17.5773
53	80	212	22	18.2536
			27.5	15.9286
	82	242	23.5	16.764
			28	15.0946
42	76	161	23	17.5606
			27	16.712
	81	213	23	16.357
			27.5	14.676

for the presence of peat water increases, resulting in lower resistivity in the peat environment. Since the humification processes can increase the quantity of humus particles, the higher degree of humification would decrease the resistivity (Asadi *et al*, 2009f, Asadi, 2010).

3.13 CORRELATIONS BETWEEN INDEX PARAMETERS OF PEAT

As with mineral soils, correlations between various index parameters have also proved to be useful for peat. Edil (2003) emphasized the importance of characterizing peat and organic soils by certain index parameters to provide a basis for comparison of results of mechanical tests. Hobbs (1986) also suggested that it is convenient to relate the basic geotechnical properties of organic soils to some of the easily determined index parameters, such as water content, organic content or liquid limits.

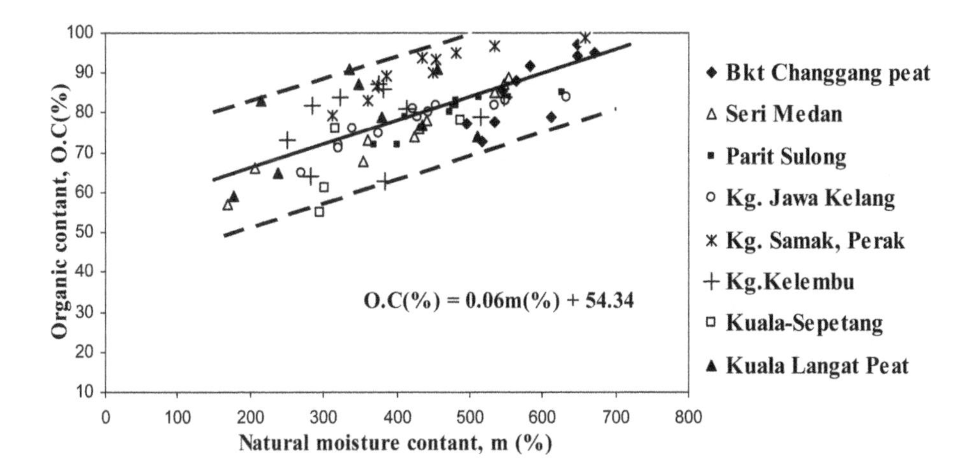

Figure 3.19 Water content vs. organic content (*after* Kazemian *et al.*, 2009).

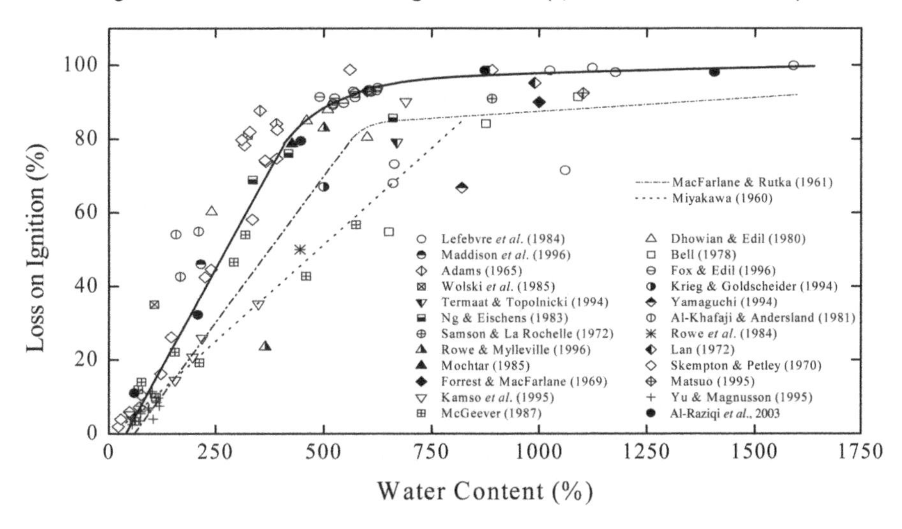

Figure 3.20 Correlation of water content with loss on ignition (*after* O'Loughlin and Lehane, 2003).

3.13.1 Water content vs. organic content

Figure 3.19 is a graph of natural water content (m) versus organic content (OC) of the tropical hemic peat of West Malaysia. The natural water contents of these soils ranged from 200 to 700%, with organic content in the range 50 to 95%. The organic content of the peat increases with increasing natural water content.

Figure 3.20 shows the plot of water content versus loss on ignition (O'Loughlin and Lehane, 2003) for Irish peat and organic soils. The relationship shown is linear but only up to loss on ignition $N = 80\%$. A high degree of scatter between loss on ignition (N) and water content (w_o) exists for soils with higher organic contents, which they attributed to the degree of humification of the organic matter.

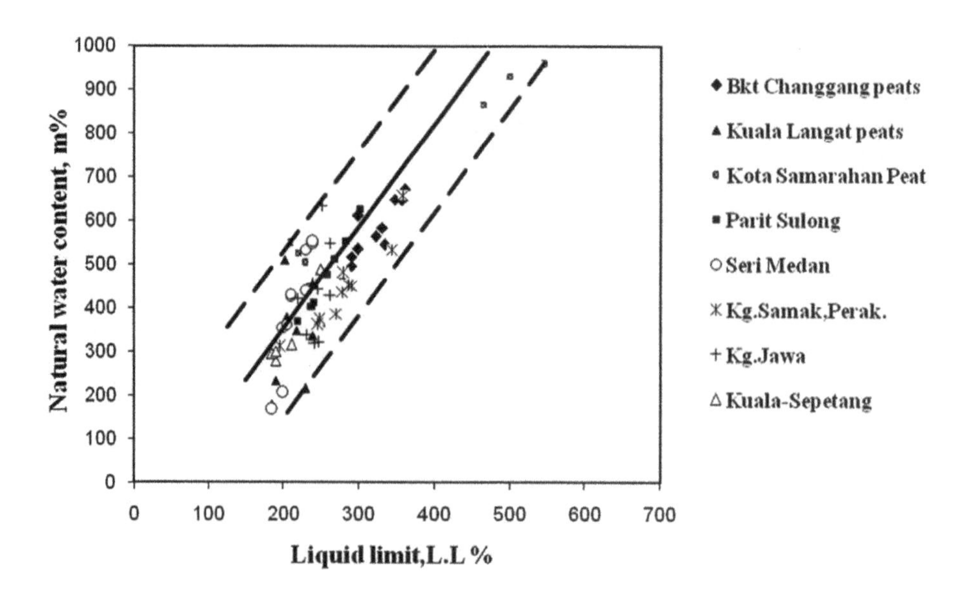

Figure 3.21 Natural water content vs. liquid limit (*after* Kazemian *et al.*, 2009).

3.13.2 Water content vs. liquid limit

Figure 3.21 shows the graph of natural water content (w) versus liquid limit (LL). The liquid limit of the peat increases with increasing natural water content. For the tropical hemic peat of Malaysia, the liquid limit is in the range 200–500%, about the same range as that of temperate (fen) peat.

3.13.3 Organic content vs. liquid limit

Figure 3.22 shows the graph of organic content versus liquid limit. In general, the liquid limit of peat increases with increasing organic content.

Skempton and Petley (1970) derived the following equation which relates LL and N for temperate peat as

$$LL = 0.5 + 5.0N \tag{3.13}$$

where LL is liquid limit and N is ignition loss, both expressed as ratio. However, the above equation does not seem to fit well for the case of tropical peat.

3.13.4 Natural water content vs. dry density

Figure 3.23 shows the graph of dry density (ρ_d) and water content (w) for soils. The best fit is given by the equation

$$\rho_d = 0.872(w + 0.317)^{-0.982} \tag{3.14}$$

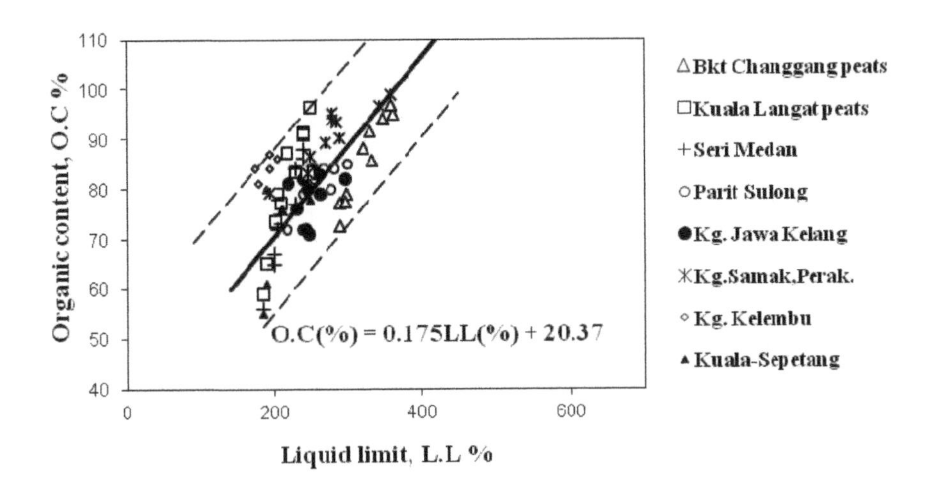

Figure 3.22 Organic content vs. liquid limit (*after* Kazemian *et al.*, 2009).

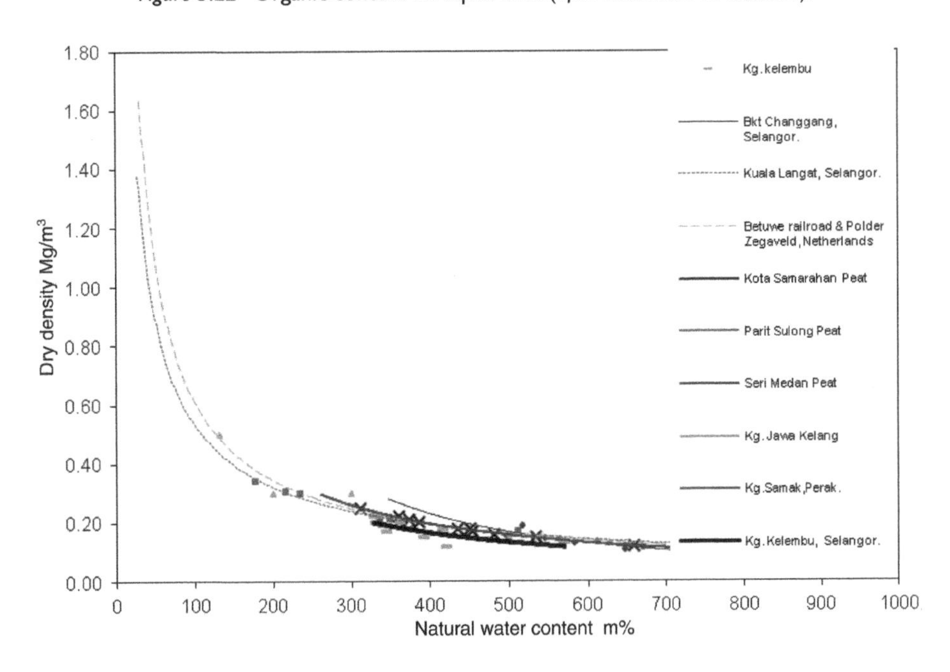

Figure 3.23 Natural water content vs. dry density (*after* Al-Raziqi *et al.*, 2003).

As shown in the graph, the densities of peat soils are significantly lower compared with those of mineral soils because of the presence of organic matter. In this case both tropical and temperate peat show a very close fit.

3.13.5 Specific gravity vs. organic content (loss of ignition)

Figure 3.24 shows the graph of specific gravity versus loss on ignition for peat soils.

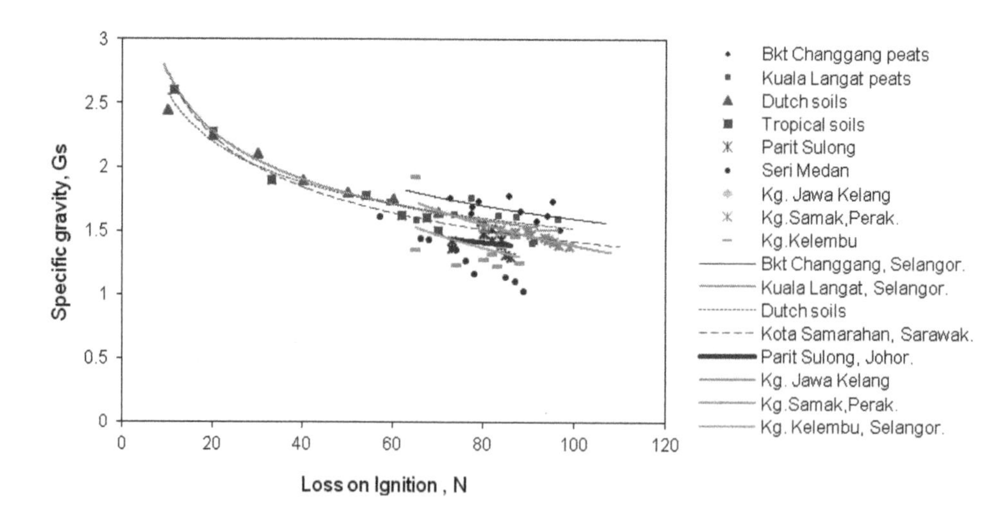

Figure 3.24 Specific gravity vs. loss on ignition (*after* Al-Raziqi *et al.*, 2003).

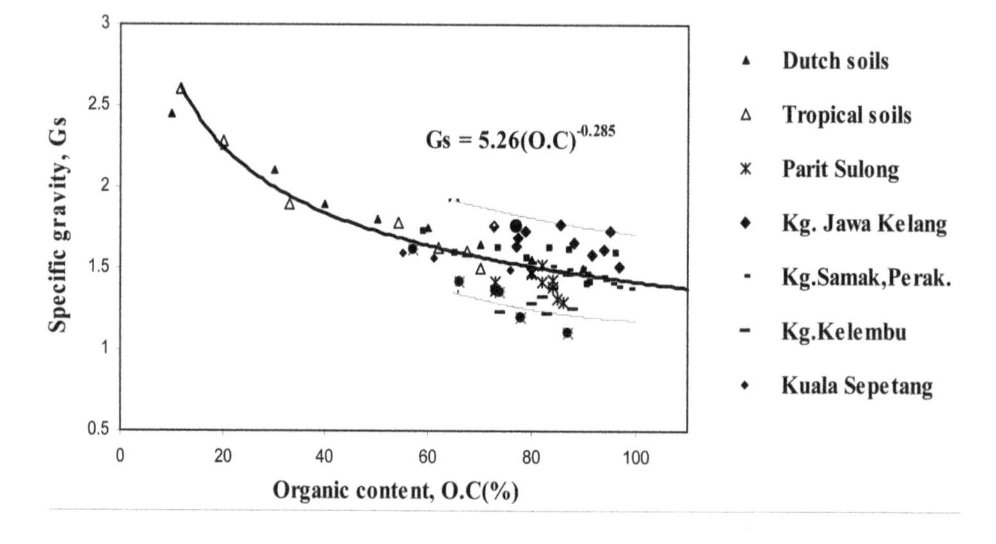

Figure 3.25 Specific gravity vs. organic content (*after* Kazemian *et al.*, 2009).

Figure 3.25 shows another plot of specific gravity (G_s) versus organic content (OC) for Malaysian peat, Dutch soils and tropical soils.

Skempton and Petley (1970) proposed the following relationship for the above two parameters.

$$\frac{1}{G_s} = \frac{1 - 1.04(1 - N)}{1.4} + \frac{1.04(1 - N)}{2.7} \tag{3.15}$$

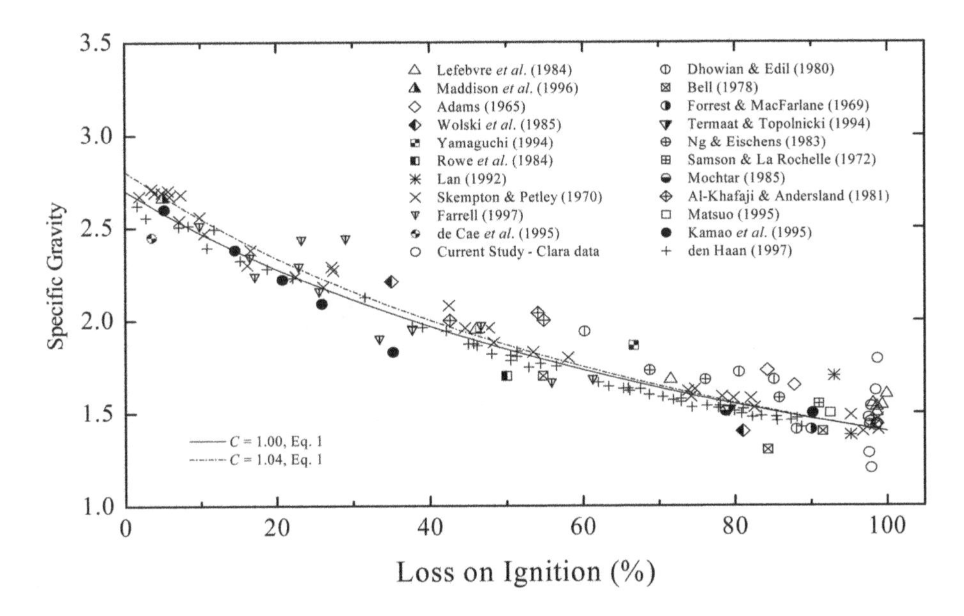

Figure 3.26 Correlation of specific gravity with loss on ignition for Irish peat (*after* O'Loughlin and Lehane, 2003).

den Haan (1997) simplified the above equation as follows.

$$\frac{1}{G_s} = \frac{N}{1.365} + \frac{(1-N)}{2.695} \tag{3.16}$$

Figure 3.26 shows the plot of specific gravity versus loss on ignition of Irish peat.

Figure 3.27 shows a similar plot for Dutch peat.

The values of specific gravity obtained from equation (3.17) are somewhat lower than given by Skempton and Petley (1970).

In general, the high lignin content of tropical peat gives it a slightly lower specific gravity compared with that of temperate peat, with G_s in the range 1.07–1.70, and an average of about 1.40.

3.13.6 Bulk density vs. loss of ignition

Figure 3.28 shows a graph of bulk densities and ignition loss for peat soils. The bulk densities of peat are in the range 0.8–1.2 Mg/m^3 compared with bulk densities of mineral soils, which are in the range 1.80–2.00 Mg/m^3. This is due to the lower specific gravity of solids found and higher water-holding capacity in peat compared with inorganic soils.

Based on extensive tests performed on Dutch peat, den Haan and El Amir (1994) proposed the following empirical relationship:

$$\gamma_{sat}(kN/m^3) = 12.266 - 3.156OC \tag{3.17}$$

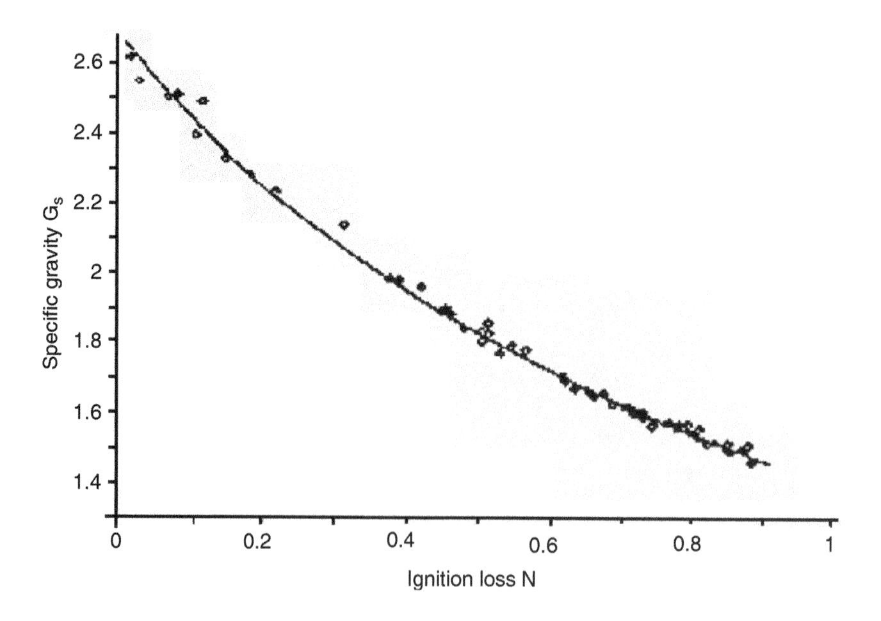

Figure 3.27 Correlation of specific gravity and ignition loss of Dutch peat (*after* den Haan, 1997).

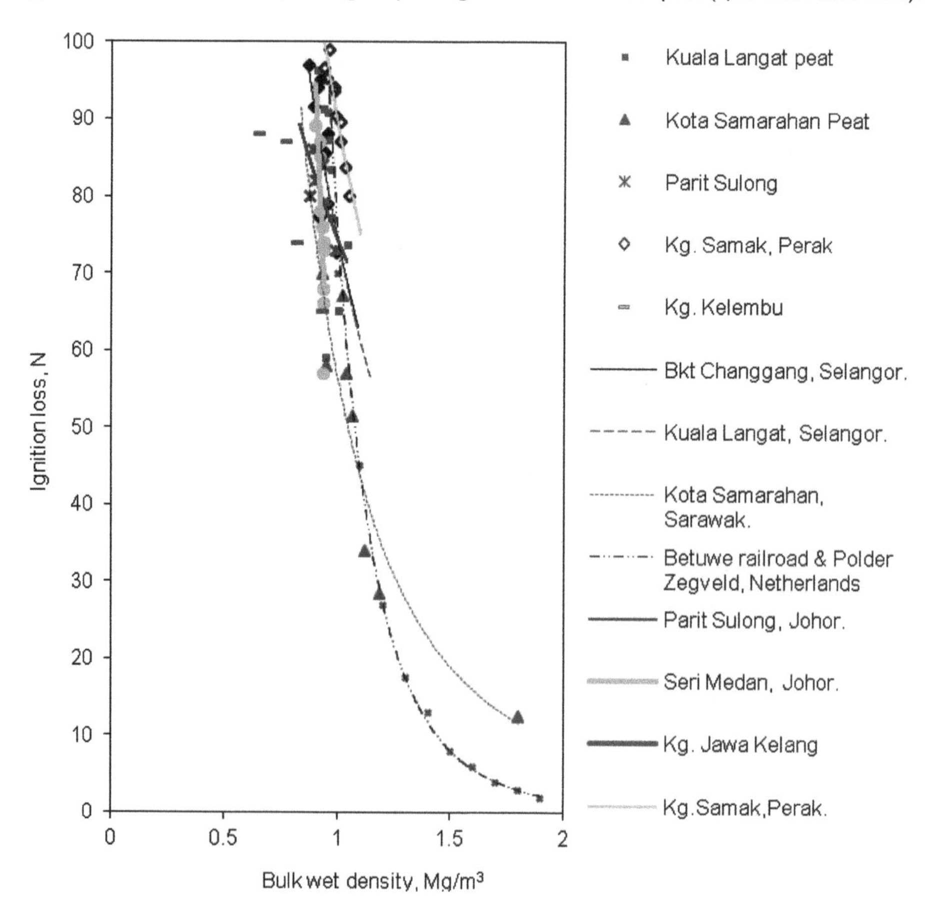

Figure 3.28 Bulk density vs. loss on ignition (*after* Al-Raziqi *et al.*, 2003).

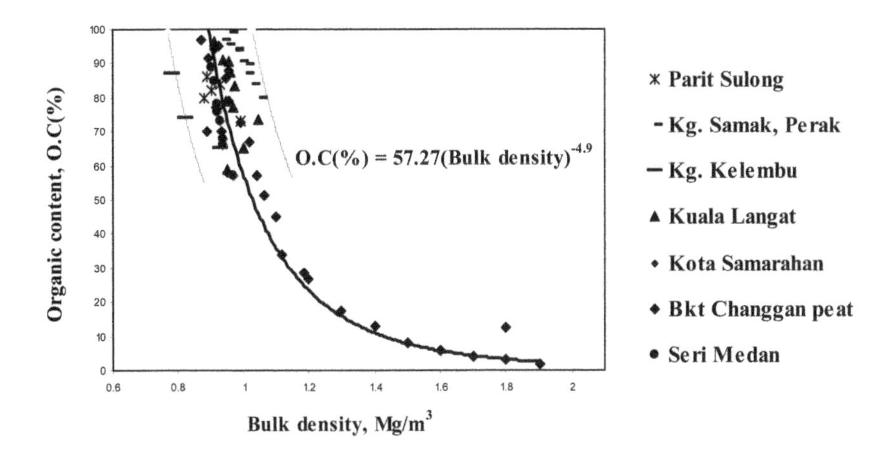

Figure 3.29 Bulk density vs. organic content (*after* Kazemian *et al.*, 2009).

Figure 3.29 shows another plot of organic content versus bulk density of Malaysian peat.

3.13.7 Bulk density vs. water content

Figure 3.30 shows a graph of bulk density vs. water content of peat.

3.13.8 Compression index vs. liquid limit

Figure 3.31 shows a plot of compression index (C_c) vs. liquid limit for Malaysian peat.

3.14 SUMMARY OF ENGINEERING PROPERTIES OF PEAT

Peat's engineering properties are summarized in this section.

Water content

The water content of peat can range from 500% to 2,000%, even reaching as high as 2,500% for some fibrous peats. Water content values of less than 500% are usually an indicator of high mineral fractions within the peat sample.

Ash content

The ash content (or non-organic content) of a peat sample is the percentage of dry material that remains as ash after controlled combustion. Peat that has grown *in situ* normally has an ash content of somewhere between 2% and 20% of its volume.

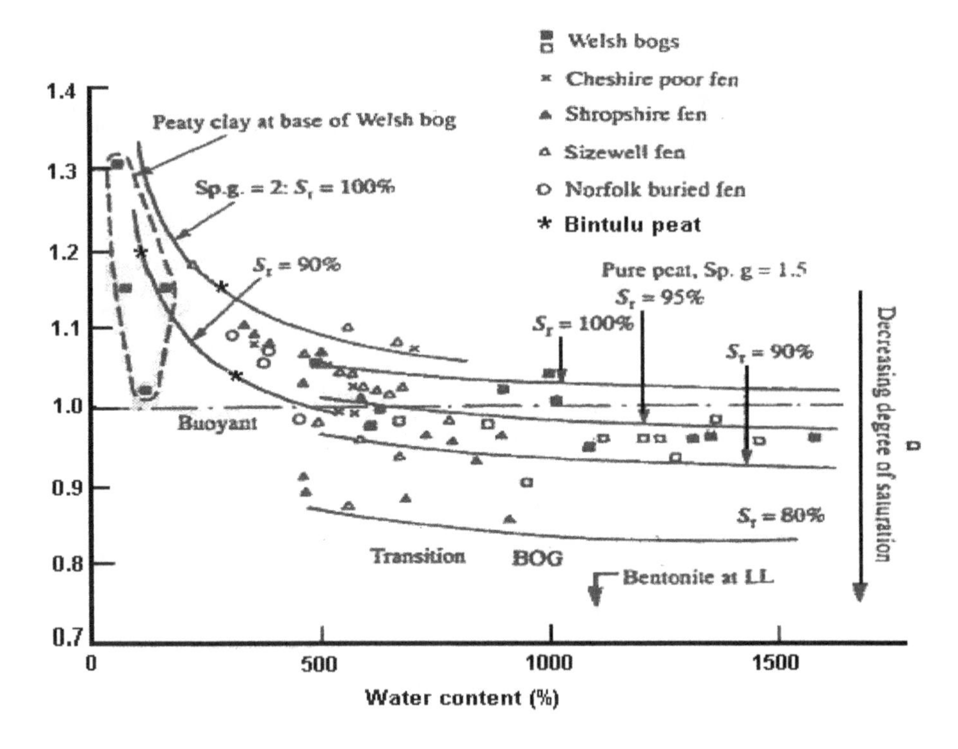

Figure 3.30 Graph of bulk density vs. water content.

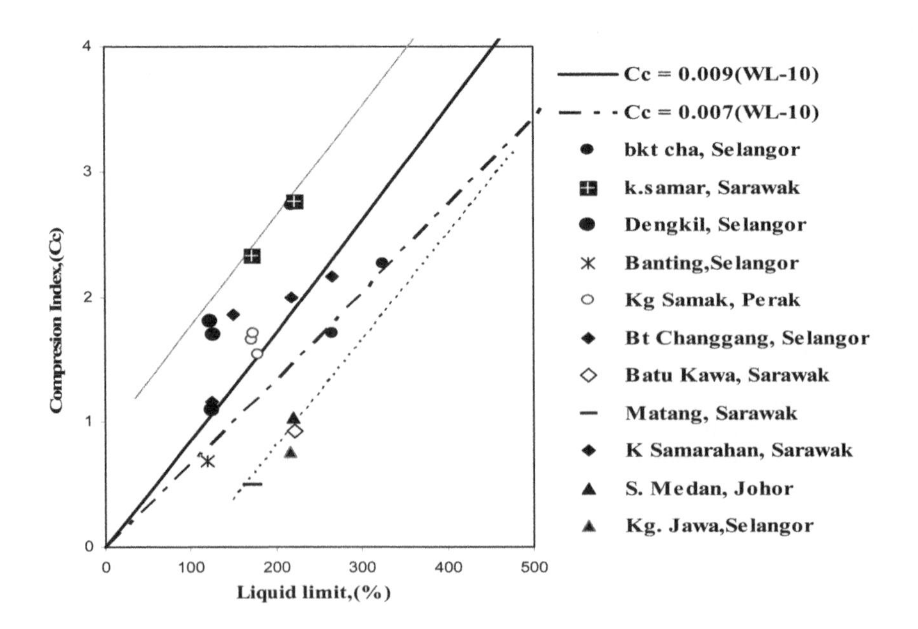

Figure 3.31 Compression index vs. liquid limit of Malaysian peat (*after* Kazemian *et al.*, 2009).

Table 3.12 Physical and chemical properties of peat (*after* Kazemian *et al.*, 2011b).

Peat type	*Natural water content (w, %)*	*Bulk density ($Mg\ m^{-3}$)*	*Specific gravity (G_s)*	*Acidity (pH)*	*Ash content (%)*	*Reference*
Fibrous-woody	484909	–	–	–	17	Colley (1950)
Fibrous	850	0.95–1.03	1.1–1.8	–	–	Hanrahan (1954)
Peat	520	–	–	–	–	Lewis (1956)
Amorphous and fibrous	500–1500	0.88–1.22	1.5–1.6	–	–	Lea and Brawner (1963)
	200–600	–	1.62	4.8–6.3	12.2–22.5	Adams (1965)
	355–425	–	1.73	6.7	15.9	
Amorphous to fibrous	850	–	1.5	–	14	Keene and Zawodniak (1968)
Fibrous	605–1290	0.87–1.04	1.41–1.7	–	4.6–15.8	Samson and LaRochell (1972)
Coarse fibrous	613–886	1.04	1.5	4.1	9.4	Berry and Vickers (1975)
Fibrous sedge	350	–	–	4.3	4.8	Levesqe *et al.* (1980)
Fibrous sphagnum	778	–	–	3.3	1	
Coarse fibrous	202–1159	1.05	1.5	4.17	14.3	Berry (1983)
Fine fibrous	660	1.05	1.58	6.9	23.9	Ng and Eischens (1983)
Fine fibrous	418	1.05	1.73	6.9	9.4	
Amorphous granular	336	1.05	1.72	7.3	19.5	
Peat portage	600	0.96	1.72	7.3	19.5	Edil and Mochtar (1984)
Peat Waupaca	460	0.96	1.68	6.2	15	
Fibrous peat Middleton	510	0.91	1.41	7	12	
Fibrous peat Noblesville	173–757	0.84	1.56	6.4	6.9–8.4	
Fibrous	660–1590	–	1.53–1.68	–	0.1–32.0	Lefebvre *et al.* (1984)
Fibrous	660–890	0.94–1.15	–	–	–	Olson and Mesri (1970)
Amorphous Peat	200–875	1.04–1.23	–	–	–	
Peat	125–375	0	1.55–1.63	5–7	22–45	Yamaguchi *et al.* (1985)
Peat	419	1	1.61	–	22–45	Jones *et al.* (1986)
Peat	490–1250	–	1.45	–	20–33	Yamaguchi *et al.* (1987)
Peat	630–1200	–	1.58–1.71	–	22–35	Nakayama *et al.* (1990)
Peat	400–1100	0.99–1.1	1.47	4.2	5–15	Yamaguchi 1990
Fibrous	700–800	~1.00	–	–	–	Hansbo (1991)
Peat (Netherlands)	669	0.97	1.52	–	20.8	Termatt and Topolnicki (1994)
Fibrous (Middleton)	510–850	0.99–1.1	1.47–1.64	4.2	5–7	Ajlouni (2000)
Fibrous (James Bay)	1000–1340	0.85–1.02	1.37–1.55	5.3	4.1	

Table 3.13 Engineering properties of peat (http://www.roadex.org/, 2004).

	Types of peat		
Property	*Fibrous*	*Medium decayed*	*Amorphous granular*
Water content (%)	700–2000	500–1200	500–900
Ash content (%)	1.5–3.0	3–8	8–30
In situ bulk density (kg m^{-3})	900–1100	900–1100	900–1100
Dry density (kg m^{-3})	40–70	70–100	100–140
Void ratio	10–25	8–17	7–13
Permeability (m s^{-1})	10^{-5}–10^{-6}	10^{-6}–10^{-7}	10^{-7}–10^{-8}

In situ bulk density

The *in situ* bulk density of a peat bog depends predominantly on its moisture content. Amorphous granular peats can have *in situ* undrained bulk densities of up to 1,200 kg/m^3, whilst very woody fibrous peats can have *in situ* densities as low as 900 kg/m^3 in unsaturated conditions.

Specific gravity

The specific gravity of peat typically varies from 1.4 to 1.8, with the higher ranges again reflecting a higher mineral content.

Void ratio

The initial void ratio of peat varies with the type of peat and moisture content, ranging from 7 to 25.

Permeability

The permeability of peat in the field can be highly variable depending on its morphology and can reduce dramatically when subjected to loading. The permeability of virgin peat, for example, can range from 10^{-2} to 10^{-5} cm s^{-1}, but when loaded, such as with a low road embankment, it can quickly reduce to 10^{-6} cm s^{-1}, or to as low as 10^{-8} to 10^{-9} cm s^{-1} under a higher embankment.

Tables 3.12 and 3.13 show some typical engineering properties of peat.

Chapter 4

Shear strength of natural peat

4.1 INTRODUCTION

As mentioned before, peat and organic soils commonly occur as extremely soft, wet, unconsolidated surficial deposits that are integral parts of wetland systems. They may also occur as strata beneath other surficial deposits. These problematic soils are known for their high compressibility and low shear strength. Access to these surficial deposits is usually very difficult, as the water table will be at, near or above the ground surface. Undoubtedly, this is the cause of the tendency to either avoid construction and building on these soils, or when this is not possible, to simply remove, replace or displace them. In some instances this may lead to possibly uneconomical design and construction alternatives. However, in many countries this material covers a substantial area. The pressure on land use from industry, housing and infrastructure is leading to more frequent utilization of such marginal ground. It is therefore necessary to be able to obtain suitable design parameters for strength and compressibility, as well as to find suitable construction techniques on these materials.

Shear strength is one of the most important parameters in engineering design when dealing with soil, especially during the pre- and post-construction periods, since it used to evaluate the foundation and slope stability of soil. If the ultimate shear strength is exceeded, the soil will fail or deform. The failure criterion is developed using the stress–strain relationship and elasticity theory. The magnitude of the strain in soil depends on parameters such as the magnitude of the applied load, the composition of the soil, past stress history and void ratio, and also on the manner in which the stress is applied (Anggraini, 2006). Peat usually has very low shear strength and the determination of shear strength is a difficult job in geotechnical engineering because it depends on factors such as the origin of the soil, its water content, organic matter and the degree of humification. During the sampling stage, disturbance of the sample will also affect the evaluation of shear strength.

The increase in shear strength upon consolidation of peat may be significant. According to Munro (2004), the greater the moisture content and decomposition, the lower the shear strength; in addition, a higher mineral content causes higher shear strength. In general, shallow peat, due to its more fibrous nature, is likely to have greater strength than more humified peat at depth (Culloch, 2006).

Peat is mostly considered a frictional or non-cohesive material due to its fibre content and the spatial orientation of the fibres. The high friction angle of peat will not actually reflect a high shear strength due to the fact that the fibres are not always solid and may be filled with water and gas. The presence of fibres will modify the strength behaviour of peat, since the fibres can be considered as reinforcement and can provide effective stress where there is none, inducing anisotropy.

The shear strength of peat can generally be found out in many ways. *In situ* methods such as the field vane shear test and the cone penetration test are very useful, and these tests can be used to avoid many of the problems associated with soil sampling. However, these methods have some inherent limitations since the shear strength can only be determined indirectly through correlation with laboratory results and also from back calculation from the results of actual failures. Further, the variable nature of peat and the difficulties in obtaining good representative samples from the field mean that laboratory testing can only give indicative results (Culloch, 2006).

The most common laboratory test is a direct shear test to determine the drained shear strength of fibrous peat. A triaxial test is frequently used to evaluate the shear strength of peat in the laboratory under consolidated–undrained (CU) conditions. This is due to the fact that the results of a triaxial test on fibrous peats are difficult to interpret because the fibres often act as horizontal reinforcement; hence failure is seldom obtained in a drained test because the triaxial test for peat with low permeability, if performed under drained conditions, may take several days to complete.

Edil and Dhowian (1981) and Landva and La Rochelle (1983) showed that the effective internal friction φ' of peat is generally higher than that of inorganic soil (e.g. undrained friction angle of amorphous peat and fibrous peat is in the range 27–32° under a normal pressure of 3–50 kPa; on the other hand, for amorphous granular peat the effective internal friction is 50° and for fibrous peat it is in the range 53–57°). The undrained friction angle of peat in West Malaysia is in the range 3–25° (Huat, 2004).

The determination of undrained shear strength is also important because peat is always present below the groundwater level. This is usually done in situ because sampling of fibrous peat for laboratory evaluation of undrained shear strength is almost impossible. Some approaches that have been used for in situ testing for peat are: cone penetration test, dilatometer test, vane shear test, pressure-meter test, plate load test and screw plate load test (Edil, 2001). The vane shear test is the most commonly used; however, the interpretation of the test results must be approached with caution. A range of 3–15 kPa is obtained for the undrained shear strength of peat, which is much lower than that of mineral soils. A correction factor of 0.5 is suggested by Noto (1991) and Hartlen and Wolski (1996) for the test results on organic soil for which the liquid limit is more than 200%.

The evaluation of shear strength of peat under undrained conditions is greatly affected by disturbance of the soil sample. Shogaki and Kaneko (1994) stated that to obtain and minimize the effect of sample disturbance, there is no need for sampling if the undrained shear strength is obtained by field tests. The undrained shear strength of the soil can also be measured by performing laboratory tests on specimens trimmed from blocks or large undisturbed samples (Kazemian and Huat, 2009a; Kazemian *et al.*, 2012). In this condition, there could be nearly zero effective confining pressure or even tension at failure in fibrous peat (Edil, 1997).

In fibrous peat, the force is taken by the fibres, which act as reinforcement if the direction of the load is the same as that of the fibres. As a result of the sedimentation process and compaction, the main direction is usually horizontal, but it is possible for a section of peat to have a vertical orientation. The effect of organic matter and stiffness of soils depends largely on whether the organic matter is decomposed or consists of fibres which can act as reinforcement (Arman, 1969; Landva *et al.*, 1983). In general, fibrous peat has higher shear strength than other groups of peat, such as hemic and sapric peats. The shear strength behaviour of peat is observed to be highly anisotropic (Hanzawa *et al.*, 1994). The shear strength of a soil is not only a function of the material itself, but also of the stress applied and the manner in which it is applied.

Shear strength parameters always play a vital role when engineering decisions have to be made about any soils that include peat. Shear strength is a concern both during construction (to support construction equipment) and at the end of construction (to support the structure). The low shear strength and high compressibility of peat soils confines them to the 'problematic' category. Accuracy in determining the shear strength of these soils is associated with several variables, as described earlier.

For peat, the presence of fibres modifies our concepts of strength behaviour in several ways. It can provide effective stress where there is none and it induces anisotropy. It also results in reduced K_o values compared to clays, as will be explained later. Finally, shear resistance may continue to develop at high strains without a significant peak behaviour.

Early research on peat strength indicated some confusion as to whether peat should be treated as a frictional material like sand or cohesive like clay. Commonly, surficial peats are encountered as submerged surficial deposits. Because of their low unit weight and submergence, such deposits develop very low vertical effective stresses for consolidation and the associated peats exhibit high porosities and hydraulic conductivities comparable to those of fine sand or silty sand (Dhowian and Edil, 1980). Such a material can be expected to behave 'drained' like sand when subjected to shear loading. However, with consolidation porosity decreases rapidly and hydraulic conductivity becomes comparable to that of clay. For example, the time for the end of primary consolidation as defined by full dissipation of the base pore pressure in a singly drained consolidation test on Middleton peat increased 10 times under the second load increment, from about 2 min to 20 min, and to 200 min after several load increments (Edil *et al.* 1991, 1994). There is a rapid transition immediately from a well-drained material to an 'undrained' material.

4.2 LABORATORY TESTING

Several methods can be used to determine the drained and undrained shear strength in the laboratory, namely the triaxial test, shear box test, ring shear or direct shear, and vane shear test. Figure 4.1 illustrates some of the equipment normally used for shear strength testing in the laboratory.

As mentioned above, unlike mineral soils, the presence of fibres in peat and their varying interaction within the shearing mode imposed by the particular testing procedure creates difficulties in assessing the true operating strength value. The use of drained strength (usually obtained from a triaxial test) with estimated pore pressure

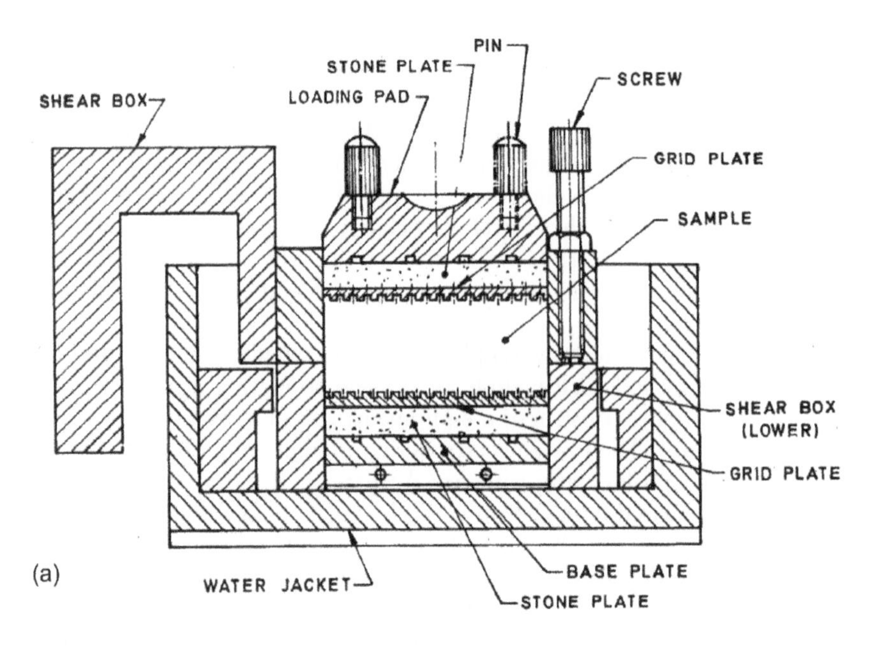

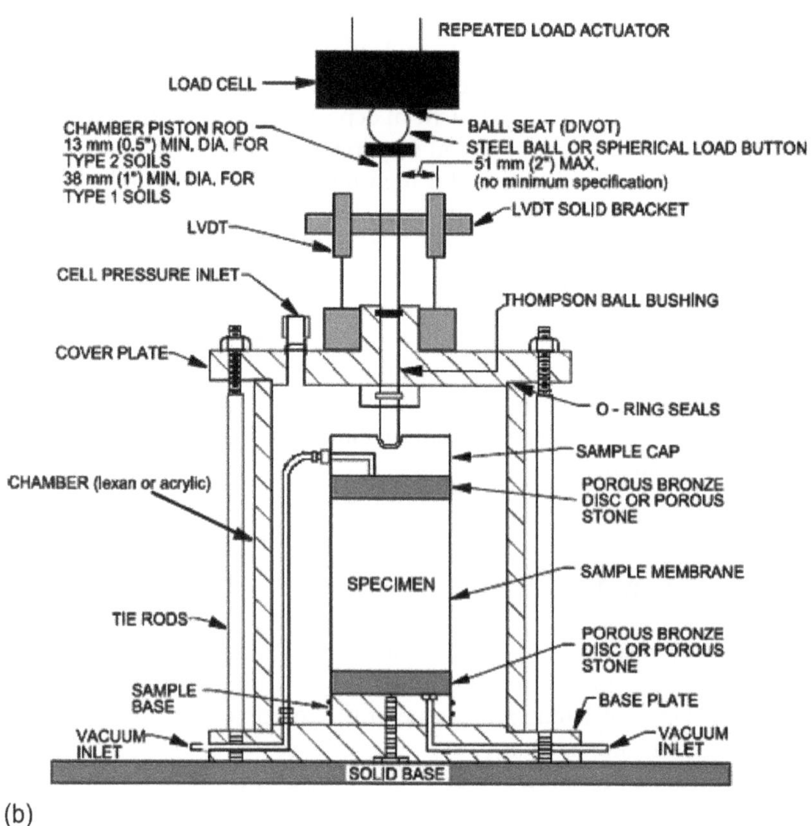

Figure 4.1b (a) Shear box (*Source*: BIS: SP 36 (Part-1) 2006; (b) triaxial test (*Source*: http://www.fhwa.dot.gov/); (c) vane shear (*Source*: BIS: SP 36 (Part-1) 2006.

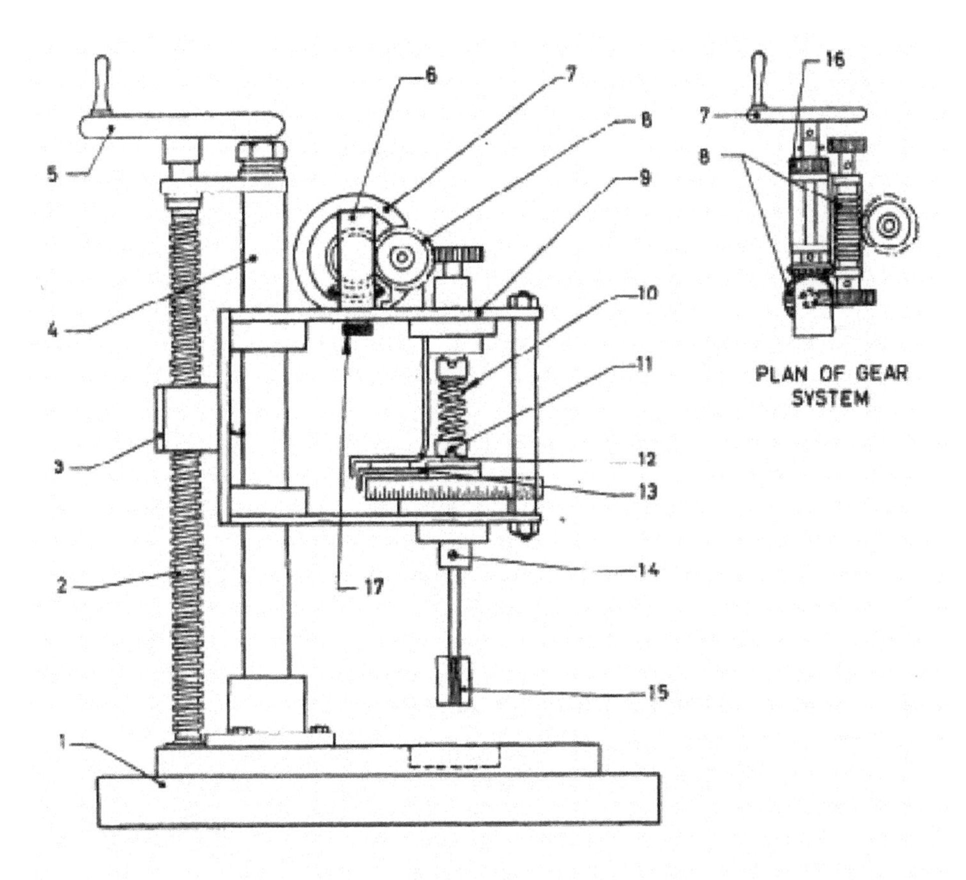

This is only a typical example and any design of appearance satisfying the requirements specified in 2 may be used.

1. Base
2. Lead screw
3. Nut
4. Support pillar
5. Lead screw handle
6. Gear bracket
7. Torque applicator handle
8. Slow motion level & work gears
9. Bracket
10. Torque spring
11. Locating pins
12. Strain indicating pointer
13. Maximum pointer
14. Vane fixing screw
15. Shear vanes
16. Normal speed gear
17. Gear bracket clamp screws

(c)

Figure 4.1 (*continued*).

is more common in stability analysis in Europe because drained parameters are considered to be more fundamental and less variable than undrained strength parameters. Both approaches, however, have their share of difficulties when applied to peat. For instance in fibrous peat there are complications in the effective strength parameters due to fibre content. Edil (1997) reported that one could have nearly zero effective confining pressure or even tension at failure in fibrous peat.

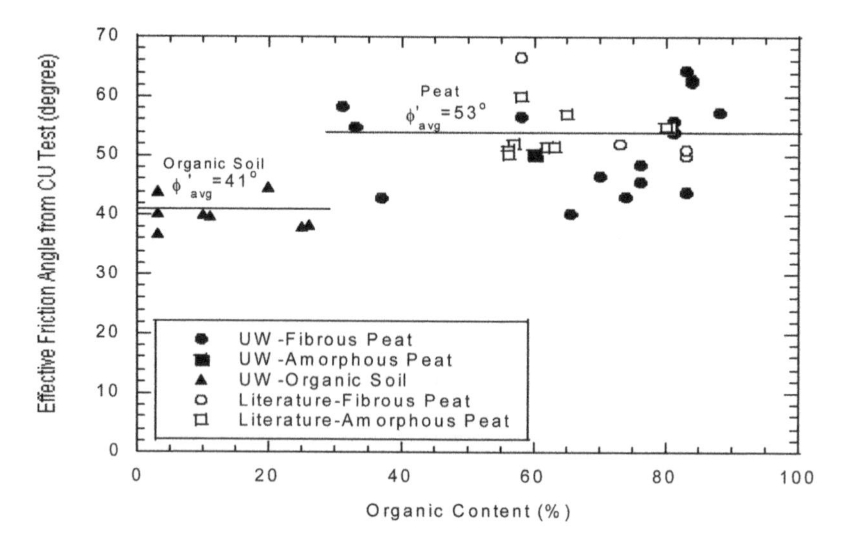

Figure 4.2 Effective friction angles vs. organic content (*after* Edil, 2003).

4.2.1 Drained shear strength parameters

The *effective friction angle* of peats is typically determined in consolidated undrained (CU) triaxial compression tests and occasionally in drained direct, ring or simple shear tests. Drained triaxial tests are seldom performed due to the gross change in specimen dimensions and shape during the test. Normally, consolidated peats exhibit zero or small effective cohesion and generally high effective friction angles.

Figure 4.2 gives the reported effective friction angle as a function of organic content compiled by Edil (2003). In this figure, materials with an organic content less than approximately 25% are called 'organic soils'. The average effective angle of friction is 53° for the peats and clearly above the average angle of 41° for the organic soils.

Yamaguchi *et al.* (1985) and Farrell and Hebib (1998) report lower friction angles in triaxial extension tests compared to triaxial compression tests.

4.2.2 Undrained shear strength parameters

Table 4.1 shows shear box test results of various tropical peats from West Malaysia. The shear strength of peat, as mentioned earlier, is low, with cohesion (c) values in the range 6–17 kPa and angle of internal friction (ϕ) in the range 3–25°. The shear strength parameters are generally lower with increasing degree of humification (less fibre content). The angle of friction is also generally higher for the more fibrous (fibric) peat.

Kazemian *et al.* (2012) have also reported a much higher friction angle for all three peats; fibrous, hemic and sapric (Figure 4.3). To carry out the consolidated undrained triaxial test, the authors collected peat from various locations in Selangor, Malaysia,

Table 4.1 Laboratory shear box test results of peats (*after* Al-Raziqi *et al.*, 2003).

Location	*Moisture content (%)*	*Organic content (%)*	*Liquid limit (%)*	*von Post scale*	*Cohesion (kPa)*	*Angle of internal friction (degree)*
Banting, Selangor	211	85	294	H1	9-11	9-20
	195	79	219	H2	6–11	9–16
	832	84	361	H5	8–10	7–10
	219	94	316	H6	11–12	9–12
	225	85	166	H8	8–12	6–11
Kg. Jawa, Selangor	215	78	180	H3	10–12	6–14
	209	89	325	H6	12–14	7–25
	786	85	368	H8	7–11	8–13
Kg. Jawa, Selangor	680	85	298	H3	11–12	10–15
	747	93	352	H5	10–12	5–10
	720	83	282	H7	7–9	9–12
Dengkil, N. Sembilan (west Malaysia)	246	98	305	H2	13–17	3–12
	301	98	335	H5	11	13–15
	786	83	377	H8	8–9	12–20

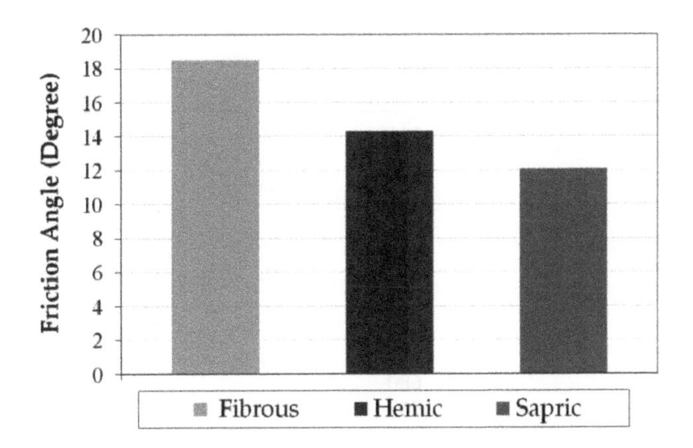

Figure 4.3 Friction angle of fibrous, hemic and sapric peats (*after* Kazemian *et al.*, 2012(a)).

so as to have all three varieties: fibrous, hemic and sapric peats. The organic contents were reported to be 94%, 81% and 75% for fibrous, hemic and sapric peats, respectively. Similarly, the fibre contents were reported to be 79%, 53% and 31% respectively. Further, Kazemian *et al.* (2012) have also presented the cohesion of various peats (Figure 4.4). The strength parameters of fibrous peat are higher than those of hemic and sapric peats, as these are affected by the content of fibres, as stated by Edil (1997).

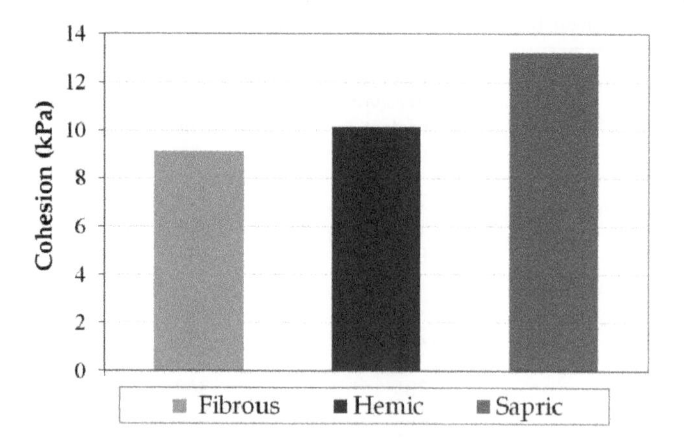

Figure 4.4 Cohesion of fibrous, hemic and sapric peats (*after* Kazemian *et al.*, 2012(b)).

Table 4.2 Vane shear strength of peat (*after* Al-Raziqi *et al.*, 2003).

Location	*Moisture content (%)*	*Organic content (%)*	*Liquid limit (%)*	*von Post scale*	*Field vane shear strength (kPa)*
Banting, Selangor	211	85	294	H1	10–12
	195	79	219	H2	11
	832	84	361	H5	10
	219	94	316	H6	7–9
	225	85	166	H8	4
	802	83	362	H10	4–6
Kg. Jawa, Klang	214	79	180	H3	11
	225	84	325	H6	8
	618	88	368	H8	5
Kg. Jawa, Klang	680	85	298	H3	10–15
	747	93	352	H5	5–10
	720	83	282	H7	9–12
Dengkil, N. Sembilan (West Malaysia)	246	98	305	H2	9–13
	301	98	335	H5	6–10
	786	83	377	H8	3–6
Berengbengkel, Central Kalimantan	467–1224	41–99	–	H2–H5	6–17

4.3 VANE SHEAR STRENGTH

Table 4.2 shows the vane shear strength of various tropical peats obtained from West Malaysia using the small field vane shear test. The undrained shear strength of the peat soil (S_u) is in the range 3–15 kPa, which is generally much lower than that of mineral soils (soft clays).

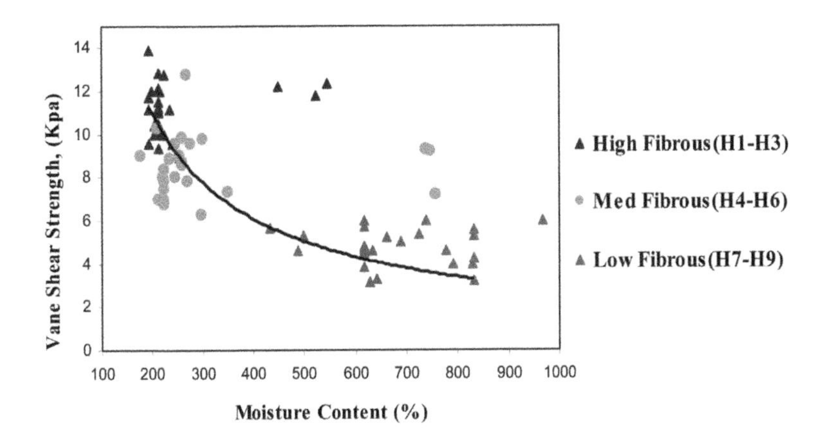

Figure 4.5 Vane shear strength vs. moisture content (*after* Al-Raziqi *et al.*, 2003).

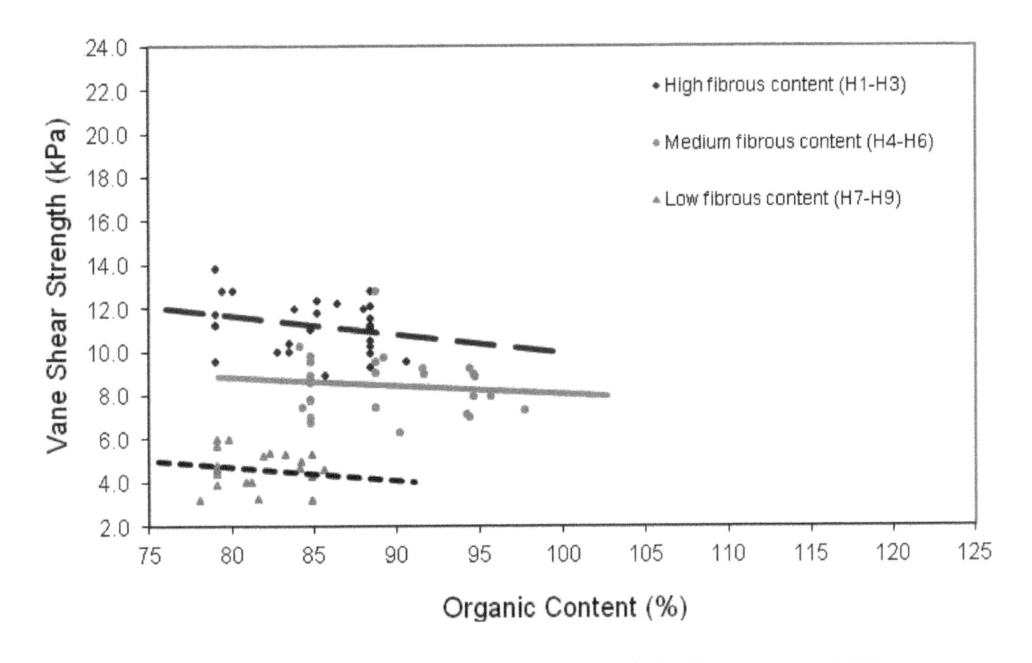

Figure 4.6 Vane shear strength vs. organic content (*after* Al-Raziqi *et al.*, 2003).

Figures 4.5 and 4.6 show graphs of vane shear strength vs. water content and organic content of peat, which shows the decreasing behaviour of peat shear strength with increasing moisture content and organic content.

According to Mitchell (1993), the effect of organic matter on the stiffness of soils depends largely on whether the organic matter is decomposed or consists of fibres, which can act as reinforcement. He investigated peat soil to study the effect of organic content on the unconfined compressive strength of soil. The result is presented in

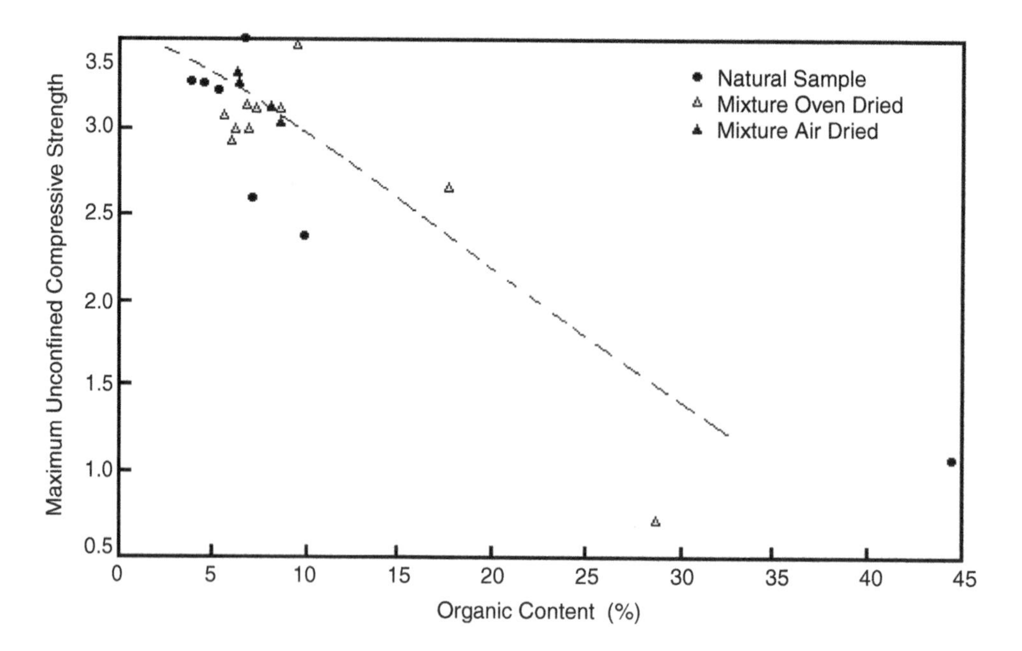

Figure 4.7 Unconfined compressive strength as a function of organic content for a natural soil and soil-peat mixture (*after* Mitchell, 1993. This material is reproduced with permission of John Wiley & Sons, Inc.).

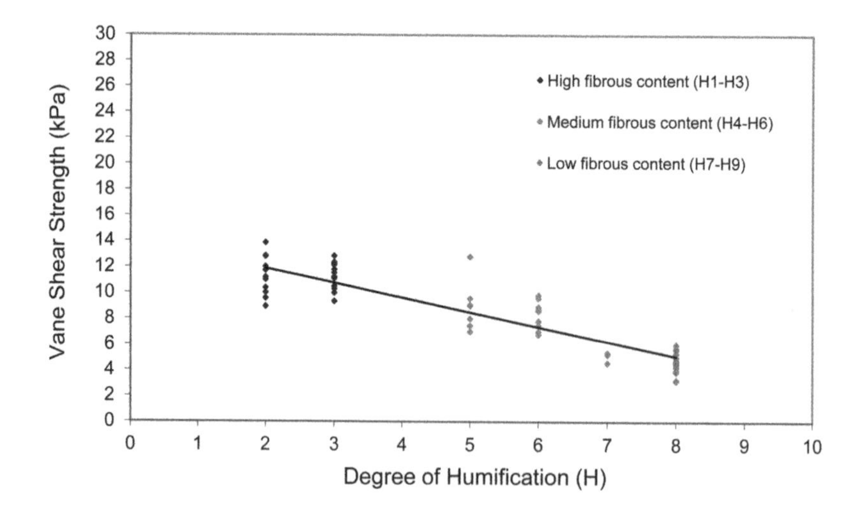

Figure 4.8 Vane shear strength vs. degree of humification (*after* Al-Raziqi *et al.*, 2003).

Figure 4.7. It can be seen that the unconfined compressive strength decreases significantly with increasing organic content.

Figure 4.8 shows the decreasing trend between the degree of humification and the vane shear strength of peat. In general, fibric peat (H1–H3) has higher shear strength

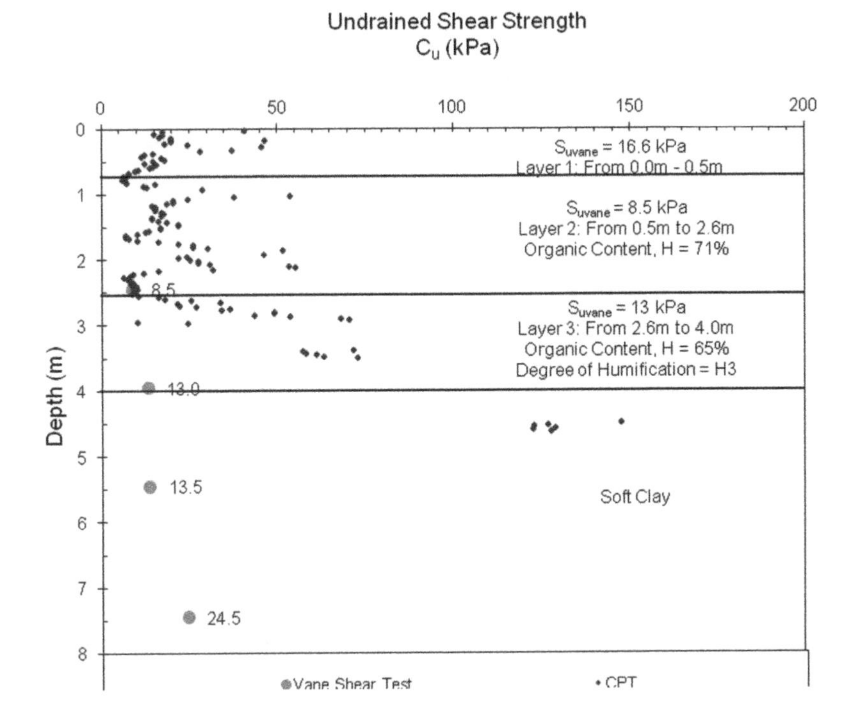

Figure 4.9 Vane shear strength profile of subsoil for site in Putrajaya, Malaysia.

than hemic (H4–H6) and sapric (H7–H10) peat. As noted by Edil (1997), the presence of fibres affects the strength behaviour of peat.

Figure 4.9 shows the field vane shear subsoil profile of a new mosque project site at Putrajaya, Malaysia. As shown, there is only a slight tendency for an increase in strength with depth. The low bulk density of peat together with the high water table implies low effective stresses with depth. Because of this there may not be a discernible increase of strength with depth.

Yogeswaran (1995) reported the average field vane shear strength for tropical peat found in Sarawak (East Malaysia) to be only 10 kPa, while the sensitivity ranged from 2 to 11. The higher water content and low dry density gives peat exceptionally low shear strength. Because of this, and hence the low bearing capacity, a surface foundation on peat has to be generally improved by fill materials before any engineering works can commence. This will be discussed in detail in Chapter 6.

4.4 SHEAR STRENGTH INCREASE WITH CONSOLIDATION

In order to estimate and take into account the rate of increase of undrained shear strength as a function of consolidation and subsequent densification of peat, Magnan (1994) suggests that laboratory compression tests are performed, and the coefficient $\Delta\lambda_{cu} = \Delta c_u/\Delta\sigma$ is then applied to the shear strength obtained from the vane test.

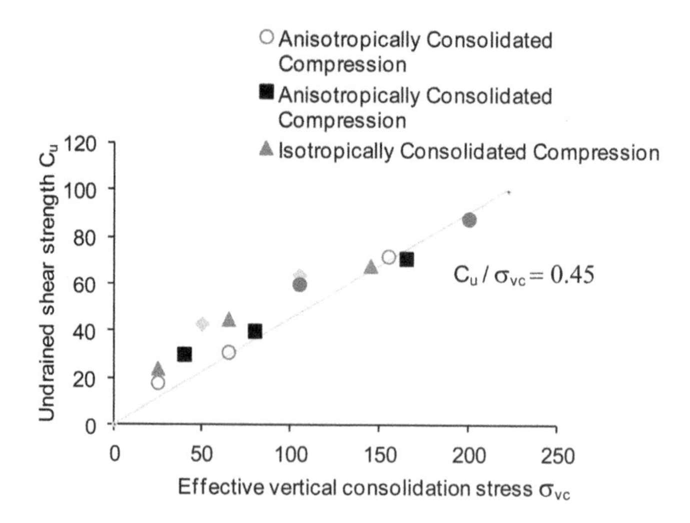

Figure 4.10 C_u vs. depth and σ_{vc} laboratory (*after* Farrel, 1997).

For peat, $\Delta\lambda_{cu}$ is generally higher than clay, with values close to 0.50. The $\Delta\lambda_{cu}$ for Malaysia West Coast clay from Klang, for example, is only 0.25 (Chen and Tan, 2003).

In the field, undrained strength is typically determined by vane shear. Plate load, screw plate and cone penetration tests have been used, but to a lesser extent.

Figure 4.10 shows a plot of undrained strength C_u versus consolidation stress, σ'_c. As shown, the ratio of C_u and σ'_c is 0.45.

$$C_u/\sigma'_c = 0.45 \tag{4.1}$$

Figure 4.11 presents the normalized undrained strength compiled by Edil and Wang (2000). Fibrous and amorphous peats show no perceptible differences and give an average normalized undrained strength of 0.59, with most of the data falling between 0.5 and 0.7. The organic soils (those with an organic content less than 20%) seem to have lower normalized undrained strength compared to the peats. These c_u/σ'_{vc} values appear unusually high compared to the typical values for inorganic clays (0.20 to 0.25).

4.5 EFFECT OF pH ON UNDRAINED SHEAR STRENGTH

Recent investigations have shown that the shear strength is not constant for a given soil and it changes when the environmental conditions change (Fang, 1997; Fang and Daniels, 2006). The effects of pore fluid pH on the vane shear strength of various tropical peaty soils were investigated by Asadi *et al.* (2011a) using the small vane shear test. In this study, to measure S_u, undisturbed samples in small acrylic moulds were cured under different pH gradients of peat water (Asadi, 2010; Asadi et al, 2011e). The peat water pH was adjusted using 0.1 M HCl or 0.1 M NaOH solution for 20 days. Then S_u was measured with a standard vane shear apparatus according to BS

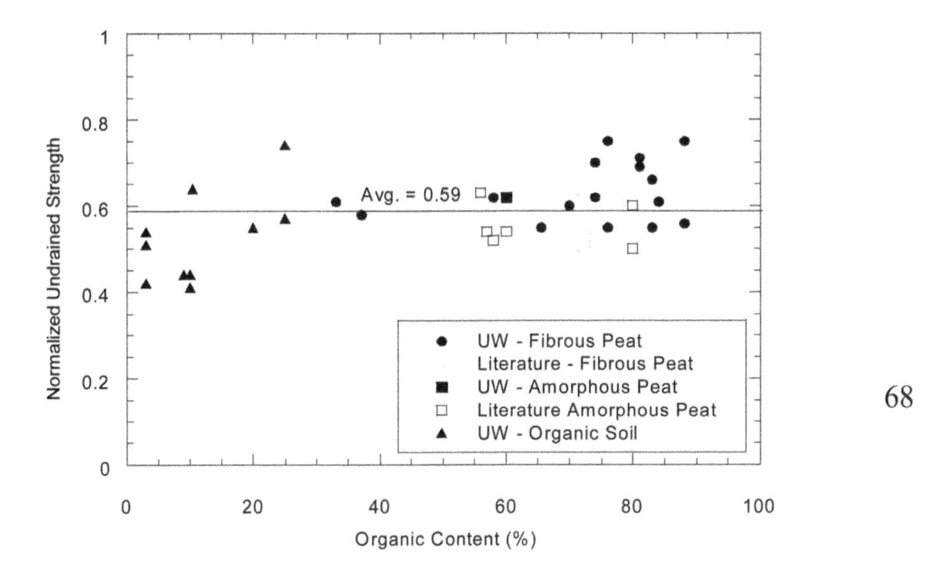

Figure 4.11 Normalized undrained strength versus organic content (*after* Edil and Wang, 2000).

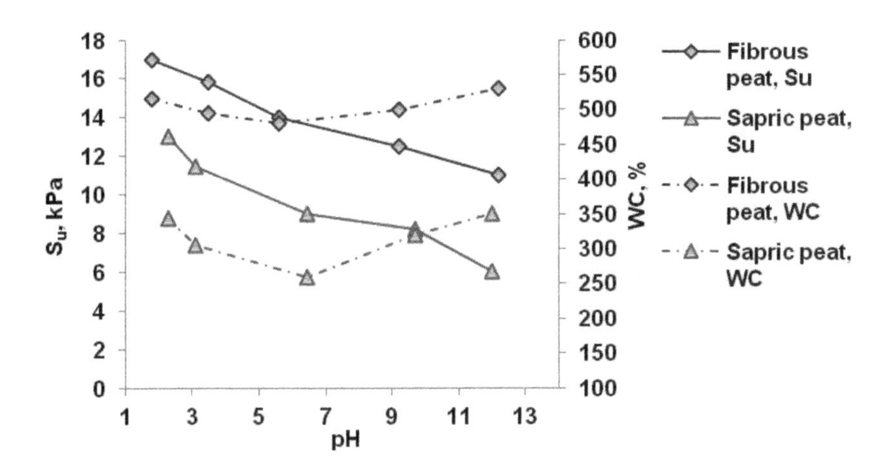

Figure 4.12 Effect of pH on undrained shear strength (*after* Asadi *et al.*, 2011a) .

1377-7-3: 1990. The tests confirmed the increase of the shear strength of both fibrous and sapric samples in the presence of H^+ (Figure 4.12).

4.6 EFFECT OF CYCLIC LOADING

Yasuhara *et. al.* (1994) studied the effect of cyclic loading on peat, which can be introduced by earthquakes, and concluded that peat loses its undrained strength when subjected to undrained cyclic loading. Even if drainage is allowed, due to the dissipation

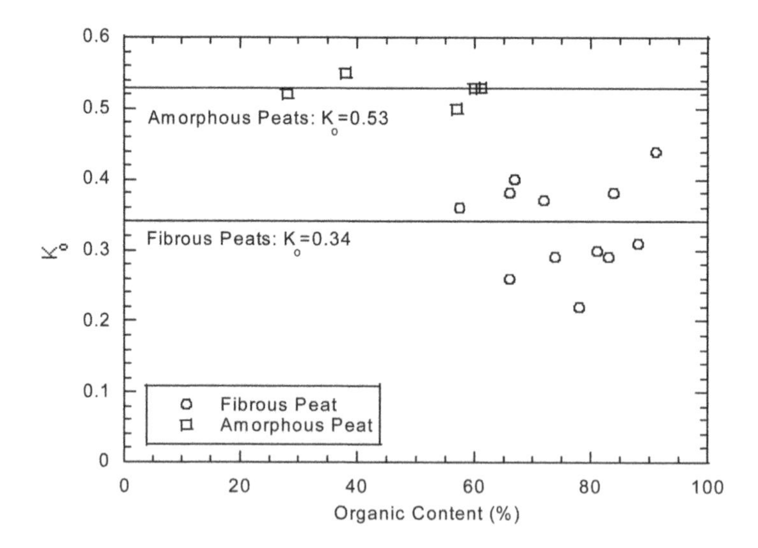

Figure 4.13 K_o vs. organic content (*after* Edil and Wang, 2000).

of excess pore pressure generated by cyclic loading undrained strength is not improved very much.

4.7 K_o BEHAVIOUR

K_o represents the one-dimensional lateral earth pressure coefficient under confined conditions in which no lateral strain is permitted, in other words, the at-rest condition. Several reports of laboratory measurements of $K_o$are available in the literature. The K_o data collected by Edil and Wang (2000) are plotted in Figure 4.13 as a function of N (loss on ignition), i.e. organic content. Figure 4.13 does not reveal a trend with respect to N; however, it is clear that the average K_o for amorphous peats (0.53) is higher than the average value for fibrous peats (0.34).

The K_o data presented above pertains to normally loaded specimens. During unloading, $K_o$becomes greater than that during loading, as would be expected of overconsolidated soils. Schmidt (1966) proposed the following formula to relate $K_{o(rb)}$(during rebound) to $K_{o(nc)}$ (during normal loading):

$$K_{\mathrm{o(rb)}} = K_{\mathrm{o(nc)}}\,(\mathrm{OCR})^{\alpha} \tag{4.2}$$

where OCR is the overconsolidation ratio (the ratio of the maximum past stress to the current stress) and α is an empirical coefficient. Several values have been proposed for α including $\sin\phi'$ (Mayne and Kulhawy, 1982), based on an extensive review of inorganic soils. Another commonly used value for α is 0.5 for inorganic soils. Similar data for peats and organic soils are limited. Kanmuri *et al.* (1998) report a value of 0.5 for α for a fibrous peat based on numerous tests in a K_o-consolidation triaxial apparatus. Edil and Dhowian (1981) reported much lower values for α (0.09 to 0.18) for a range of peats. However, their tests were in a K_o-test tube and they were concerned about the elastic rebound response of the tube wall influencing their data during unloading.

4.8 SUMMARY

The shear strength of a peat deposit depends on degree of humification, water content and mineral content. The higher the moisture content the lower will be the shear strength; the higher the degree of humification the lower the shear strength; and the higher the mineral content the higher the shear strength.

Measuring the *in situ* shear strength of a peat deposit in the laboratory is not easy, due to the difficulty in obtaining a good representative sample in the field, getting it to the laboratory quickly and trimming it to size without disturbance. Because of this, simple *in situ* field tests, such as the vane test, have been used to give an indication of *in situ* shear strengths. These have their limitations, however, and should not be relied upon without supporting evidence.

Chapter 5

Deformation characteristics of peat

5.1 INTRODUCTION

When a soil is subjected to an increase in compressive stress due to foundation load the resulting soil compression (generally called settlement) generally consists of elastic compression (immediate settlement), primary compression (consolidation settlement) and secondary compression. Compared with mineral soils, peat soils are highly organic and highly compressible. Their compression or settlement process may take a considerably longer amount of time. Peat generally possesses low undrained strength and high compressibility. Buildings on peat are usually supported on piles, but the ground around them may still settle, creating a scenario as depicted in Figure 5.1.

Fibrous peat has high organic and fibre content with a low degree of humification. The behaviour of fibrous peat is different from mineral soil because of different phase properties and microstructure (Edil, 2003). Landva and Pheeney (1980) described fibrous peat particles as consisting of fragments of long stems, thin leaves, rootlets, cell walls and fibres, often very large.

Since fibrous peat has hollow perforated cellular structures (Figure 5.2) and a network of fibrous elements in vertical and horizontal sections, it is most likely to have very low shear strength and high compressibility, and the long-term consolidation of peat results in a large deformation. Peat lands have presented difficult subsurface conditions for the construction of roads, dikes, housing developments, storage facilities, industrial parks and so on, including high initial costs and/or continued maintenance operations for years (Colley, 1950; Hanrahan and Rogers, 1981). Peat soils are geotechnically problematic soils as the compression and settlement processes may take a considerably longer time, which increases when the ground water level decreases. Besides settlement, stability problems during construction, such as localized bearing failures and slip failures need to be considered (Duraisamy *et al.*, 2009).

One of the towns in Malaysia that is badly affected by land subsidence on peat is Sibuin Sarawak. The problem is mainly caused by either uncontrolled land filling or ground water lowering due to over-drainage (or both). A substantial part of the town centre is currently in a state of disrepair, with rows of residential housing units being abandoned altogether, due to excessive ground subsidence. Ground subsidence has also resulted in negative gradients to drainage, resulting in unhealthy water stagnation in many parts of the town. Much of the town is also prone to flooding, both locally as well as regionally. The regular occurrence of flooding is, however, the very basis on which the peat soils are geologically formed and sustained (Figure 5.3).

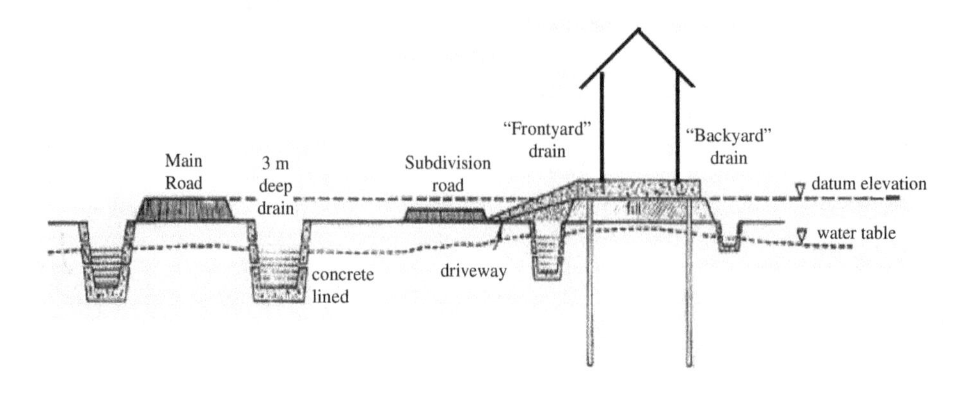

(a)

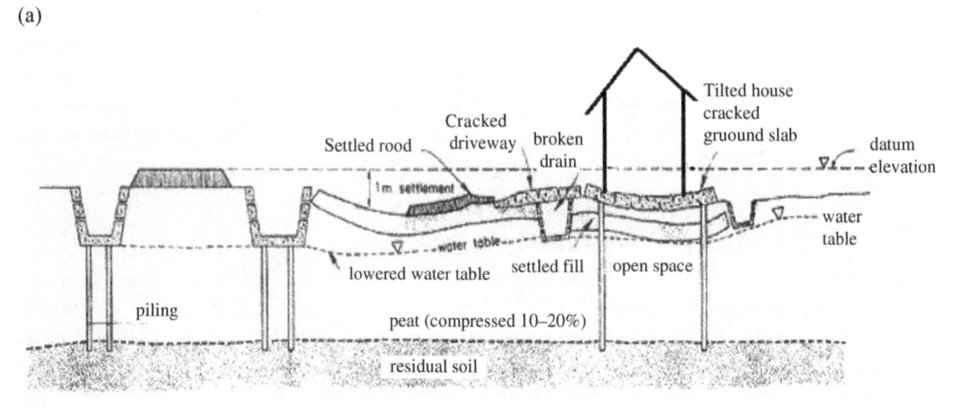

(b)

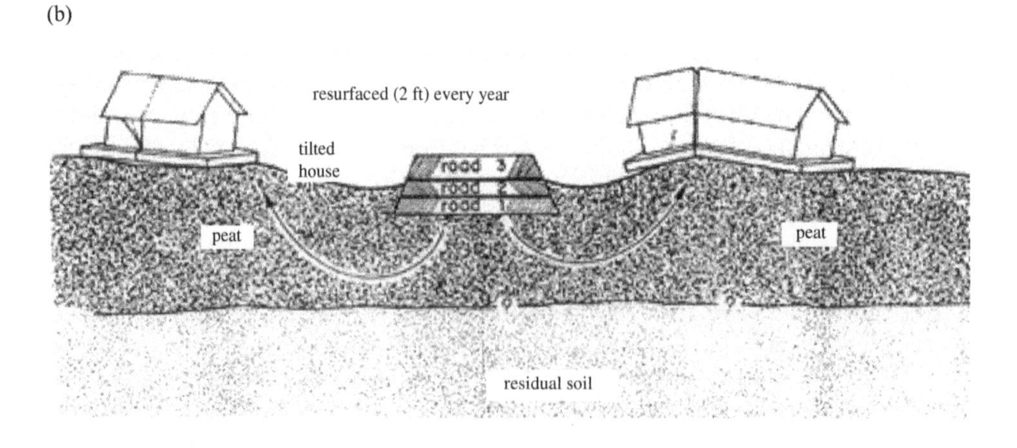

(c)

Figure 5.1 (a) Typical drain layout for housing estate (immediately after completion of construction); (b) several years after completion of construction (scale exaggerated slightly); (c) Sketch of problem due to the bulging of peat near roads when they settle (vertical scale and damage exaggerated).

Figure 5.2 Scanning electron microphotograph of peat: (a) vertical section and (b) horizontal section (*after* Mesri and Ajlouni, 2007).

Field experience on earth-filling projects on several sites on peat shows that peat exhibits a short rapid primary consolidation stage, followed by a prolonged and slow secondary/creep compression (Wong, 2003).

The calculation of settlement requires evaluation of soil parameters from the compression curves usually obtained in oedometer tests. The results of incremental loading oedometer tests are usually presented as the relationship between void ratio, e, and effective vertical stress, σ_v' σ. The vertical effective stress may be plotted on a linear scale to determine the coefficient of volume change, m_v, and oedometer modulus, M, or on a logarithmic scale to determine the compression index c_c. Figure 5.4 shows a typical sketch of an oedometer.

Kirov (2003) found that when testing soft organic soils, the loading schedule used in the oedometer tests has a considerable influence on the results obtained. He suggested that it is better to use an individual loading schedule, as close as possible to the loading conditions expected *in situ*. If other (i.e. different from *in situ* and hence, unsuitable) schedules are used, misleading values for the coefficient of volume compressibility may be obtained. This can subsequently lead to mistakes in the calculation of the settlement of project structures designed for site in question.

5.2 COMPRESSIBILITY PARAMETERS OF PEAT

A peat can consolidate, compress and settle in two ways when loaded:

(a) Slowly, with gradual consolidation and compression allowing time for the peat mass to respond to the load. This is the desired method for constructing a road on peat and allows time for the peat to improve its strength and bearing capacity.
(b) Rapidly, without a change in volume, with rapid spread and shear of the peat causing failure. Peat is highly vulnerable to 'shear overstress' and loadings need to be carefully controlled to keep stresses within the available strength.

In the normal course of events the consolidation, compression and settlement of peat may be considered as taking the form of three phases, instantaneous settlement

(a)

(b)

(c)

Figure 5.3 (a) Ground subsidence in Sibu (Malaysia); (b) damage to infrastructure; (c) February 2001 flood in Sibu (Malaysia) (*after* Tai and Lee, 2003).

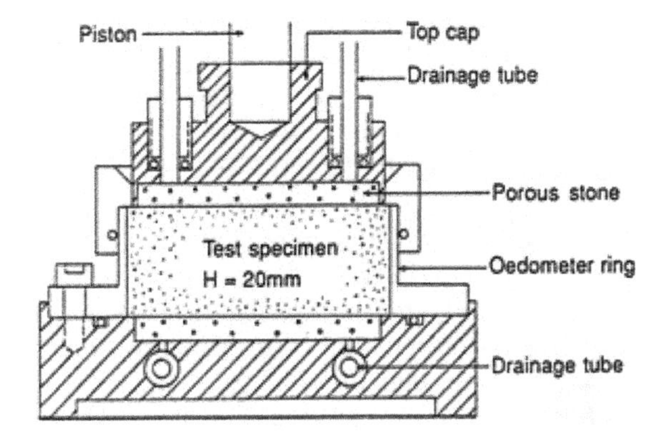

Figure 5.4 Sketch of an oedometer (*Source*: http://www.ngi.no).

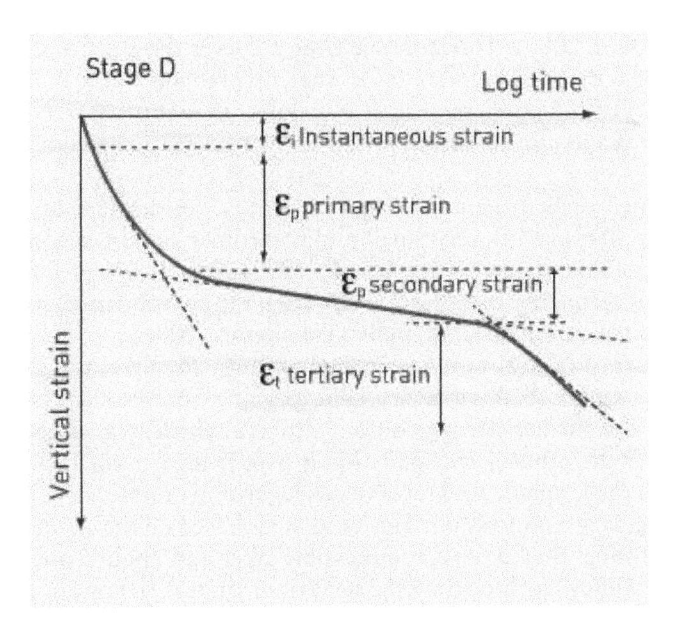

Figure 5.5 Consolidation behaviour of peat (*Source*: http://www.roadex.org/).

or initial compression, primary consolidation and secondary compression. A fourth phase, 'tertiary settlement' is also mentioned by some sources (Figure 5.5).

Initial compression occurs instantaneously after the load is applied, whereas primary and secondary compression depend upon the length of time the load is applied. The initial compression occurs mainly due to the compression of gas within the pore spaces and also due to the elastic compression of soil grains. Primary consolidation observed during the increase in effective vertical stress causes the dissipation of excess pore water pressure. After this dissipation has completed, secondary compression takes place at constant effective vertical stress.

The compression behaviour of peat varies from the compression behaviour of other types of soil in two ways. First, the compression of peat is much greater than that of other soils. Secondly, the creep portion of settlement plays a more significant role in determining the total settlement of peat than for other soil types.

The primary consolidation of fibrous peat takes place very rapidly. A large secondary compression, and even tertiary compression, is also observed to take place (Kazemian and Huat, 2009a).

The dominant factors controlling the compressibility characteristics of peat include the fibre content, natural water content, void ratio, initial permeability, nature and arrangement of soil particles, and inter-particle chemical bonding in some of the soils (Mesri and Ajlouni, 2007). Determination of the compressibility of fibrous peat is usually based on the standard consolidation test.

The *in situ* void ratio of fibrous peats is very high because of the fact that very compressible and bendable hollow cellular fibres form an open entangled network of particles, giving a high initial water content. During both primary and secondary compression, water is expelled simultaneously from within and among the peat particles (Mesri and Ajlouni, 2007). Therefore the e-log p curves show a steep slope indicating a high value of compression index (C_c). The compression index of peat soil ranges from 2 to 15. Furthermore, there is a possibility that secondary compression will start before the dissipation of excess pore water pressure is completed (Leonards and Girault, 1961).

As mentioned earlier, the unit weight of peat is close to that of water; thus the *in situ* effective stress (σ') is very small and sometimes cannot be detected from the results of a consolidation test (Mesri *et al.*, 1997). It is also very difficult to obtain the beginning of secondary compression (tp) from the consolidation curve because the preliminary consolidation occurs rapidly (Yulindasari, 2006).

Compression of fibrous peat continues at a gradually decreasing rate under constant effective stress, called secondary compression. The secondary compression of peat is due to the further decomposition of fibers which is assumed to occur at a slower rate after the primary consolidation is over (Mesri *et al.*, 1997). The slope of the final part of the graph of void ratio versus logarithm of time curve (C) is defined as the rate of secondary compression. This estimate is based on the assumptions that C is independent of time, thickness of compressible layer and applied pressure. The ratio C/C_c has been widely used to study the behaviour of peat (Dhowian and Edil, 1980) and a range between 0.05 and 0.07 for C/C_c is reported by Mesri *et al.* (1997).

Although the rate of primary consolidation of fibrous peat is very high, it decreases with the application of consolidation pressure. According to Lea and Brawner (1963), there will be a significant decrease in the rate of coefficient of consolidation (C_v) during application of pressure from 10 to 100 kPa. The significant reduction factor of 5–100 is attributed to the reduction of permeability due to the increasing pressure.

As with mineral soils (silt and clay), the settlement parameters of peat (i.e. consolidation settlement) may also be determined from standard incremental oedometer (one-dimensional compression) tests, as shown in Figure 5.4. The parameters are interpreted from traditional e-log and σ_v plots. There may be differences in the magnitudes of various quantities measured, but the general shape of the consolidation curves appears reasonably similar and the formulation developed for clay compression can be used to predict the magnitude and rate of settlement (Edil, 1997).

There is, however, a certain class of peats – typically high organic and fibre content materials with low degree of humification – that do not conform to the concept of conventional clay compression because of their different solid phase properties and microstructures. Analysis of the compression of such materials presents certain difficulties when conventional methods are applied because the curves obtained from conventional oedometer tests and the behaviour exhibited by them show little similarity to the clay behaviour. The difference becomes particularly apparent at low vertical stresses, i.e. for surficial peat deposits in early load increments in the laboratory. Primary consolidation is very rapid, and large secondary and even tertiary compressions are observed. Analysis of the compression of such materials presents certain difficulties when conventional methods are applied because the curves obtained from conventional oedometer tests and the behaviour exhibited by them show little resemblance to the clay behaviour. Furthermore, such materials also are more prone to decomposition during oedometer testing. Gas content and additional gas generation may also complicate the interpretation of odometer tests. The behaviour of such peats and the recent advances in formulating their behaviour have been presented by Edil (1997) and den Haan (1997).

5.2.1 Compression index, c_c and void ratio

Figure 5.6 shows the graph of compression index, c_c, versus liquid limit (LL) for various tropical peats. As shown, c_c measured in laboratory tests increases with the liquid limit. Farrell *et al.* (1994) considered that the empirical relationship between the compression index and the liquid limit suggested by Skempton for organic soils

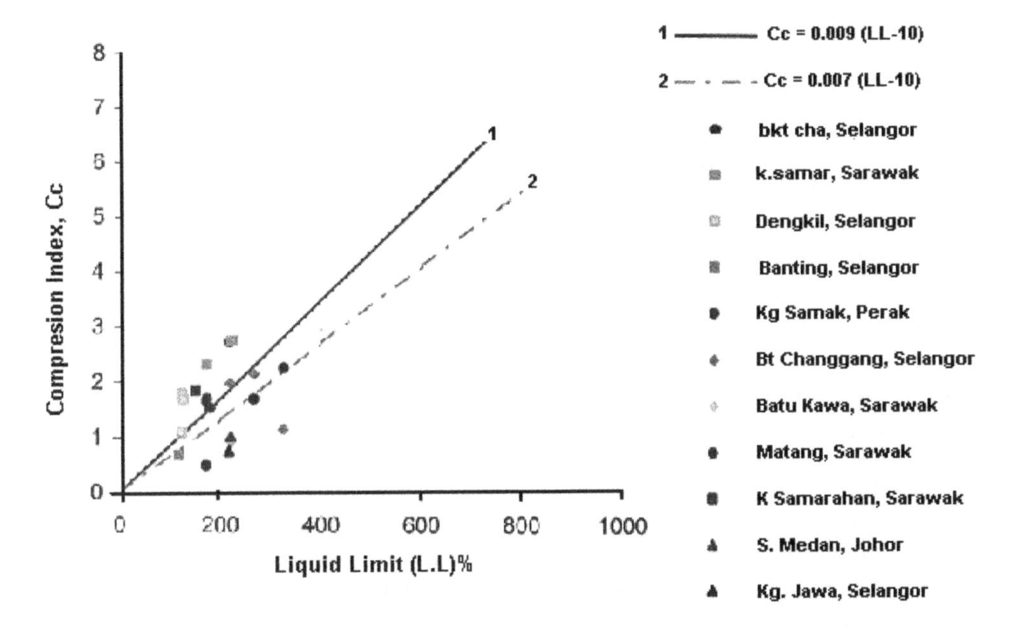

Figure 5.6 Compression index (c_c) vs. liquid limit (LL) (*after* Al-Raziqi *et al.*, 2003).

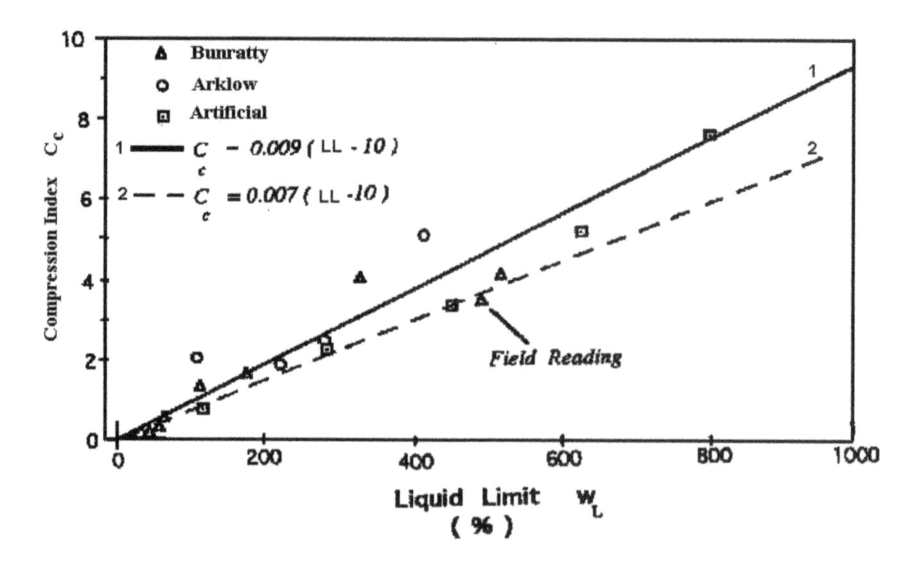

Figure 5.7 Compression index (c_c) vs. liquid limit (LL) (*after* Farrel *et al.*, 1994).

(Eqn. 5.1) gave a reasonable approximation of this parameter. The plot is shown in Figure 5.7.

$$c_c = 0.009(LL - 10) \tag{5.1}$$

Hobbs (1986) estimated the compression index of temperate (fen) peat to be about

$$c_c = 0.007(LL - 10) \tag{5.2}$$

which gives a slightly lower value of c_c. Average values of c_c for tropical peat, however, are a little higher than the above two relationships. The c_c values for peat can be as high as 5 to 10 compared with clay of only 0.2 to 0.8.

Azzouz *et al.* (1976) gave the following relationship for organic soil and peat:

$$c_c = 0.0115\, w \tag{5.3}$$

where w is soil natural water content (%).

Figure 5.8 shows a plot of the initial voids ratio versus liquid limit obtained for tropical peats from Malaysia, together with those for normally consolidated peat found by Miyakawa (1960) and Skempton and Petley (1970). The general trend is that the void ratio increases with an increase in the liquid limit.

The void ratio of peat ranges from 9 for amorphous peat up to 25 for fibrous peat. Such high void ratios give rise to phenomenally high water contents. For comparison, Malaysian marine clay, for instance, has an initial void ratio in the range 1.5 to 2.5. The natural void ratios of peats indicate their higher capacity for compression.

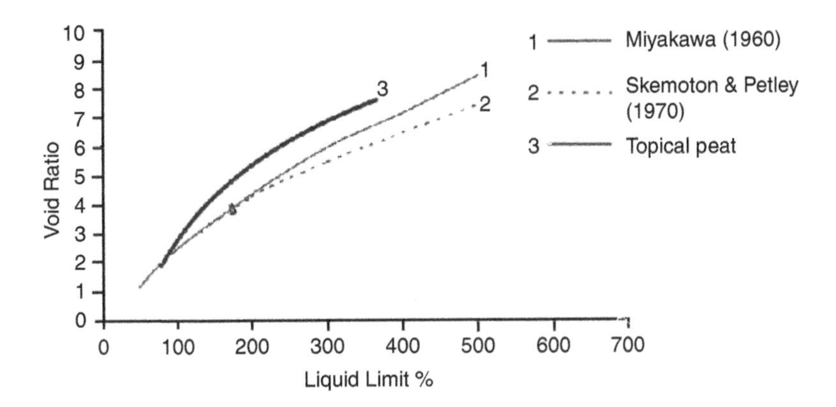

Figure 5.8 Liquid limit vs. void ratio (*after* Al-Raziqi *et al.*, 2003).

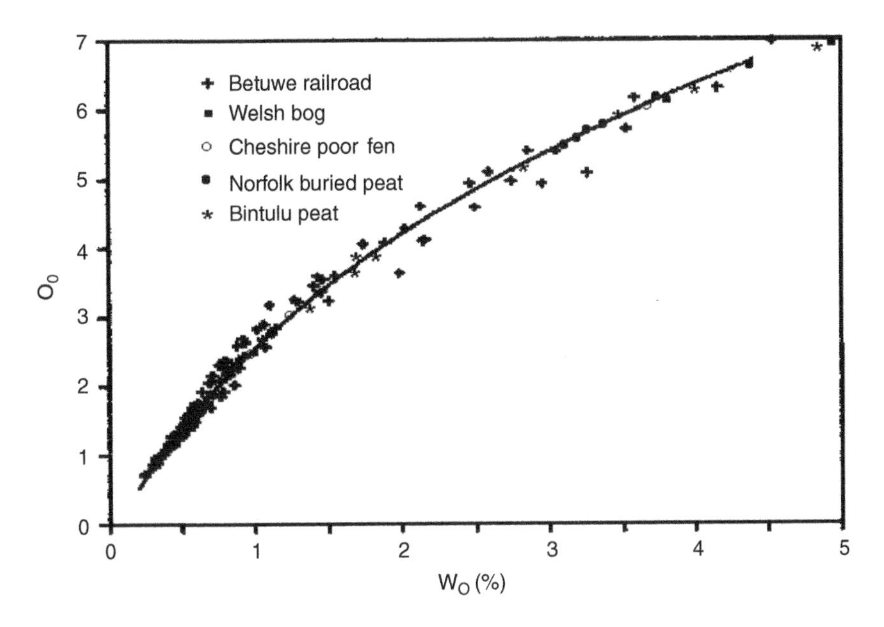

Figure 5.9 Initial water content vs. void ratio for Dutch peat (*after* den Haan, 1997).

Figure 5.9 shows the graph of initial void ratio (e_o) and natural water content (w_o) for Dutch peat. The best-fit line in the above figure is expressed by:

$$e_o = 30.65\frac{(w_o + 0.88)^{0.116}}{1.12} - 30 \tag{5.4}$$

As for the liquid limit, the void ratio increases with increasing natural water content.

Hobbs (1986) found that despite the large variations that occur within peat, the variation in the ratio of $c_c/(1 + e_o)$ is relatively small. This is shown in Figure 5.10.

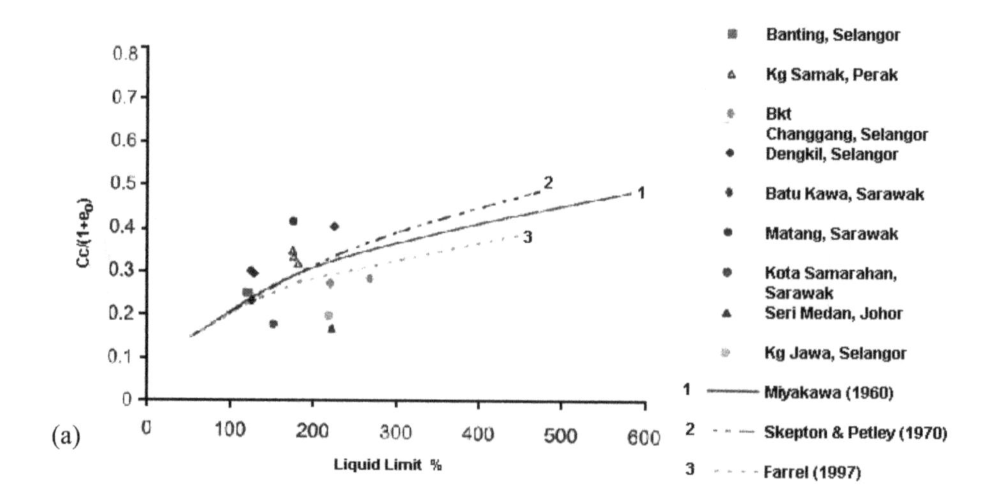

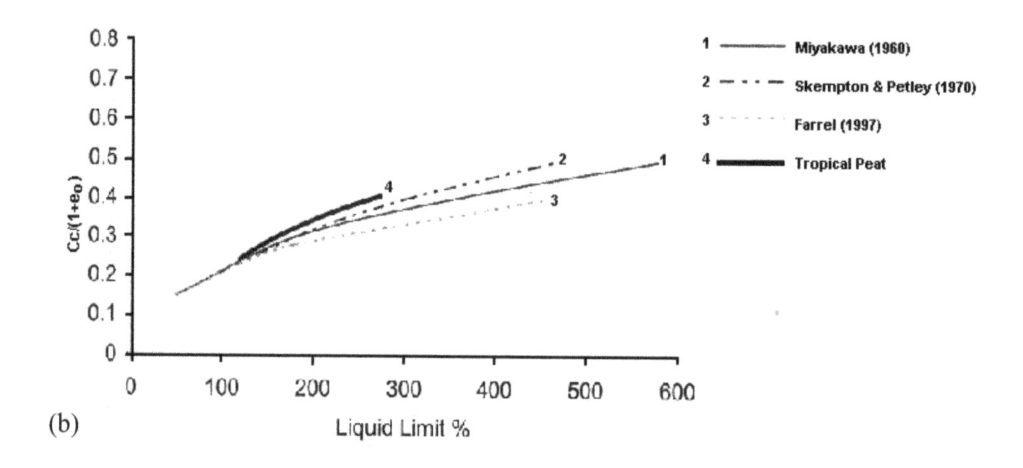

Figure 5.10 Liquid limit vs. $c_c/(1+e_o)$ (*after* Al-Raziqi *et al.*, 2003).

Figure 5.11 shows the settlement parameters for some French bog peat. The trend of the relationship between the settlement parameters $c_c/(1+e_o)$ is to increase with increasing natural water content.

5.2.2 Coefficient of consolidation, c_v

There is a significant decrease in the coefficient of consolidation of the undisturbed samples with an increase in effective stress above the apparent preconsolidation pressure (σ'_c), as can be seen in Figure 5.12. This decrease is more marked in the samples that had higher organic contents (Farrell *et al.*, 1994). The coefficient of consolidation, c_v, generally decreases as the liquid limit of the soil increases.

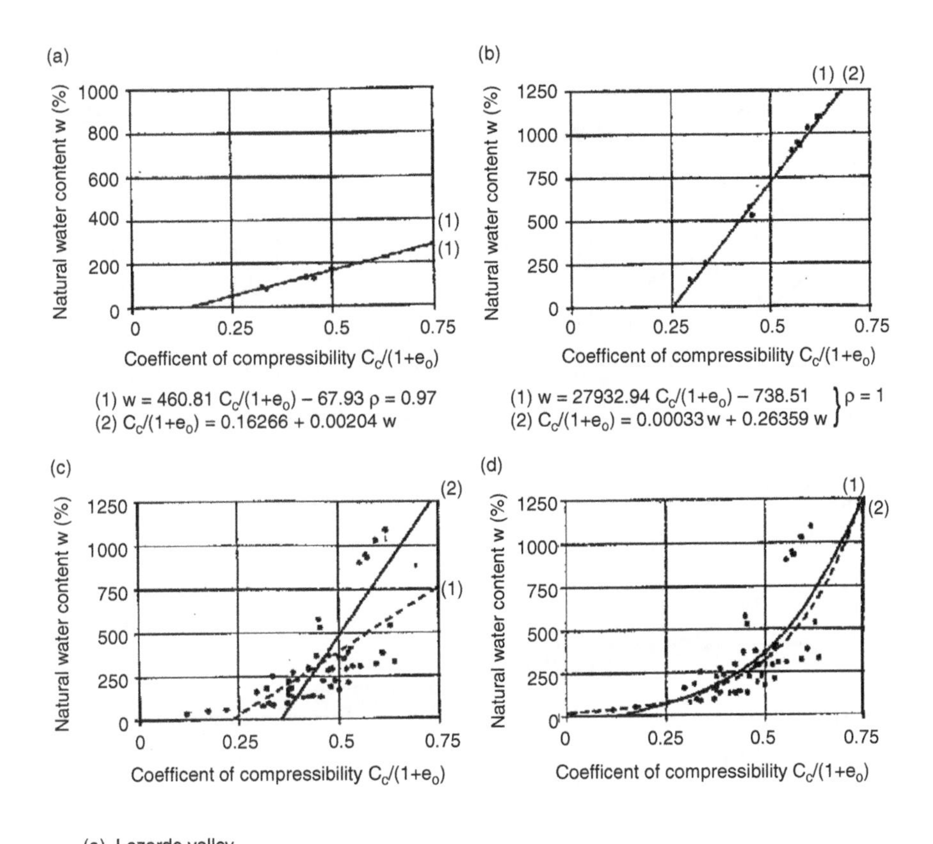

Figure 5.11 Relationship between water content and $c_c/(1+e_o)$ of French bog peat (*after* Magnan, 1994).

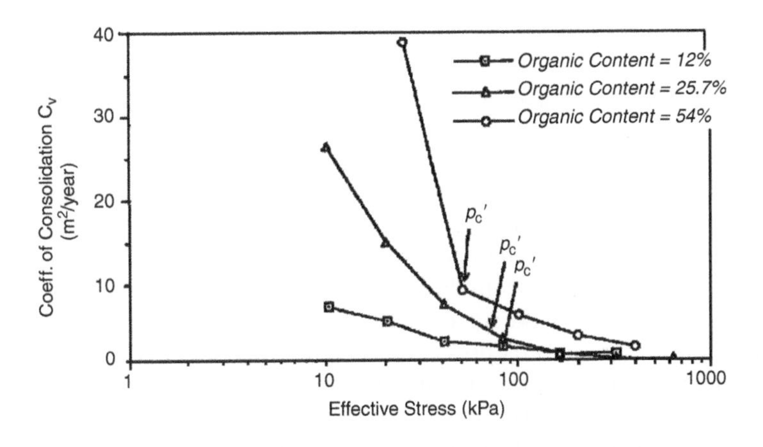

Figure 5.12 Effective stress vs. c_v (*after* Farrell *et al.*, 1994).

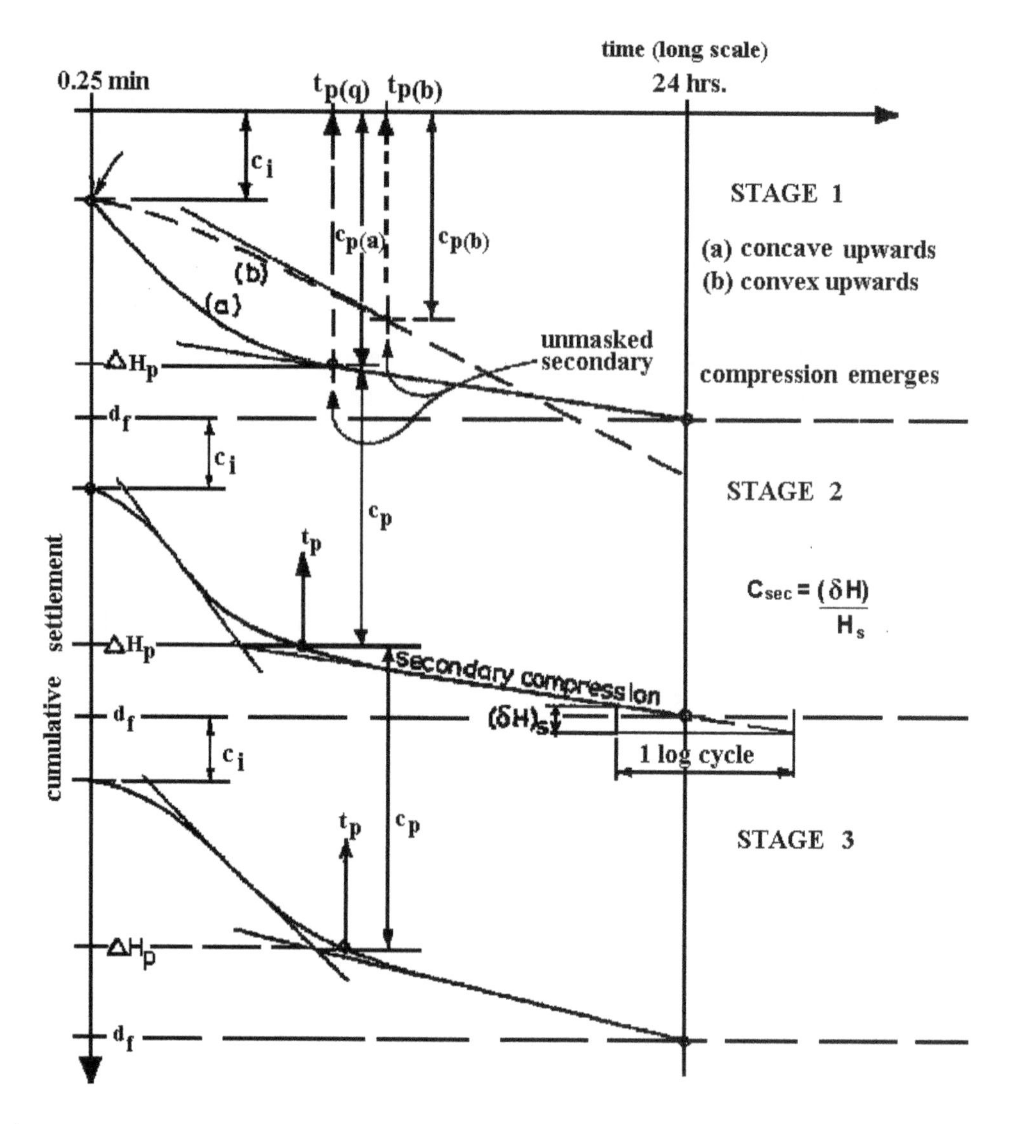

Figure 5.13 Illustration of symbols used in consolidation analysis for peat.

For clay, the log time method proposed by Casagrande and Fadum in 1940 or the square root time method as suggested by Taylor in 1942 is normally used to evaluate c_v. For peat, the method illustrated in Figure 5.13 is usually applied. Primary compression (consolidation, C_p) is a stage from the end of consolidation of the previous stage to the end of the 'primary' stage of the stage considered – that is, compression accompanied by dissipation of excess pore water pressure as defined above. The time elapsed is denoted t_p. Initial compression (C_i) is the amount of compression which occurs from the instant of loading ($t = 0$) to the arbitrarily selected time $t = 15$ s (0.25 min), being the time at which the first sensible settlement reading can usually be observed (d_o). Compression

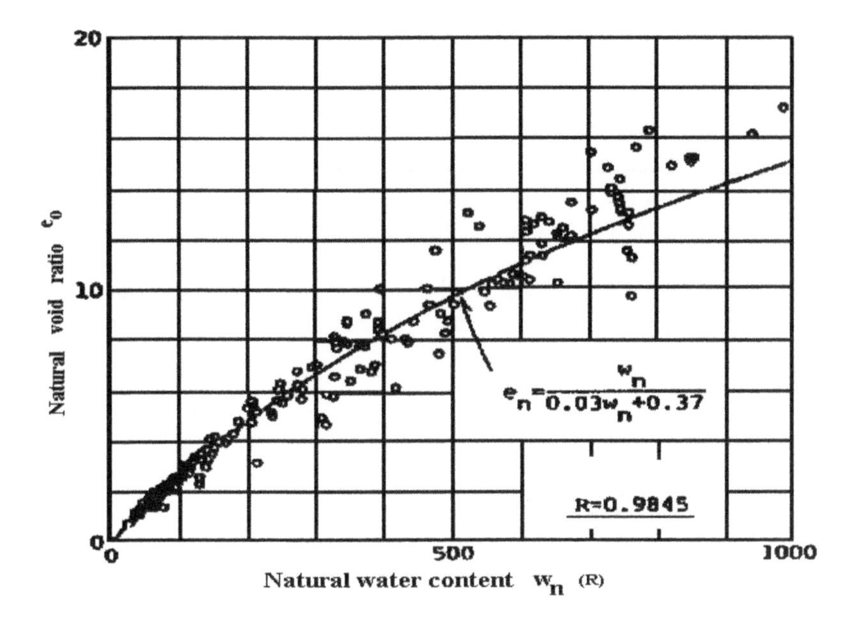

Figure 5.14 Relationship between natural water content w_n and void ratio e_o (*after* Oikawa and Igarashi, 1997).

ΔH_p is the accumulated compression of the specimen to time t_p; normally when ΔH_p extends a horizontal line this will give d_{100}, where 100% consolidation occurs.

Oikawa and Igarashi (1997) propose the following equations (5.5 and 5.6) for calculating c_v. The only input needed for these is the natural water content, without having to do the consolidation test.

$$e_f = 2.47\left[1 - \frac{1}{\exp\left(\frac{2.91}{p}\right)^{0.39}}\right] w_n^{0.85\left[1-\frac{1}{\exp(1.85p)^{0.45}}\right]} \tag{5.5}$$

$$\log c_v = \left[\frac{1.51}{W_n} + 0.20\right]\left[\frac{e_o + e_f}{2}\right] - \frac{1.12}{(W_n - 0.21)^{0.68}} + \log\frac{(1+e_o)}{e_o - e_f} + \log(p - p_o) - 1.06 \tag{5.6}$$

where e_f and e_o are the final and initial void ratios, w_n is the water content, and p is the consolidation pressure in kg cm^{-2}. c_v is in cm s^{-1}. Figure 5.14 shows the plot of water content and void ratio from Oikawa and Igarashi.

Kazemian and Huat (2009a) have studied the compressibility parameters of fibrous peat. High levels of organic matter are an indicator of high compressibility and swell characteristics of the soils (Mesri and Ajlouni, 2007; Puppala *et al.*, 2007; Kazemian *et al.*, 2009). Fibrous peat undergoes large settlements in comparison to clays when subjected to loading. The compression behaviour of fibrous peat varies from the compression behaviour of other types of soil in two ways. First, the compression of peat is much greater than of other soils. Second, the creep portion of settlement plays a more significant role in determining the total settlement of peat than of other soil types. The primary consolidation of fibrous peat is very rapid, and large secondary compression, and even tertiary compression, is observed. Secondary compression (discussed in the next section) is generally the more significant part of compression because the time rate is much slower than for primary consolidation (Yulindasari, 2006).

Figure 5.15 shows the changes in coefficient of volume compressibility (m_v) with consolidation pressure. It was observed by the authors that the m_v decreased with increasing consolidation pressure.

Kazemian and Huat (2009a) have also studied the variation of the coefficient of consolidation with consolidation pressure. Figure 5.16 shows this for fibrous peat.

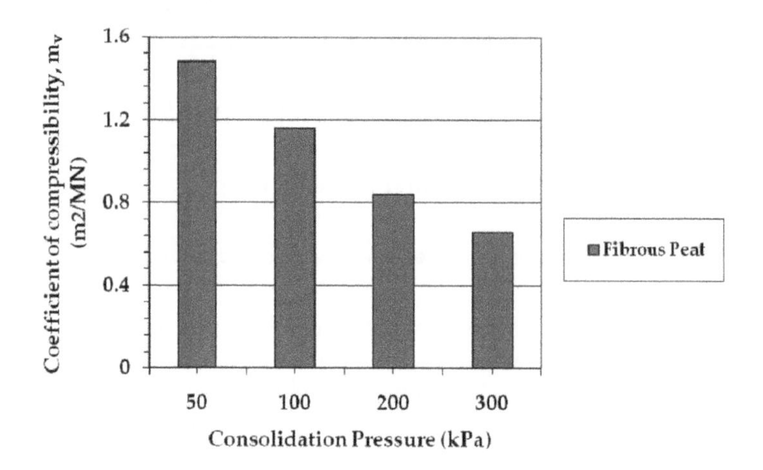

Figure 5.15 Coefficient of compressibility vs. consolidation pressure for fibrous peat (*after* Kazemian and Huat, 2009a).

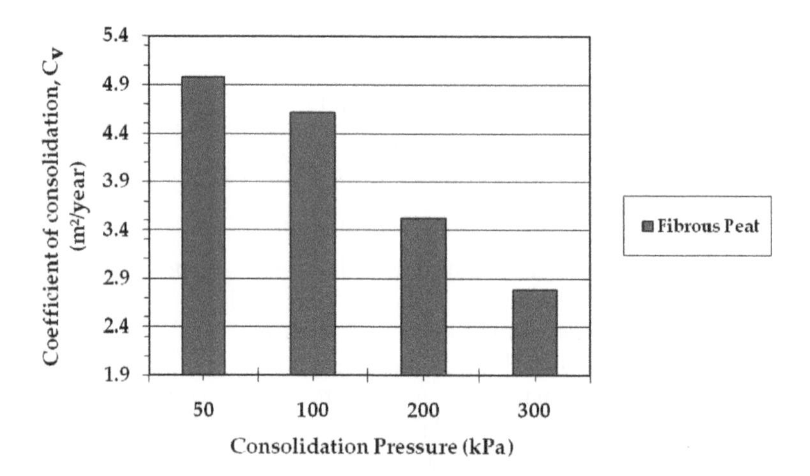

Figure 5.16 Coefficient of consolidation vs. consolidation pressure for fibrous peat (*after* Kazemian and Huat, 2009a).

5.2.3 Secondary compression

In the 'secondary compression' period the load on the peat is further transferred from the water within the peat to the internal peat skeleton as the peat continues to respond to the applied load. This secondary phase is now generally accepted to be the result of the loaded vegetable fragments within the peat mass slipping and reorganizing to form a denser matrix. As the peat fragments come together and pore sizes close up, the permeability through the peat reduces in response.

This simple scenario does not of course give a full picture of the complex processes at work in peat consolidation and strength improvement. The descriptions of 'primary consolidation' and 'secondary compression' are two parts of a continuous dynamic consolidation process at work within the loaded peat mass. The amount of primary consolidation incurred in any location will vary with the type of peat, but it can be generally approximated to be around 50 per cent of the total settlement over time. Secondary compression is normally accepted to take place over a period of 30 years.

Primary compression (also commonly known as consolidation) is the process of compression due to the extrusion of water from soil voids as a result of increased loading. For mineral soils (clay), this process may takes a very long time (several years) to complete. The associated settlement, as mentioned above, is referred to as consolidation settlement. Peat and organic soils of high water, organic and fibre content, however, tend to show a rather rapid primary compression/consolidation (a few minutes in the laboratory) under the applied stress, followed by significant secondary compression that continues for a long time (well beyond 24 hours in the laboratory). Thus secondary compression is severe in such materials and cannot be as easily ignored, as has often been done when dealing with firmer inorganic soils (Edil, 1997).

Secondary compression, or creep, as defined for one-dimensional compression of soils, is the continuing volumetric compression under constant vertical effective stress. This time-dependent component of total settlement is typically taken to occur after essentially all of the excess pore pressure has dissipated: a stage that is considered to occur after consolidation has ended. The associated settlement is called secondary compression or creep.

Referring to Figure 5.13, secondary compression is the final portion of the time settlement curve.

The estimation of secondary consolidation or creep of peat is complex, and the lack of consensus on the best approach is apparent. Several researchers have proposed analytical methods that attempt to model this complex area. However, the estimation of secondary consolidation in practice is frequently carried out using the simplistic assumption that secondary settlement varies linearly with the logarithm of time. Holtz and Kovacs (1981) have illustrated the limitation implied by using this method, but it has nevertheless been widely used both in practice and in research publication (Farrell, 1997).

The definition of the coefficient of secondary consolidation (c_α) in this book is

$$c_{\mathrm{a}} = \frac{\Delta H}{H_o}(\log(tp + \Delta t)/t_p)) \tag{5.7}$$

where H_{o} is the layer thickness, Δt the time increment producing secondary compression ΔH, and t_{p} is the time of completion of consolidation (see Figure 5.17).

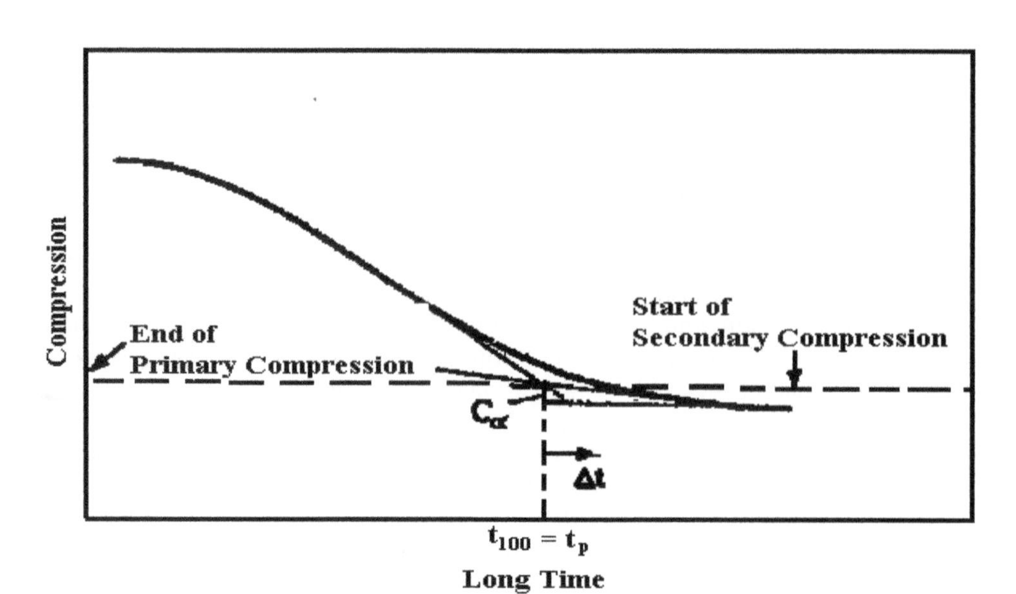

Figure 5.17 Start of secondary compression.

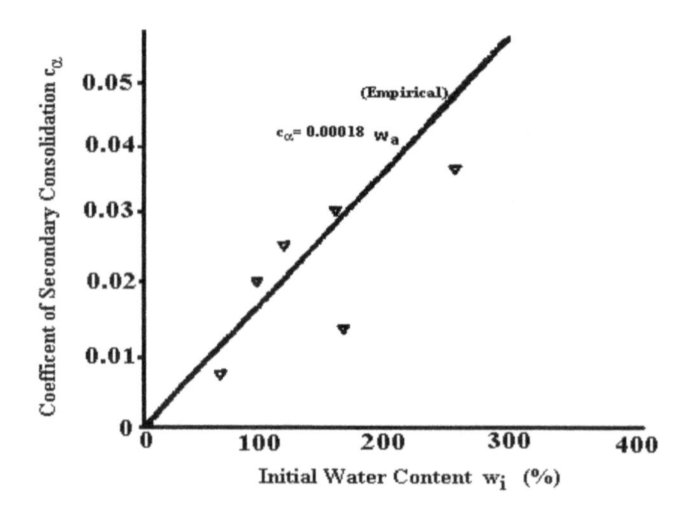

Figure 5.18 Coefficient of secondary consolidation c_α vs. initial moisture content (*after* Farrell *et al.* 1994).

Farrell *et al.* (1994) indicated that the relationship between the laboratory-determined coefficient of secondary consolidation c_α and the initial moisture content up to a moisture content w_n of about 250%, as shown in Figure 5.18, was generally close to the empirical relationship.

$$c_\alpha = 0.00018\, w_\mathrm{n} \tag{5.8}$$

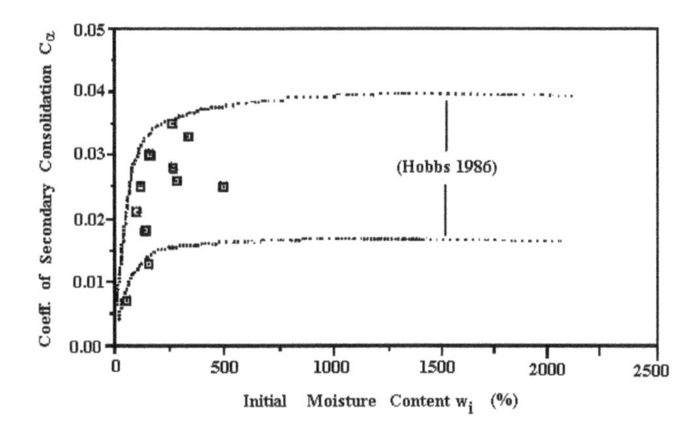

Figure 5.19 c_α vs. initial moisture content (*after* Hobbs, 1986).

There are insufficient results to confirm the trend noted by Hobbs (1986) of a relatively constant value of c_α above a moisture content of 250%, as shown in Figure 5.19. Hobbs also noted that c_α would reach limiting values of between 0.02 and 0.04 for highly organic soils.

O'Loughlin and Lehane (2003) found that c_α increases with increasing organic content. c_α evidently also depends on w_o and consequently e_o. They also suggested a relationship between c_c and c_α as shown in Figure 5.20.

According to Mesri *et al.* (1997), reliable data suggest that c_α/c_c for peat lies within the range 0.06 ± 0.01. However, studies carried out by Paikowsky *et al.* (2003) on Cranberry bog peat, Massachusetts, showed that c_α/c_c is not constant, but varies as shown in Figure 5.21.

Compression of fibrous peat continues at a gradually decreasing rate under constant effective stress, and this is termed secondary compression. The secondary compression of peat is thought to be due to further decomposition of the fibres, which is conveniently assumed to occur at a slower rate after the end of primary consolidation (Mesri *et al.*, 1997). Kazemian and Huat (2009a) presented the results of the ratio of compressibility (c_α/c_c) versus consolidation pressure of fibrous peat (Figure 5.22). As expected, this ratio increased with increasing consolidation pressure.

5.2.4 Tertiary compression

Secondary compression carries with it the widespread connotation of constant rate with the logarithm of time, which means a decreasing true settlement rate with time. There is some evidence that the coefficient of secondary compression (c_α) may change with time, even for clays (Mesri and Godlewski, 1977). For fibrous peat, there is a significant deviation from constant c_α in most cases of stress application (Edil, 2003).

Edil and Dhowin (1979) reported the steepening of the log t versus compression curve in oedometer tests on Wisconsin peat (Figure 5.23), and defined a tertiary compression phase after the secondary compression phase. Dhowian and Edil (1980) first introduced the term, finding rather strong effects in fibrous and hemic Wisconsin peat. They also proposed to redefine tertiary compression as the steepening of natural strain

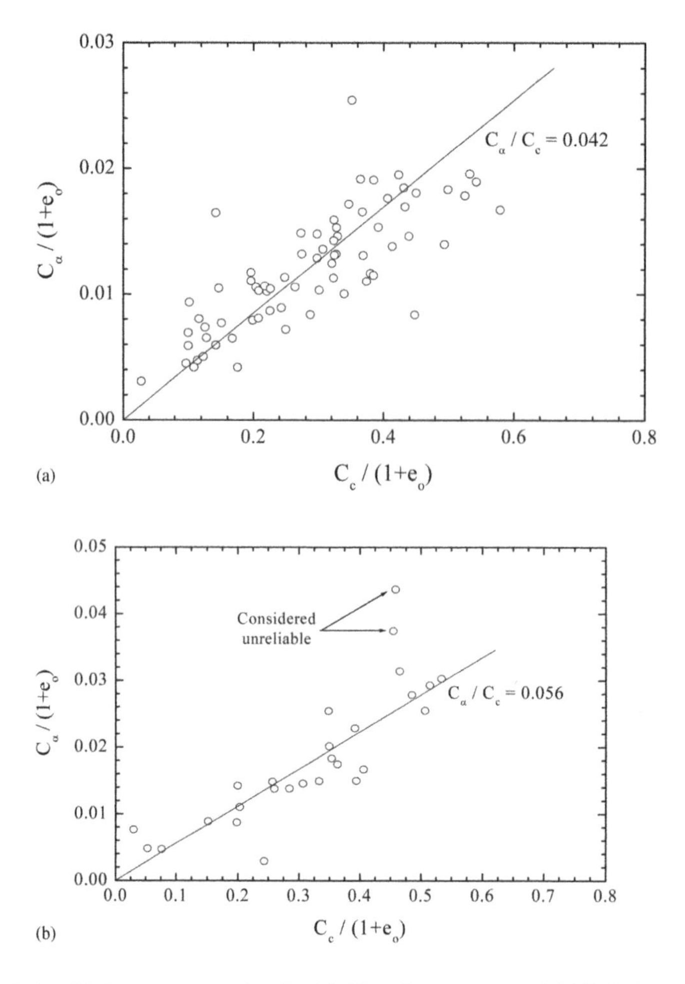

Figure 5.20 Relationship between c_α and c_c for (a) Clara fibrous peat and (b) Ballydermot fibrous peat, Irish Midland (*after* O'Loughlin and Lehane, 2003).

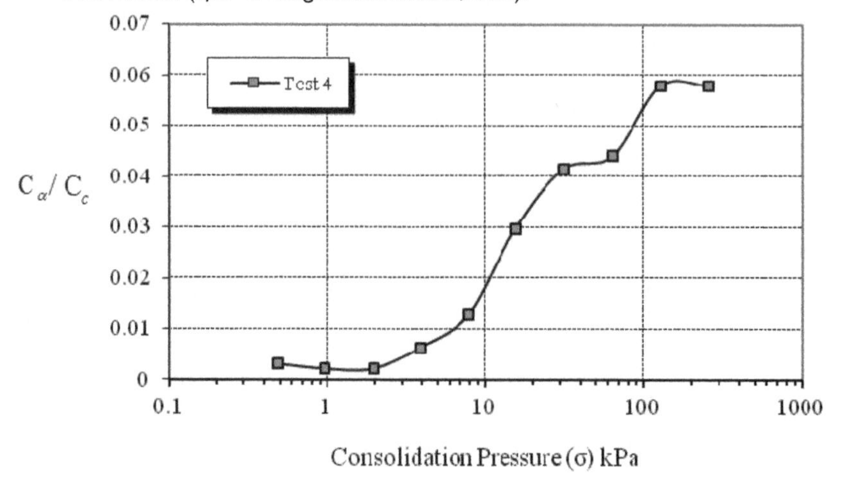

Figure 5.21 c_α/c_c vs. consolidation pressure (*after* Paikowsky *et al.*, 2003).

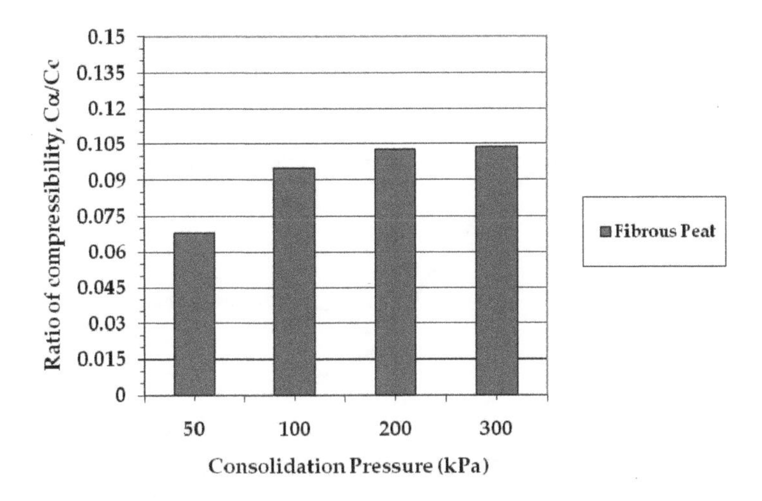

Figure 5.22 Ratio of compressibility (c_α/c_c) versus consolidation pressure for fibrous peat (*after* Kazemian and Huat, 2009a).

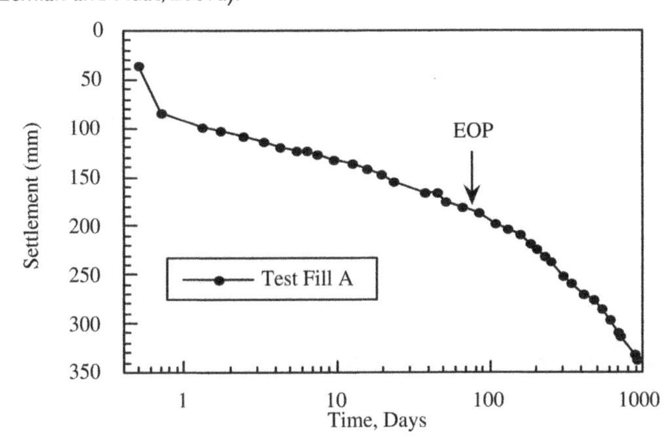

Figure 5.23 Settlement vs. time curve at the centre of Middleton Test Fill.

log (creep rate) curves, and asserted that the false effects of resetting of time-zero are circumvented in this manner. However, at large time the different is small, so strain-log t curves that steepen after long periods probably represent tertiary compression. The term 'tertiary' compression was introduced as a convenience device to designate increasing c_α with time.

Figure 5.24 illustrates the tertiary phase, following the secondary and primary phases. Figure 5.25 shows a plot of log strain rate versus log time.

5.3 HYDRAULIC CONDUCTIVITY

The physical structure and arrangement of constituent particles, e.g. fibres and granules, in peat greatly affect the size and continuity of pores, resulting in a wide range

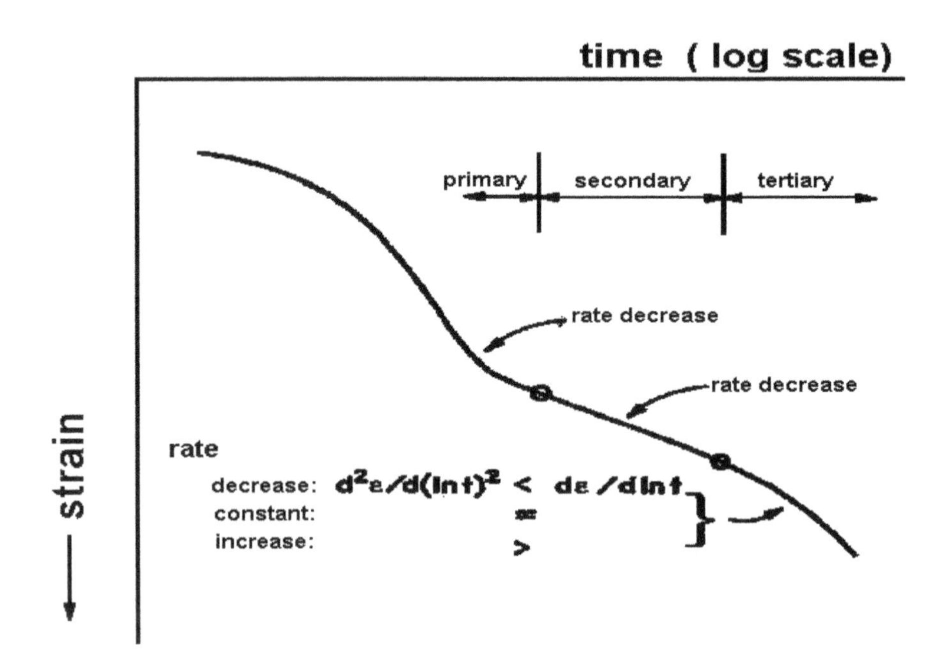

Figure 5.24 Primary, secondary and tertiary phases in oedometer compression.

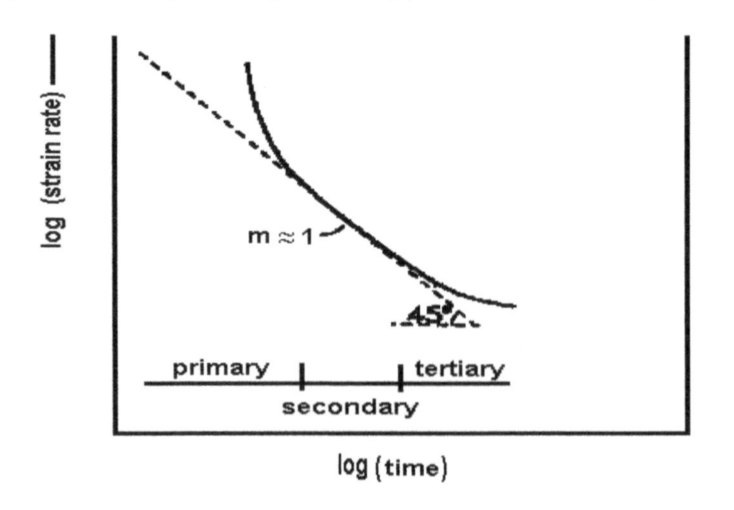

Figure 5.25 Primary, secondary and tertiary phases in oedometer compression (strain vs. time plot on a log-log scale).

of hydraulic conductivities. The highly colloidal amorphous peat tends to have lower hydraulic conductivity compared to fibrous peats. The dominant factors, in addition to the original structure and material characteristics that control hydraulic conductivity, are density (or degree of consolidation) and extent of decomposition. These factors can change with time and result in a change in hydraulic conductivity. In its natural

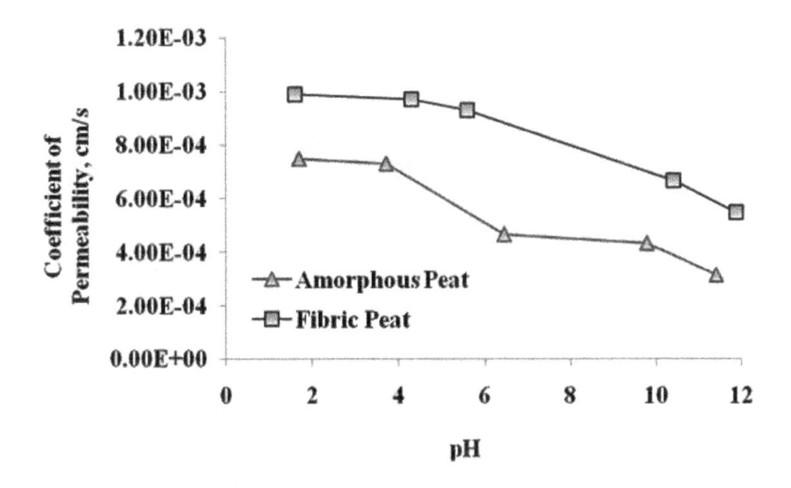

Figure 5.26 Effect of pH on permeability (*after* Asadi *et al.*, 2011a).

state, peat can have a hydraulic conductivity as high as sand, i.e., 10^{-3} to 10^{-2} cm s^{-1}. Hydraulic conductivity decreases markedly under load, down to the level of silt or clay hydraulic conductivity, i.e. 10^{-6} to 10^{-7} cm s^{-1} or even lower (Dhowian and Edil, 1980).

In general, horizontal hydraulic conductivity, especially for predominantly fibrous peats, is greater than that in the vertical direction by an order of magnitude or more (Dhowian and Edil, 1980).

5.3.1 Effect of pH on permeability

Asadi *et al.* (2011a) investigated the effect of pH on permeability of tropical peaty soils under laboratory conditions. Initially the pH of the pore peat water was adjusted by drop-wise addition of a 0.1 M HCl or 0.1 M NaOH solution. Then the undisturbed sample in the mould was saturated under pore peat water pressure. The pore peat water was allowed to flow through the specimen and the permeability value calculated. It was found that in both amorphous and fibrous peats, the permeability increased as the pH decreased (Figure 5.26). The alkaline conditions had contrary effects, i.e. as the pH increased the permeability of both amorphous and fibrous decreased. The results showed that effects of acids and bases on the permeability of the fibrous peat were lower than those of amorphous peats (Asadi, 2010).

5.4 FINAL SETTLEMENT DUE TO SURFACE LOAD

A conventional one-dimensional formula may be used with the above compressibility parameters to calculate the total settlement in peat. Alternatively, a simpler formula, as proposed by den Haan and El Amir (1994), may also be used. According to them,

$$\varepsilon = \frac{w_o - w}{w_o + 0.371 + 0.362\,N} \tag{5.9}$$

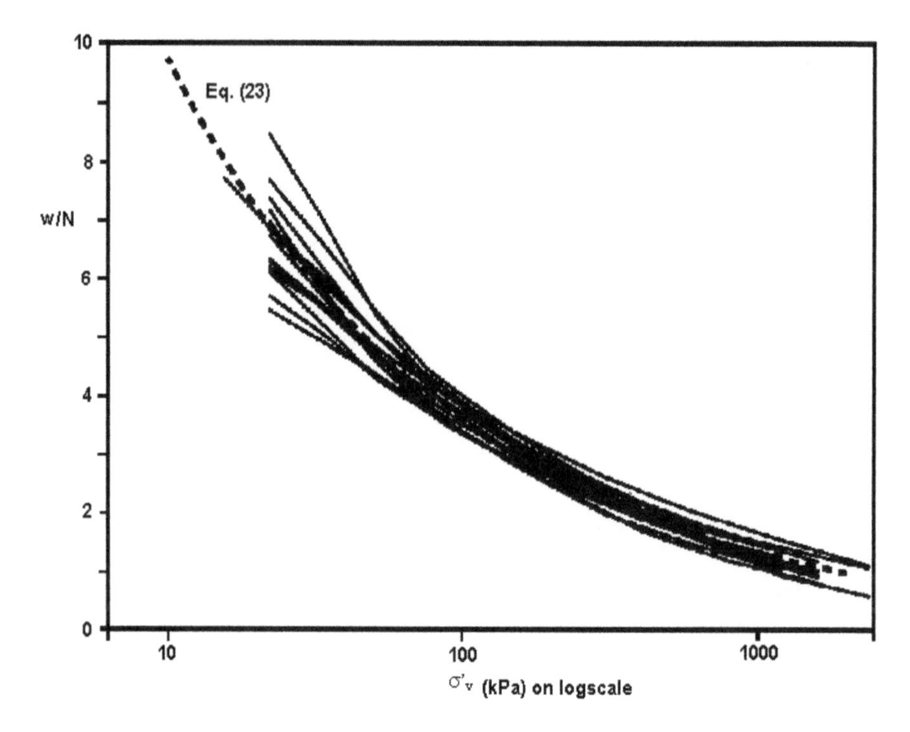

Figure 5.27 $w/N - \log\sigma_v'\ \sigma$ (*after* den Haan and El Amir, 1994).

where ε = compression strain
w_o = initial water content
N = loss of ignition
w = water content, given by the following equation:

$$w/N = 26.7\sigma_v'\ \sigma^{-0.437}$$

(See Figure 5.27) σ_v' is the final stress in kPa.

5.5 OBSERVATIONAL METHODS

Because of the difficulties associated with correlating laboratory behaviour with the field, it is desirable to use field observations to determine the mechanical properties governing the settlement time behaviour. Asaoka's (1978) observational procedures use early field settlement data to predict the end of primary settlement and the *in situ* coefficient of consolidation. This method has enjoyed increasing popularity, especially for consolidation of clays. In the Asaoka method, the one-dimensional consolidation settlement S_1, S_2, S_3, S_{n-1}, S_n etc. at times 0, dt, 2dt, etc. (i.e. at equal time increments) can be expressed, as a first-order approximation, by:

$$S = \alpha + \beta\, S_{n-1} \tag{5.10}$$

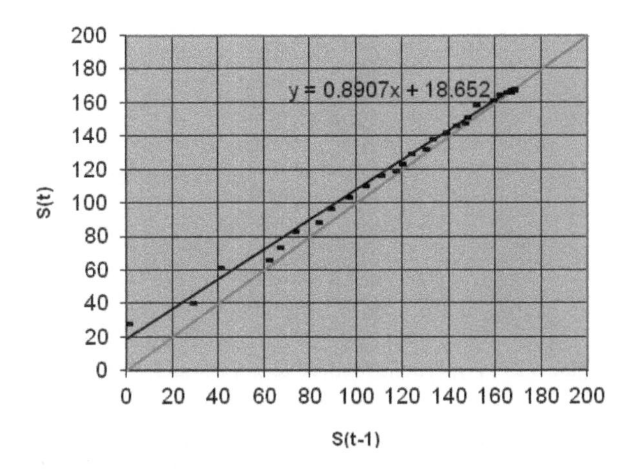

Figure 5.28 Typical Asaoka plot.

Equation (5.10) represents a straight line when the values of S_n are plotted on the vertical axis and values of S_{n-1} are plotted on the horizontal axis. Hence α represents the intercept on the vertical axis and β the gradient.

A typical straight-line plot that emerges when value of S_{n-1} are plotted against S_n for a series of equal time intervals is illustrated in Figure 5.28.

From the settlement-time curve in Figure 5.23, when settlement is complete, $S_n = S_{n-1}$. The equilibrium line $S_n = S_{n-1}$ is the straight line drawn at 45°, i.e. $\beta = 1$.

The ultimate (100%) settlement, S_{100}, can be obtained by substituting $S_n = S_{n-1} = S_{100}$ into Equation (5.10):

$$S_{100} = a + \beta S_{100}$$

and

$$S_{100} = a/(1 - \beta) \tag{5.11}$$

The 90% settlement S_{90} is thus given by:

$$S_{90} = 0.9a/(1 - \beta) \tag{5.12}$$

The number of time increments j_{90} needed to achieve 90% settlement is given by:

$$j_{90} = \ln(1 - U_{90})/\ln\beta \tag{5.13}$$

The above provides the basis for utilizing settlement data to make assessments of the degree of settlement that is occurring within the monitored area. Referring to the straight-line Asaoka plot shown in Figure 5.28, the value S_{100} is obtained when the best fit straight line through the site data allows increasingly refined predictions of the magnitude and rate of total settlement to be made.

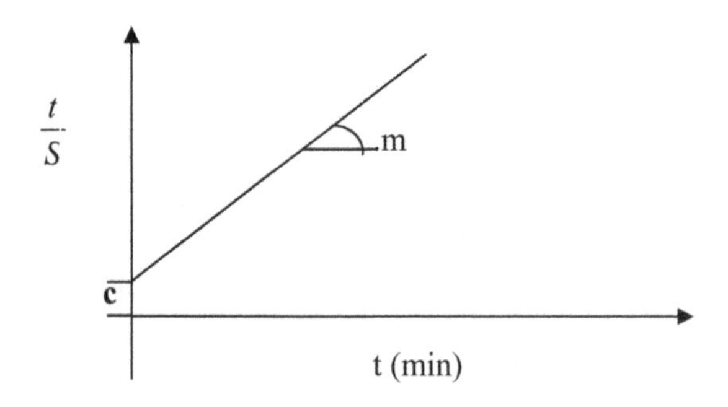

Figure 5.29 Hyperbolic method to predict future settlement.

For clay soils this method seems to work rather well (Huat, 1996, 2002; Rahim and George, 2004). For peat soils, Cartier *et al.* (1989) used Asaoka's method for the analysis of a test embankment on peat and reported a reasonable prediction of settlement and time of 98% primary consolidation. Edil *et al.* (1991), on the other hand, applied the procedure to a variety of clay and peat cases and questioned its applicability to peat settlement.

Another useful method is the hyperbolic method (Tan, 1971; Chin, 1975). This method is based on the assumption that the settlement-time curve is similar to a hyperbolic curve and can be represented by the equation:

$$S = \frac{t}{(c + mt)} \tag{5.14}$$

Where S is the total settlement at any time after the excess pore water pressure has dissipated and m and c are empirical constants. Figure 5.29 shows a plot with the ratio t/S on the ordinate and time t on the abscissa. A straight line gives the intercept c with slope m, and the significance of m is seen by writing the equation as follows:

$$\frac{1}{S} = m + \frac{c}{t} \tag{5.15}$$

When t becomes very large, i.e. $\rightarrow \propto$, then $1/t \rightarrow 0$ and $1/S = m$, which means the ultimate settlement, $S_{ult} = 1/m$.

Studies carried out by Al-Raziqi *et al.* (2003) showed that this method may be used for predicting the primary phase of peat settlement. However, for the secondary (creep) phase, the settlement prediction could be misleading.

Chapter 6

Soil improvement and construction methods in peat

6.1 INTRODUCTION

Peat represents the extreme form of soft soil. It is subject to instability, such as localized sinking and slip failure, and massive primary and long-term settlement when subjected to even moderate load increases. In addition, there is discomfort and difficulty of access to the sites, tremendous variability in material properties and difficulty in sampling. Peat may also change chemically and biologically with time. For example, further humification of the organic constituents will alter the soil's mechanical properties, such as compressibility, shear strength and hydraulic conductivity. Lowering of ground water may cause shrinking and oxidation, leading to humification with a consequent increase in permeability and compressibility.

Buildings on peat are usually suspended on piles, but the ground around it may still settle, creating a scenario as depicted in Figure 5.1. Much of this chapter is therefore devoted to methods of construction for line structures, such as for road embankments, which are subject not only to instability and settlement, but also to the tremendous variability in peat soil material properties, as illustrated in Figure 6.1.

Even if failure is avoided, it is inevitable that soft waterlogged soil and peat takes a long time to settle when loaded due to embankment or soil fill. Under these conditions, the embankment will settle continually into the ground below, even if the soil does not fail by displacement. This is illustrated in Figure 6.2. To make matters worse, the addition of more fill to make up the depression in the embankment will ensure further settlement.

Where possible, engineers seek to avoid building on this problematic ground. Avoidance is sometimes possible by changing the location of the construction, as illustrated in Figure 6.3. However this is not always possible. In cases where there is no alternative due to the high demand for land for development, even this less desirable land has to be developed.

The main purpose of this chapter is to describe some of the methods and recent technology that have been used to mitigate the challenges faced in construction (in particular those for line structures such as road embankments) on peat and organic soils. Some technical details will be presented for the methods described. In must also be noted that in practice these technologies overlap and are commonly used concurrently. This chapter gives some insights on the construction methods currently employed where yet there is no consensus on the best methods of analysis. Undoubtedly construction on peat and organic soils is not easy, but with better understanding it can be more manageable.

(a)

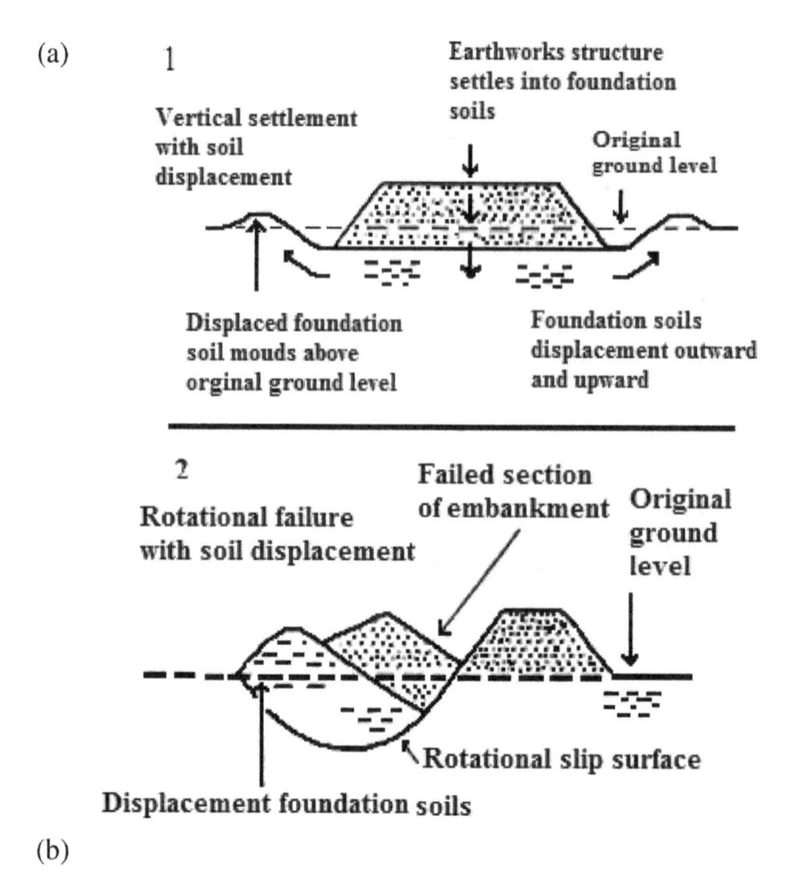

(b)

Figure 6.1 (a) Bearing capacity failure in peat; (b) embankment crack (*after* Hussein and Mustapha, 2003).

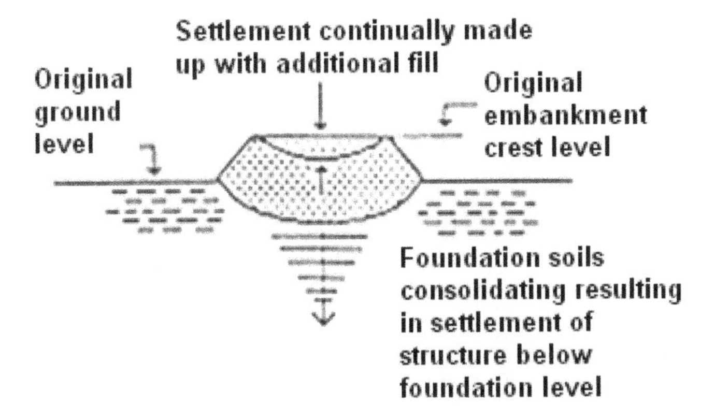

Figure 6.2 The settlement problem in peat.

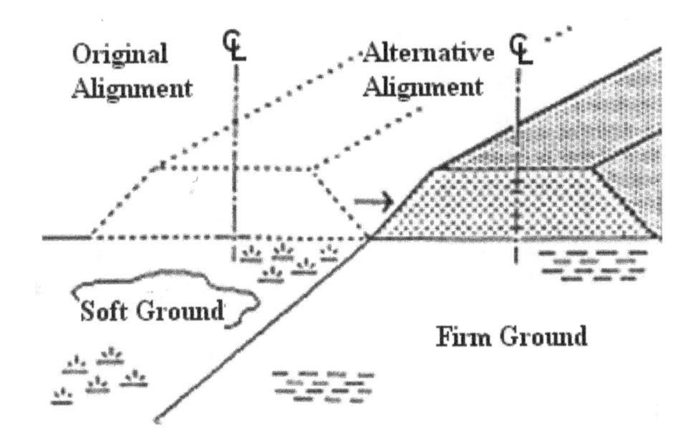

Figure 6.3 Avoidance technique.

A number of construction options that can be applied to peat are stated by Edil (2003):

- **Avoidance:** consider marginal lands; as mentioned above, this is typically the first choice if feasible.
- **Excavation – displacement/replacement:** practical typically up to 5 m depth.
- **Ground improvement and/or reinforcement to enhance soil strength and stiffness:** this refers to any method or technique that improves the engineering properties of soil, such as shear strength, compressibility, stiffness or permeability. Raju (2009) classified soil improvement methods according to the following principles: (i) consolidation (e.g. prefabricated vertical drains and surcharge, vacuum consolidation, stone columns), (ii) chemical modification (e.g. deep soil mixing, jet grouting, injection grouting), (iii) densification (e.g. vibro-compaction, dynamic compaction, compaction grouting) and (iv) reinforcement (e.g. stone columns, geosynthetic reinforcement). The majority of companies and geoengineers working in these fields agree that soil improvement methods are usually done based on

Table 6.1 Various soil improvement techniques (*After* Hausman, 1990).

Mechanical modifications	Shallow Surface Compaction	Static Roller	Smooth Steel Rollers and Pneumatic-tired Roller	
			Sheepsfoot Rollers	
			Grid Roller	
		Impact and Vibratory Equipment	Tamper, Rammers, and Plate Compactors	
			Vibrating Rollers	
			Impact Rollers	
	Deep Compaction Technique	Pre-compression		
		Explosion		
		Heavy Tamping (Dynamic compaction)		
		Vibration (Vibro-compaction and Vibro-replacement)		
	Hydro-mechanicl Compaction	Hydraulic Fill		
		Dry Fill with Subsequent Spraying or Flooding		
		Compaction of Rock Fill with Water Jets		
Hydraulic modify cations	Open Sumps and Ditches			
	Gravity Flow Wells			
	Pre-loading and Use of Vertical Drains			
	Vacuum Dewatering Wells and Preloading			
	Electrokinetic Dewatering and Stabilization			
Physical and chemical modifications	Mixing method	Shallow Mixing Method (SMM)		
		Deep Mixing Method (DMM)		
	Injection and grouting	Permeation Grouting	Jet Grouting	Single Phase
		Compaction Grouting		Dual Phase
		Hydro fracture Grouting		Triple Phase
		Compensation		Super Jet Grouting
Modifications by inclusions and confinement	Flexible Geosynthetic Sheet Reinforcement			
	In Situ Ground Reinforcement	Ground Anchorage		
		Rock Bolting		
		Soil Nailing		
	Soil Confinement by Formwork	Crib Walls		
		Gabion and Mattress		
		Fabric Formwork		

the following classifications: (i) mechanical modifications, (ii) hydraulic modifications, (iii) physical and chemical modifications and (iv) modifications by inclusions and confinement (Hausman, 1990). This classification is explained in detail in Table 6.1. The methods that are related to peat will be discussed later.

- **Stage construction and preloading:** used to overcome problems of instability in fills constructed over weak deposits. It takes time but can be accelerated by the use of vertical drains (wick drains) and stability can be enhanced by geosynthetic reinforcement. Loading can be achieved by the placement of loads on the surface or vacuum consolidation.
- **Deep *in situ* mixing (lime-cement columns):** forced mixing of lime, cement or both with soft deposits to form stabilized soil columns (application to peat is still under development).
- **Stone columns:** filling water jetted holes in soft ground with compacted gravel.
- **Piles:** expensive, but reliable for building foundations if used with suspended floors. Also used for embankment support in some countries.
- **Thermal precompression:** moderate ground heating (15–25°C) is used to accelerate settlement and reduce long-term compression upon cooling (field tested but no commercial applications yet).
- **Preload piers:** preload piers or geopiers involve densely packing stone in layers in a hole; i.e. an intermediate foundation system. They are intended for radial precompression of the ground and are currently under development.
- **Reduce driving forces by lightweight fill:** use of woodchips, sawdust, tire chips, geofoam or expanded shale. Lighter but sufficiently strong and stiff fill materials.

The methods for improving the engineering performance of peat are described in detail in the following sections.

6.2 EXCAVATION – DISPLACEMENT AND REPLACEMENT

One solution is to replace the poor soil by excavation or by dumping suitable imported fill materials if the soils are of very high liquid type, as illustrated in Figure 6.4. This is naturally very expensive on materials. It is also difficult to control the underground movement of the material. In addition, there must be an environmentally acceptable location to waste the excavated soil within an economically acceptable haul distance and there must be a source of adequate fill, again within an economically acceptable haul distance (Jarrett, 1995). Furthermore, this method can only be effective for up to a depth of 5–6 m. It is normally used for surface peat in France to avoid maintenance work related to long-term settlement and horizontal movements (Magnan, 1994).

An example case history is the 1.4 km Samariang ring road project in the division of Kuching, Sarawak, Malaysia (Figure 6.5). The site was reported to be overlain by soft peat and organic soils 5.0 m to 7.0 m deep, with very high moisture content (1000–1700%), low specific gravity (1.38–1.54) and low bulk density (1.015–1.025 Mg m^{-3}). The method adopted in this construction is to excavate the soft deposits to a depth of 4.5 m to 5.5 m below the existing ground surface with a side slope of 1:3. Extensive dewatering had to be employed because of the high water table. Sand was used as the

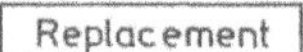

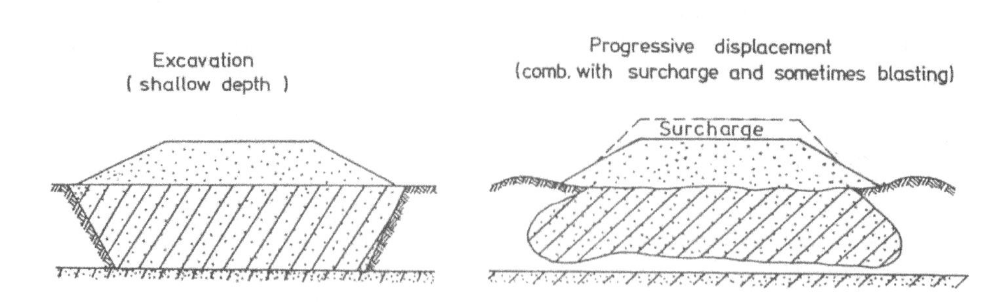

Figure 6.4 Excavation and replacement.

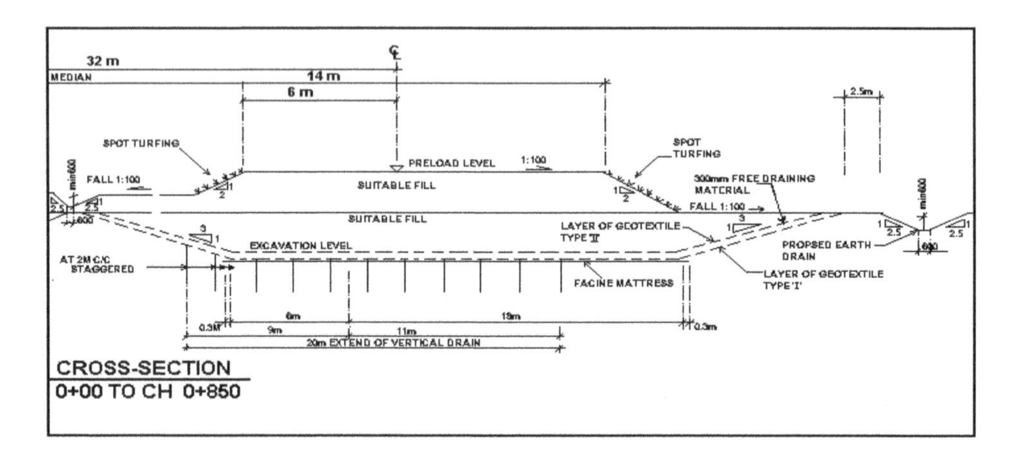

Figure 6.5 Excavation and replacement of the Samariang ring road project (*after* Toh *et al.*, 1994).

replacement material up to the original ground surface, followed by another 2 m of earth fill to the road platform.

6.3 SURFACE REINFORCEMENT, PRELOADING AND VERTICAL DRAIN

6.3.1 Surface reinforcement

Geotextile, geogrids, timber or bamboo mattresses can be used to separate granular embankment fill from the soft subsoil on which it is being placed and for tensile reinforcement to increase the overall stability of the embankment. As a separator, the geotextile prevents the fill from becoming contaminated by the weak subsoil material. thus preserving the better load-bearing properties of the fill. As the fill becomes thicker, the need for separation decreases and the geotextile then becomes a tensile reinforcement, reinforcing the base of the higher embankment. Such reinforcement assists in preventing a number of forms of failure, including lateral spreading and the normal circular arc. Case histories are given in Rowe and Mylleville (1996) and Farrell (1997).

The geotextile may also be an aid to construction. On a wet site it helps with the establishment of an initial working platform, and in more fibrous peat it helps to preserve the natural surface mat from the destructive activity of the construction traffic. It should also be noted that the natural fibres of peat are helpful in maintaining the stability of the fill.

The addition of a bamboo mattress is similar to the use of corduroy. Corduroy construction involves placing cut trees or planks, often obtained from clearing the site, at the soil-to-fill interface as a mean of distributing the load and bridging the soft spots in an organic subsoil. This still represents an excellent method of construction when timber is available and labour is economically viable to place the timber (Jarrett, 1997).

An example where the replacement and geotextile method was adopted was the construction of the Kuching Boulevard road project in the proposed new financial centre of the city (Kuching, Sarawak Malaysia). The site was underlain by some 2–4 m of very soft clayey silt with some decayed wood and peat, with a natural moisture content of 40–300% and an undrained shear strength of 10–20 kPa, overlying some 2–5 m of firm to stiff clayey silt. The superficial deposit was excavated and replaced with a suitable earth fill. Bamboo mattresses and sandy aggregates separated by non-woven geotextile were utilized as a separator-cum-drainage layer above the original ground at the excavated level (Figure 6.6).

An example of a case history where bamboo mattresses were combined with non-woven geotextile to provide a separation layer on which a working platform could be

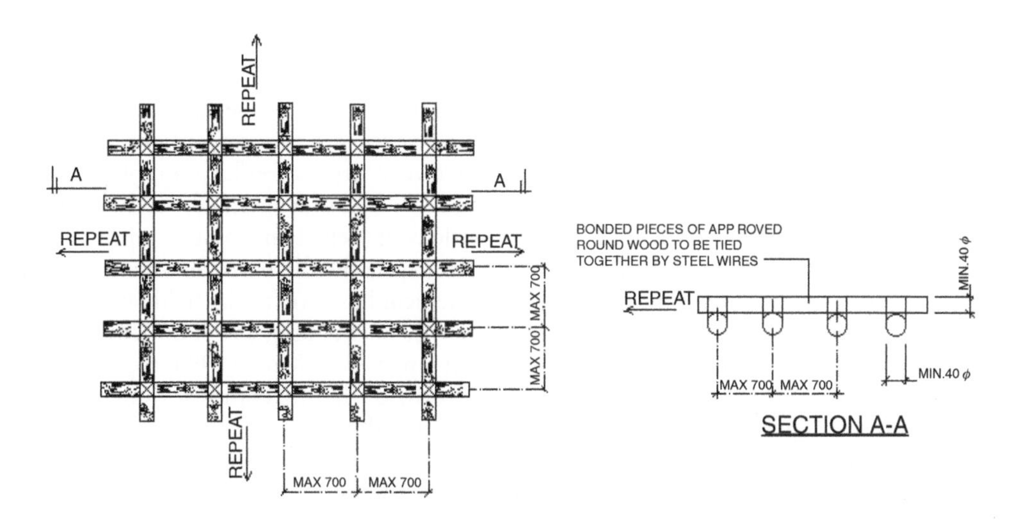

Figure 6.6 Geotextile fascine mattress (*after* Toh *et al.*, 1994).

developed is the construction of the Kuching ring road, north of Kuching city (Toh *et al.*, 1994). Bamboo was used to aid in the rolling out of the geotextile over the soft ground. Wide stability berms were constructed on both sides of the main embankment section to maintain stability. The subsoil was reported to comprise 2–4 m of very soft organic soil and peat, underlain by a 3–5 m thick stratum of firm to stiff clayey silts. The soft organic soil was reported to have the following geotechnical parameters: very high natural water content of 280–1000%, high compressibility, with compression index, C_c, in the range of 5.6–11.0, and extremely soft, with undrained shear strength less than 10 kPa.

In a number of instances, only the upper few meters of the very soft deposits were excavated and replaced with stable fill. This was one of the methods employed in the construction of some sections of the Bakun access road project in Sarawak Malaysia, where the nominal thickness of the soft subsoil is quite shallow and the embankments are of moderate height (2–5 m). Some 3 m of the very soft subsoil was excavated and replaced with sand fill. A single layer of woven geotextile (Stabilenka 200) was used as both separator and reinforcement layer.

6.3.2 Preloading

The preloading principle was also adopted in the above-mentioned example to minimize post-construction settlement.

It is interesting to note that preloading is not a recent development, but it does still represent the method of construction with the best economic and engineering returns (Carlsten, 1988). It is, however, underused because of the necessity of prior planning (Jarrett, 1987). The basis of preloading is to place a temporary fill over the construction site that is thicker than the final design fill. This causes settlement to occur more rapidly than would have occurred under the final fill design height. The preload is ideally left in place until it has settled more than the total amount that the design fill is expected to settle in its design life. Then the thickness of the preload fill is reduced to the final design thickness, with the expectation that most of the settlement has now finished. Calculations for rates of settlement, however, are notoriously difficult for all peat and organic soils as they commonly have very large secondary compressions, and the time-settlement relationship is often not of the characteristic form making it difficult to separate primary from secondary settlement (den Haan and El Amir, 1994). In fact it is common to monitor the rate of settlement of the fill after construction and to make projections of the time required for settlement based on initial field settlement (Edil and Simon-Gilles, 1986; Garga and Medeiros, 1995; Huat, 1996, 2002a).

An experience with peat preloading has been reported by Kirov (1994) for a Varna West Harbor project in Bulgaria. The subsoil profile of the site comprises some 17 m thick soft soil deposits, and alternates layers of peat and organic clays overlying marls. The site was first excavated to 3.75 m, then filled with a layer of coarse sand 0.8 m thick, preloaded with hydraulically filled sand to another 6 m, and maintained for one year.

In France, for cases of deep peat (>6 m), preloading or surcharge and stage loading are used to control the deformation and stability of structures built on peat (Magnan, 1994).

Another case history of the preloading technique was presented by Samson and La Rochelle (1972) for an expressway construction in Quebec, Canada. The original thickness of the peat varied from 3 to 6 m. The initial undrained strength of the peat was low (10 kPa), organic content was 84–95% and bulk density was 0.87–1.04 Mg m^{-3}. The preload construction was carried out in four stages:

- Stage 1: a 1.2 m thick sand working platform was left for 10 days. This created a large settlement of 1.2 m.
- Stage 2: fill was increased to 2.5–3.3 m and left to consolidate for 6.5 months. Settlement varied from 1.1 to 2.6 m.
- Stage 3: a preload surcharge of 1.0 to 1.5 m was added and left for 1 year. The total settlement varied from 1.4 to 3.3 m.
- Stage 4: the surcharge was removed, and the road was completed in about three months. The embankment heaved for a period of 200–475 days. A total rebound of 43 to 79 mm was observed. After this, slow settlement was observed at a constant rate with respect to log time for 5.5 to 8 years, after which the rate was found to increase somewhat. Total settlement over a period of 18 years after completion of the highway was 270 mm to 1.19 m (Samson, 1985).

The preloading principle, often in conjunction with vertical drains, was also adopted in the above-mentioned methods to accelerate settlement and minimize post-construction settlement (see Figure 6.7). Where it is desirable to try to speed up the dissipation of pore water pressures beneath an embankment and hence speed up the settlement process, geosynthetic vertical drains have become a modern substitute for sand drains. There are two basic reasons to wish for the quick dissipation of the pore water pressures: in a stage construction, where the strength gain of the subsoil at each stage is needed to ensure the stability of the next stage, quicker dissipation allows quicker construction; while in preloading it reduces the time necessary for the preload fill to

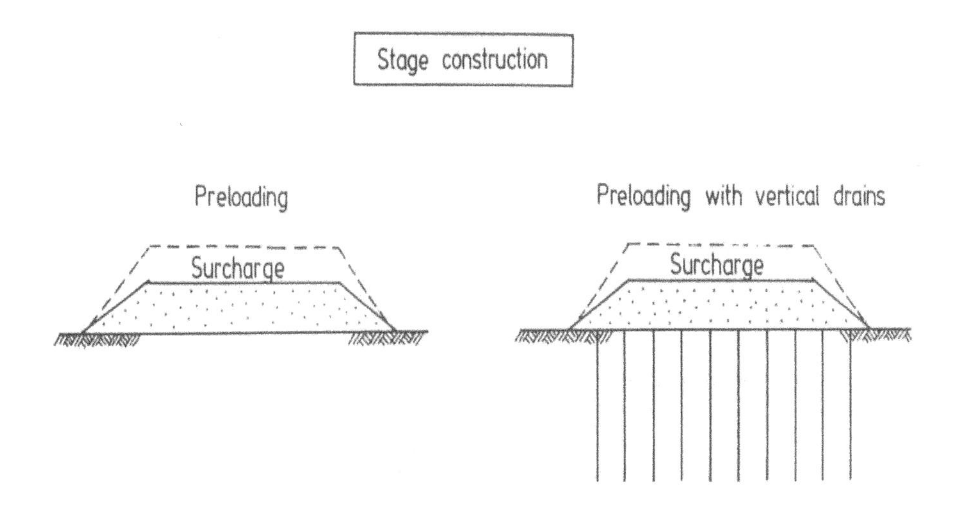

Figure 6.7 Preloading (stage construction) and vertical drain.

remain in place. Geosynthetic vertical drains are commonly 100 mm wide and 5 mm thick and are composed of a plastic drainage core wrapped in a non-woven geotextile sock. They come in long rolls and are commonly punched into the ground with a mandrel in a triangular or square configuration with a spacing of 1 m to 2 m to whatever depth is required. This punching process causes a problem for this technology, namely the displacement and disturbance of the soil by the mandrel and the smearing of the geotextile sock by the disturbed soil. This can cause a radical decrease in permeability at the drain entrance and prevent the free drainage expected. There has been considerable discussion concerning this topic and whether the drains do in fact work. An example is presented by Othman *et al.* (1994). Specifically, the high initial permeability of peat may render the drains useless. Later in the consolidation process, permeability might decrease to levels making the drains effective, but buckling and filter contamination might render them less effective before then (den Haan, 1997).

Koda and Wolski (1994) reported successful applications of this technique for construction in peat in Poland. Figure 6.8 shows the peat soil profile, construction schedule and settlement performance of their field trial.

However, as mentioned above, there is some controversy regarding the effectiveness of vertical drains in peat. Surficial peats often exhibit very high permeability until they are compressed, and a large part of their total compression takes place under constant effective stress. Vertical drains may be effective in accelerating strength gain (Kurihara *et al.*, 1994) but not total settlement. The effectiveness of the strip drains may additionally be limited by deterioration and buckling of the drain and the consequent decline in discharge capacity. However, the general consensus is that vertical drains are effective tools for construction over peat (Edil, 1994).

6.3.3 Vacuum preloading

Vacuum preloading techniques have also been developed, in which an impermeable cover is placed over the surface and a vacuum pressure is created under it to speed consolidation (Figure 6.9). This technique is often used in conjunction with vertical drains. One interesting aspect of this method is the possibility of achieving preloading or precompression without the stability problem often associated with high embankments.

The vacuum is applied by means of a tight sheet or membrane and vacuum pump. This method enables the equivalent construction of a very high embankment on very soft ground to be made over a relatively short period of time by reducing the development of shear strain in the soil (Mitachi *et al.*, 2003). An interesting observation about this technique when applied to peat was made by Hayashi *et al.* (2003): the $(\frac{\Delta C_u}{\sigma'_v})$ of the peat layer showed a higher value than that of the untreated peat layer (Figure 6.10). The $(\frac{\Delta C_u}{\sigma'_v})$ of the peat layer to which only the prefabricated vertical drain method is applied demonstrated a higher value than that of the untreated peat layer. The difference, however, is not as significant as in the vacuum preloading method.

Hayashi *et al.* (2003) also suggested that the spacing of vertical drains in peat must be less than 90 cm.

Further, vacuum dewatering (vacuum consolidation method), with or without preloading, is a soft soil improvement methods and it has been applied in a number of countries (Shang *et al.*, 1998; Chu *et al.*, 2000; Mohamedelhassan and Shang, 2002;

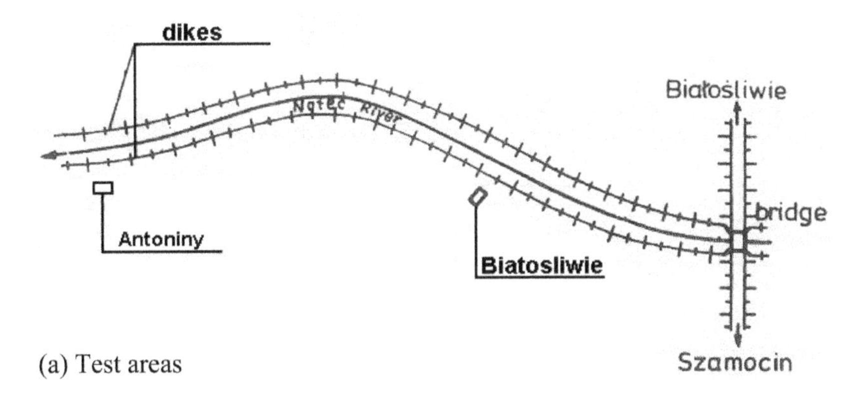

(a) Test areas

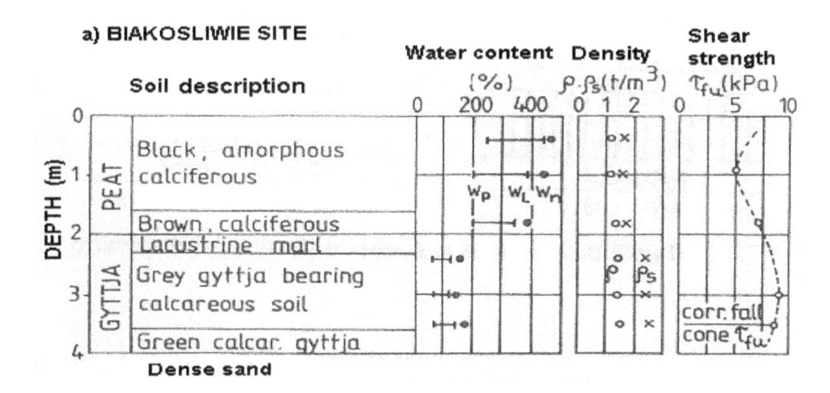

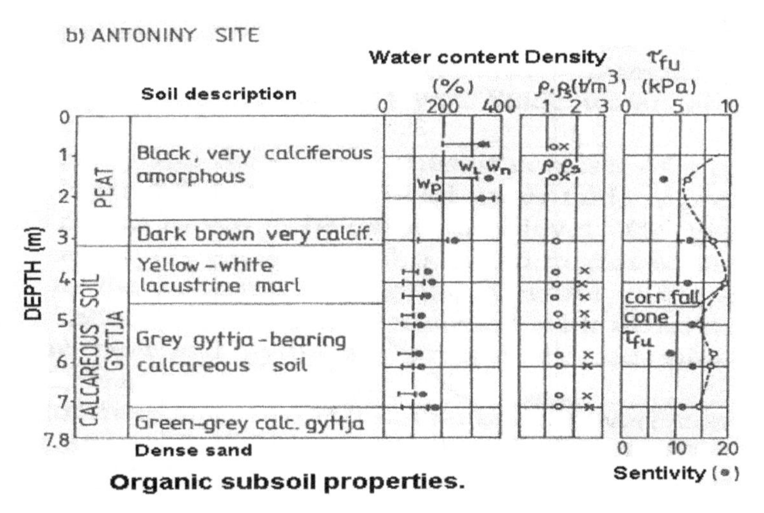

(b) Organic subsoil properties

Figure 6.8 Trial site and settlement performance using prefabricated vertical drain in Poland (*after* Koda and Wolski, 1994).

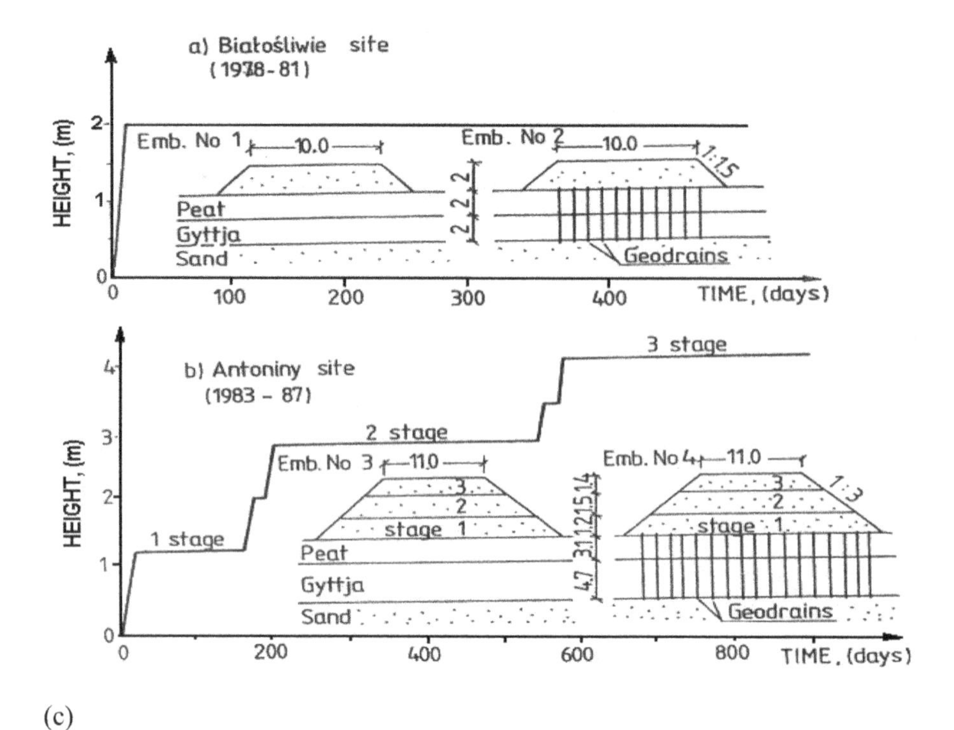

(c)

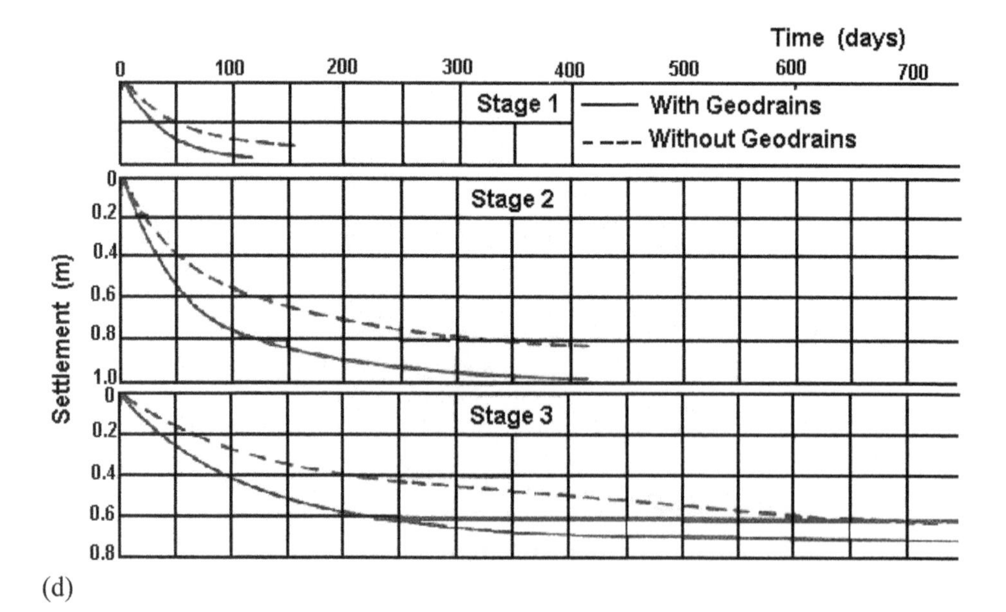

(d)

Figure 6.8 (continued).

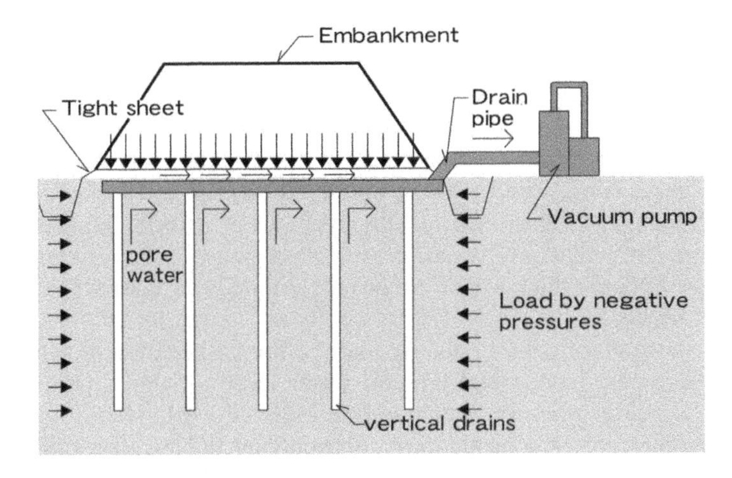

Figure 6.9 Typical setup for the vacuum consolidation method.

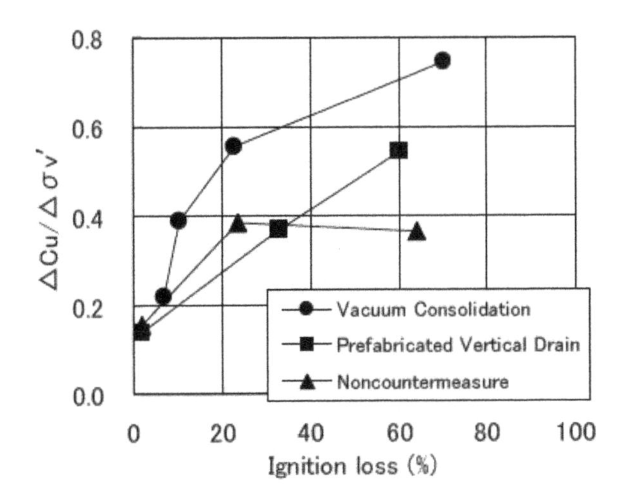

Figure 6.10 $(\frac{\Delta C_u}{\sigma'_v})$ for vacuum preloaded, vertical drains, and untreated peat.

Chai *et al.*, 2006). Kjellmann (1952) stated that when a vacuum is applied to a soil mass, it generates a negative pore water pressure. If the total stress remains constant (unchanged), the negative pore pressure results in an increase in the effective stress in the soil, which leads to consolidation. Qiu *et al.* (2007) emphasized that the principle of effective stress indicates that the soil's incremental vertical effective stress is equivalent to the pore pressure reduction and is not the vacuum pressure; i.e. the vacuum pressure is fluid pressure and is less than the atmospheric pressure, whereas the pore pressure reduction in the soil is the difference between the initial and current pore pressures and is usually considered as the vacuum pressure in vacuum preloading projects.

Recently, a new technique of applying vacuum pressure to soft clayey subsoil has been developed in which the vacuum pressure is combined with a special prefabricated vertical drain (PVD). This consists of a PVD, a drainage hose and a cap connecting the PVD and the hose, and is known as a cap-drain (CPVD) (Fujii *et al.*, 2002). Chai *et al.* (2008) explained that the method uses a surface or subsurface soil layer as a sealing layer and there is no need to place an airtight sheet on the ground surface; consequently there is no need to worry about air leakage caused by damage to the sheet. The thickness of the surface sealing layer can be determined according to the field conditions, and generally variations in this thickness will not cause additional cost. In this method, the thickness of the surface sealing layer is estimated as suggested by Chai *et al.* (2008).

Vacuum consolidation is an effective method for stabilization of soft soils in particular. Soils like peat have a high ground water level horizontal permeability (k_x), which is much greater than its vertical permeability (k_z) (Mesri and Ajlouni, 2007). This increases the rate of consolidation. By applying the vacuum below the water table, so that it is used in combination with dewatering, the equivalent preload can be increased significantly (Thevanayagam *et al.*, 1994). In comparison with pre-loading with vertical drains, the vacuum consolidation method has more advantages, such as (Cognon *et al.*, 1994; Jacob *et al.*, 1994; Shang *et al.*, 1998; Chai *et al.*, 2006):

(i) no/less fill material is required;
(ii) construction periods are generally shorter;
(iii) there is no need for heavy machinery;
(iv) a vacuum pressure up to 600 mm Hg (80 kPa) can be achieved in practice using the vacuum equipment available, which is equivalent to a fill 4.5 m in height;
(v) there is no need to control the rate of vacuum application to prevent bearing capacity failure because applying a vacuum pressure leads to an immediate increase in the effective stress in soil.

However, vacuum consolidation also has some shortcomings. For example, the applied vacuum is limited by atmospheric pressure and it may cause cracks in the surrounding surface area due to consolidation-induced inward lateral displacement of the ground (Thevanayagam *et al.*, 1994; Tang and Shang, 2000; Chai *et al.*, 2006). Furthermore, due to the complication of air-water separation and badly sealed *in situ* boundary conditions, the efficiency of the system decreases. Theoretically, the maximum vacuum pressure that may be applied is one atmosphere (about 100 kPa), but practically achievable values are normally in the range from 60 to 80 kPa (Bergado *et al.*, 1998; Tang and Shang, 2000; Qiu *et al.*, 2007). A system with 75% efficiency shows results with only 4.5 m of equivalent surcharge and to stabilize very soft soil and peaty soils, which need more negative pressure, pre-loading is necessary.

The area affected by vacuum consolidation has been investigated by many researchers (Noto, 1990; Thevanayagam *et al.*, 1994; Leong *et al.*, 2000; Hayashi *et al.*, 2003; Chu and Yan, 2005) and it has been reported that a distance of 0.7–1.5 m between the drains is the zone of effective area. However, Gabr *et al.* (1996a,b) emphasized that the zone of effective area is approximately 10 times the effective diameter of the PVDs. One of the key parameters usually addressed in PVD design is the determination of the equivalent diameter of band-shaped drains to radial drains, which dictates the size of the inflow surface (Abuel-Naga and Bouazza, 2009). PVDs are rectangular in cross-section and not round. In order to design for geotechnical or remediation purposes or some laboratory work, an equivalent circular drain diameter of PVD must be used. There are many methods and procedures for determining the equivalent diameter of a PVD with a rectangular cross-section (Welker *et al.*, 2000). Long and Covo (1994) used an electric analogue approach to develop their equation (Eqn. 6.1). Abuel-Naga and Bouazza (2009) conducted a numerical study and stated that the diameter of a PVD well under equal flow conditions shows that for such a rectangular section ($0.033 \leq t/w \leq 0.0875$), the equivalent diameter is a function of PVD width only and is in agreement with Long and Covo's (1994) findings:

$$d_{eq} = 0.5w + 0.7t \tag{6.1}$$

where d_{eq} = equivalent diameter, w = width of PVD and t = thickness of PVD.

6.4 DEEP STABILIZATION

In Sweden and Finland, deep stabilization techniques are quite popularly used for stabilization of soft soil (Åhnberg *et al.*, 1995b). Chemicals such as lime may be used to improve the strength and settlement characteristics of the soft soils. But the effectiveness of the method varies. Unslaked lime has been replaced with cement/lime mixes, usually in the ratio of 50:50, while pure cement has also been used. The strength of silt and clay can be improved up to 30-fold. In peat, however, the strength gain may not be that high. The high water content and low strength of peat require a significant gain in strength, which is inhibited by organic matter (Figure 6.11).

However, by adding enough stabilizers, such as cement, the strength gain may be adequate. Pure cement is found to be more effective in peat than cement/lime mixtures, and certain additives, such as gypsum, improve the cement's reactivity. The strength gain is mainly due to hydration products formed by cementitious reactions. It depends on the type of soil, dosage of binder, water content and curing conditions. The mechanisms whereby organic matter interferes with strength gain are not completely understood but are thought to include the following (Janz and Johansson, 2002):

- Organic matter can alter the composition and structure of calcium silica hydrate (C-S-H) gel, a cementing compound that forms bonds between particles, and also the type and amount of other hydration products, e.g. ettringite.
- Organic matter holds 10 or more times its dry weight in water and may limit the water available for hydration.
- Organic matter forms complexes with aluminosilicates and metal ions, interfering with hydration.

Figure 6.12 illustrates the application of lime/cement columns as deep stabilizers.

Huttunen *et al.* (1996) report the unconfined compressive strength of peat with different degrees of humification. They found that strength increases with increasing dosage of cement and decreases as humification increases, and the chemical and

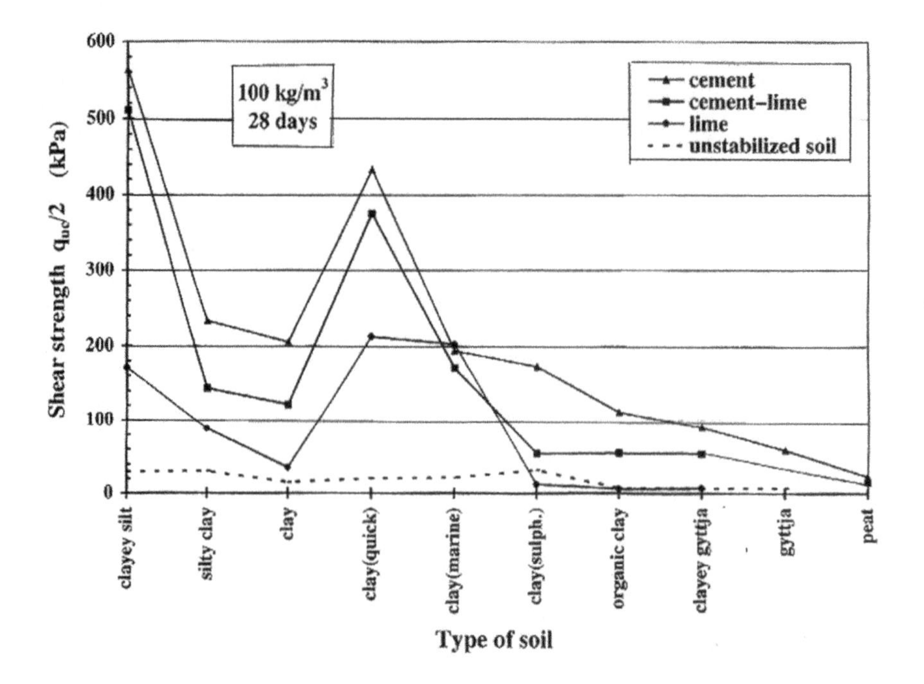

Figure 6.11 Shear strength gain (28 days) of various types of soil (*after* Åhnberg *et al.*, 1995b).

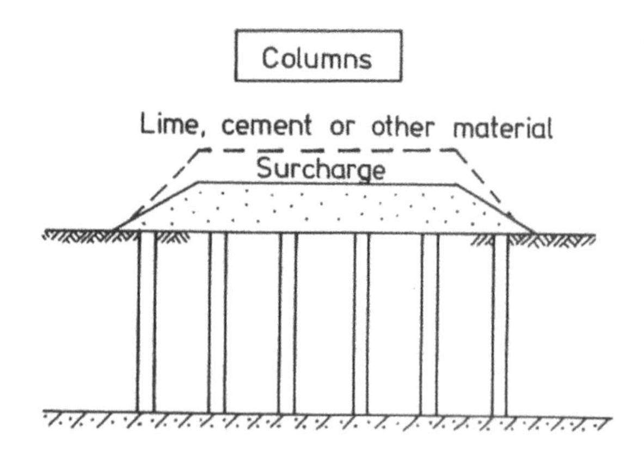

Figure 6.12 Sand or stone columns.

physical properties of the peat (water content, organic content) dictate the reaction with the binding agent.

Dry Deep Soil Mixing (DSM), also referred to as the Lime Cement Column Method, or just Cement Columns, is a variant of the deep stabilization technique invented by Kjeld Paus almost 30 years ago. It is a form of soil improvement involving the introduction and mechanical mixing of *in situ* soft and weak soils with a cementitious compound such as lime, cement or a combination of both in different proportions. The mixture is often referred to as the binder. The binder is injected into the soil in a dry form. The moisture in the soil is utilized for the binding process, resulting in an improved soil with higher shear strength and lower compressibility. The removal of the moisture from the soil also results in an improvement in the soft soil surrounding the mixed soil. Dry DSM methods have been used in Sweden and Finland since 1967 for soil improvement in soft clays and organic soils, and are mainly used to increase stability and reduce the settlements of embankments. A typical dry DSM unit consists of a track-mounted installation rig fitted with a leader and a drill motor, as shown in Figure 6.13. The binder is carried in pressurized tanks, which are mounted on the rig itself or on a separate shuttle, as shown in Figure 6.13. Soil columns up to 14 m long have been constructed using this method. Figure 6.14 shows a typical embankment section.

The Deep Mixing Method (DMM) is today accepted worldwide as a soil improvement method which is performed to improve the strength, deformation properties and permeability of the soil. It is based on mixing binders, such as cement, lime, fly ash

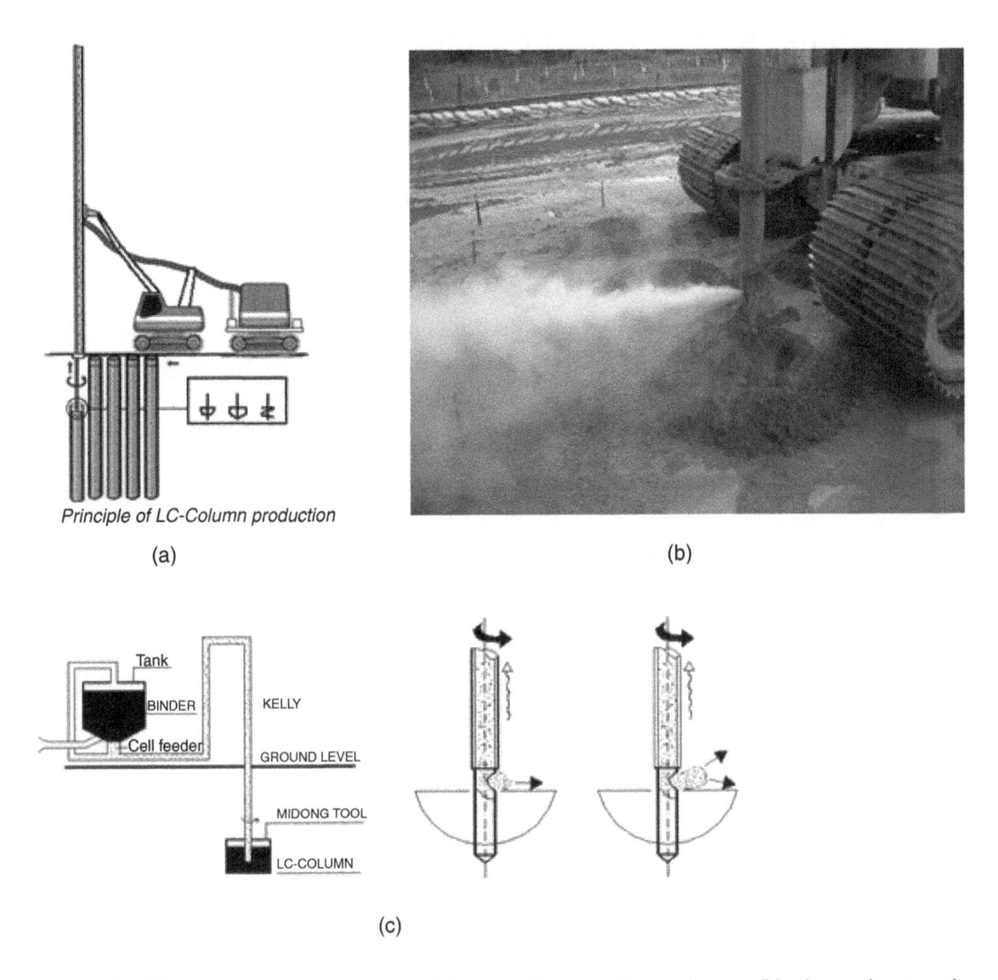

Figure 6.13 (a) Schematic showing a dry DSM machine installing columns; (b) photo showing the ejection of the binder from the mixing tool in a dry form; (c) schematic showing mixing operation in dry DSM process (*Source*: LCM Marktechnik).

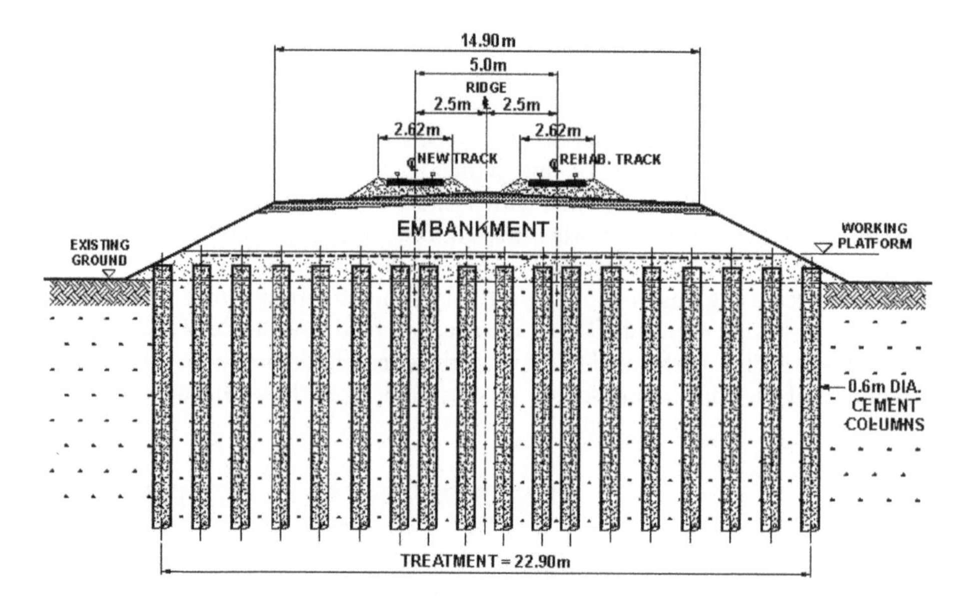

Figure 6.14 Schematic showing typical cross-section of embankment treatment with DSM (*after* Raju *et al.*, 2003).

and other additives, with the soil by the use of rotating mixing tools in order to form columns of a hardening material as a result of pozzolanic reactions between the binder and the soil grains. In Sweden and Finland, deep stabilization techniques are quite popularly used for stabilization of soft soil (Åhnberg *et al.*, 1995b).

Based on design requirements, site conditions, soil and rock layers, restraints and economics, the use of DMM is increasing. These methods have been suggested and applied for soil and rock stabilizing; slope stability; liquefaction mitigation; vibration reduction (along railways); road, railroad and bridge foundations and embankments; construction of excavation support systems or protection of structure close to excavation sites; solidification and stabilization of contaminated soils etc. (Kazemian and Huat, 2009b).

The demand for improving and stabilizing land for different purposes is expected to increase in the future and probably the best way to fulfil it is by using DMM. It is strongly suggested that, where sufficient space is unavailable, sliding and overturning stability be augmented using soil anchors. The main advantage of these methods is the long-term increase in strength, especially for some of the binders used. Pozzolanic reactions can continue for months or even years after mixing, resulting in an increase in the strength of cement-stabilized clay with increasing curing time (Bergado, 1996; Roslan and Shahidul, 2008).

According to most researchers, there are four basic jet grouting systems that are widely used nowadays (Keller Holding, 2005):

I. Single phase (grout injection only).
II. Dual phase (grout + air injection).

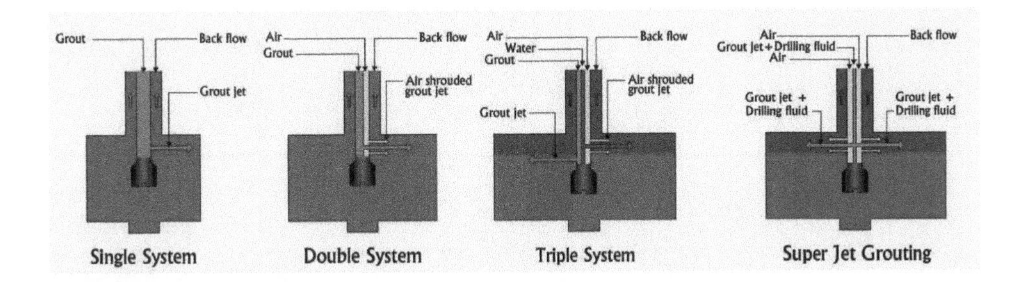

Figure 6.15 Systems in jet grouting (*after* Keller Holding, 2005; Kazemian and Huat, 2009b).

Table 6.2 Methods used in DMM (*after* Kenneth and Andromalos, 2003; Kazemian and Huat, 2009b)

Shallow soil mixing (SSM)	*Deep soil mixing (DSM)*	*Jet grouting systems*			
		Single phase	*Dual phase*	*Triple phase*	*Super jet grouting*
Ras-columns	Cement Deep Mixing System (CDM)	I. Maxperm grouting system II. Navigational drilling system III. Vacuum grouting injection	Dry jet mixing system (DJM)	Jumbo eco pile system (JEP)	Ras-jet system

III. Triple phase (water + air injection and followed by grout injection).
IV. Super Jet Grouting (air injection + drilling fluid by grout injection).

Figure 6.15 shows the procedure adopted in the jet grouting systems.

It is important to note that DMM are divided into three types, namely SSM (Shallow Soil Mixing), DSM (Deep Soil Mixing) and JGS (Jet Grouting Systems); Table 6.2 clearly names the methods which are utilized in each type (Kenneth and Andromalos, 2003).

6.4.1 Ras-columns

Ras-columns is one of the most common soil mixing methods in DDM, based on mechanical soil mixing technology. This method has been used for improving shallow soils and seldom in deep mixing (Kazemian and Huat, 2009b).

Columns of stabilized material are formed by mixing the soil in place with a 'binder'. The interaction of the binder with the soft soil leads to a material which has better engineering properties than the original soil (Hebib and Farrell, 2003).

The mixing head is combined with blades which can rotate inversely. In other words, in the bottom auger the mixing blades rotate clockwise and in the upper auger they rotate anticlockwise (Raito Kogyo, 2006).

This technique causes the cement to mix with soil homogeneously and thus produces higher quality soil-cement columns. The first step in this method is rig positioning

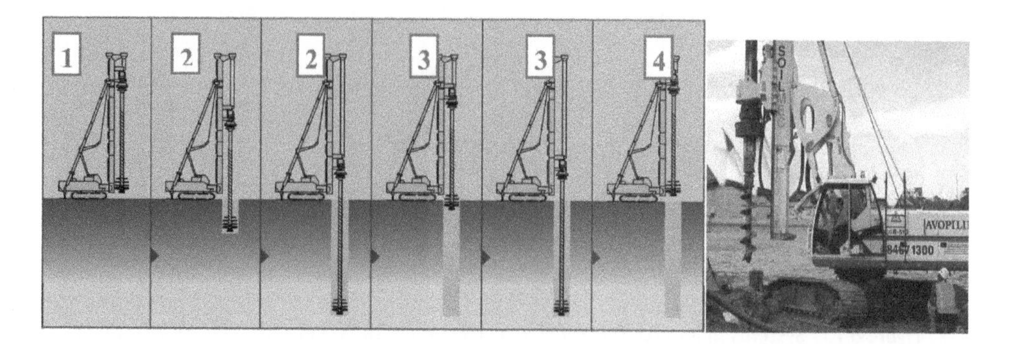

Figure 6.16 The steps in the ras-columns system (*after* Raito Kogyo, 2006) and its mixer (SDFEC, 2007; Kazemian and Huat, 2009b).

Figure 6.17 (a) CDM machine with two augers (*after* Raito Kogyo, 2006), (b) CDM machine with three augers (*after* Keller, 2005), (c) two augers which have been parted-off (*after* Balvac, 2008).

(Figure 6.16 (1)). This is followed by penetration, whereby after passing through a dry excavation zone slurry should be injected without any jetting (Figure 6.16 (2)). The third step is churning or moving the head up and down to mix the soil and cement thoroughly (Figure 6.16 (3)). Finally, in the completion step the head is withdrawn and the soil-cement column is completed (Figure 6.16 (4)).

6.4.2 Cement deep mixing system (CDM)

The second method is related to DSM and is one of the DMM methods, known as cement deep mixing (CDM). As mentioned earlier, in this method, a series of overlapping augers and mechanical mixing shafts are used. Figures 6.17(a) and 6.17(b) show CDM machines with two and three augers respectively. Two augers which have been parted-off (detached) are shown in Fig. 6.17(c).

CDM is normally used in soft soil that contains mineral soils such as clay or sand. In some conditions where mineral soils are absent, sand should be added before mixing in

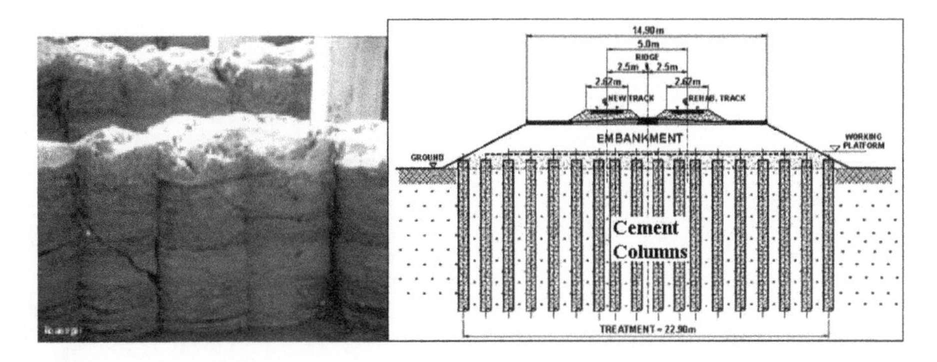

Figure 6.18 Cement mortar providing strength in CDM and the soft soil being solidified (*after* Raito Kogyo, 2006; Huat, 2004).

the cement slurry. Figure 6.18 shows the shape of cement mortar after strengthening. A series of overlapping augers ensures that the main geotechnical purpose will be achieved (Kazemian and Huat, 2009b).

CDM is a soil stabilization method which mixes cement slurry with soft soil *in situ* to attain the required strength. Soft soil is stabilized by a two-phase chemical reaction. A hydration reaction occurs and an ettringite of capillary crystals is generated when the cement mixes with water. Then a pozzolanic reaction follows, whereby the hydration product reacts with the clay minerals in the soil (Raito Kogyo, 2006).

6.4.3 Jet grouting systems

Jet grouting systems, the third type of DMM method, have some similarities with the previous methods. This method has been used for soft clays. Apart from having the same mixing tools, this method also applies the same process, whereby the *in situ* soil will be cut and broken by a high pressure jet of slurry to produce a homogeneously improved zone around the mechanically mixed core.

6.4.4 Vacuum grouting injection

It is worth pointing out that pressure injection may be less successful when the pressures needed to dispel gases and liquids from the voids are so high as to risk disrupting the structure. For instance, this may happen when the voids consist of many fine interstices which do not always interconnect (which may result in the need for a very large number of injection points), when complete filling is very difficult to achieve, or when it is difficult to confine the grout to the area to be injected.

The third step in single-phase jet grouting is vacuum grouting injection. In this technique, a partial vacuum is first established in a portion of the structure (or the whole of the structure if it is small enough), drawing off gases and liquids from the voids and interstices.

This vacuum holds the structure together, rather than exerting any potentially disruptive forces as in pressure injection. After achieving a stable vacuum, the injection

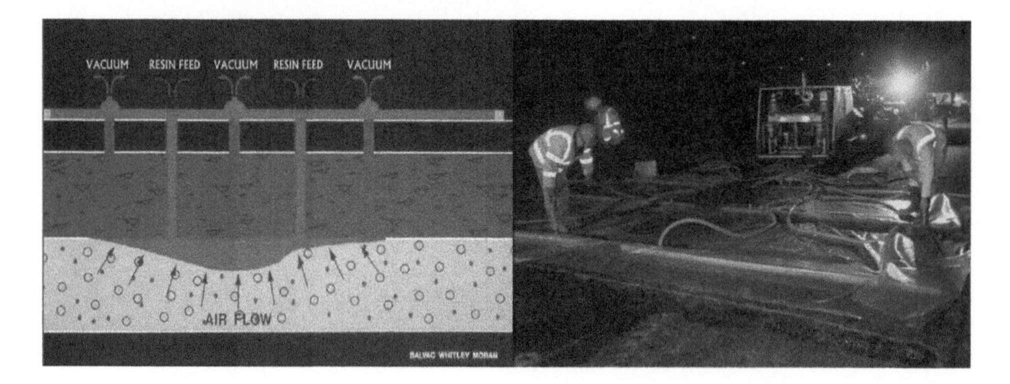

Figure 6.19 Vacuum grouting injection process and equipment. The plastic cover is used to suck in all the entrapped air so as to have no void space (*after* Balvac, 2008).

liquid is introduced either through injection pipes set at appropriate intervals and depths or over the surfaces and into the structure through cracks, fissures and porous areas. Figure 6.19 shows the process of vacuum grouting injection.

6.4.5 Dry jet mixing system (DJM)

The second type of jet grouting system is the dual phase system (DPS). One of the most commonly used methods in DPS is dry jet mixing system (DJM). This is a highly effective ground treatment system used to improve the load performance characteristics of soft clays, peats and other weak soils. The concept of using a dry binder for deep soil mixing was first presented in 1967 by Kjeld Paus from Sweden. Thirty years have passed since then and the technique has evolved considerably. The method is based on injecting dry binder carried by compressed air into soil. DJM uses mixing blades to mix dry reagents, such as cement or lime, with *in situ* soils for remediation (Kazemian and Huat, 2009b).

In this method, the process employs the effects of both hydration and the bonding of soil particles to increase the shear strength and reduce the compressibility of the soil mass (Keller Holding, 2005).

6.4.6 Dynamic replacement method

Another deep stabilization method is the dynamic replacement (DR) method. The process consists of dropping heavy weights (ranging from 10 tons to 15 tons) from heights of 10 m to 15 m to form large diameter granular columns in cohesive soil deposits. Subject to the site working conditions, the nature and consistency of the soil conditions and the required length of the DR columns, it may be required to perform pre-excavated DR columns (Sin, 2003). The size of the excavated crater may be 2 to 3 m deep over an area of 4 m^2. Very often, the actual size of this pre-excavated DR crater is governed by the ground water table condition. After the excavation, the crater is filled with granular materials and a series of pondering and ballasting phases

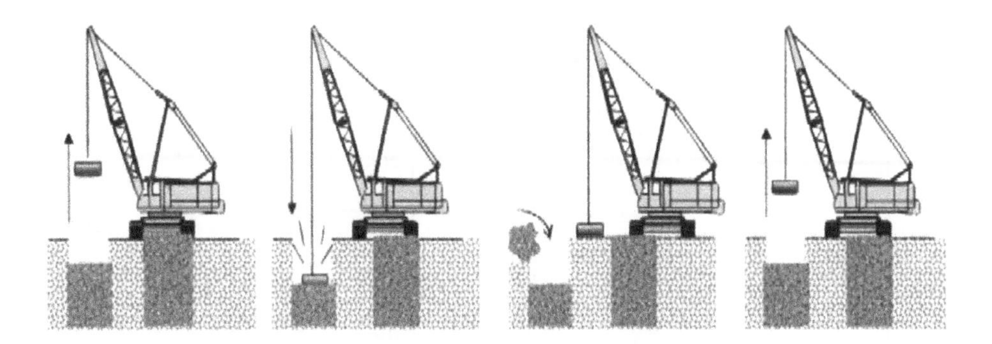

Figure 6.20 Dynamic replacement to form sand columns.

of work is carried out until completion of the installation of a DR column, as shown in Figure 6.20.

This method seems to work well for soft silty clay, but its application to peat is yet to be proven.

6.4.7 Sand drains and sand/stone columns

Kurihara *et al.* (1994) reported the favourable performance of the central Hokkaido expressway, built on peat using sand drains as shown in Figure 6.21. The sand drain-treated sections were found to experience smaller post-construction settlement over five years of operation compared with the untreated section. They also concluded that the sand compaction pile would be more expensive in this respect.

Very weak surficial zones of peat may not have adequate strength to provide lateral support to the sand or stone columns. The use of geotextile and synthetic fibres has been suggested to remedy this problem (Al-Refeai, 1992).

6.4.8 Vibrated concrete column

In this technique concrete columns are created *in situ*. A vibrator penetrates the weak subsoil until it reaches the proposed bearing stratum. Concrete is then pumped as the vibrator is withdrawn. By re-vibrating the concrete at the base and top, bulbous ends are formed which enhance base resistance and form pile caps at the surface. The embankment above the piles is reinforced with geogrid layers to promote load transfer to the columns. An example of this application is given by Maddison *et al.* (1996).

6.5 PILE SUPPORT

Piles may be described as long, slender concrete, steel or timber members that can be used in transferring loads through weak/unstable soil to ground of higher load-bearing capacity.

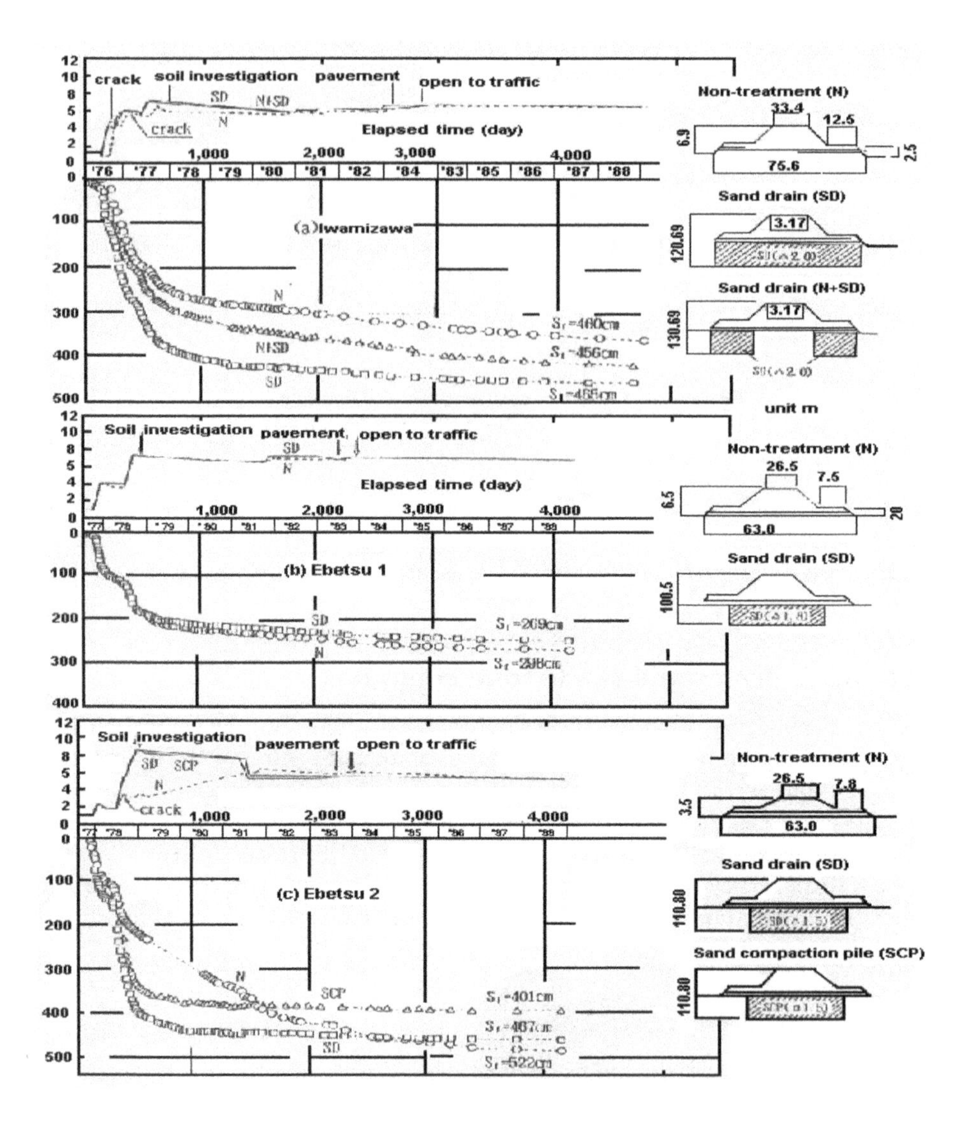

Figure 6.21 Performance of sand drain embankment (*after* Kurihara *et al.*, 1994).

6.5.1 Types of pile

These are categorized piles into six different types as follows:

- **End bearing piles**: these terminate in hard, relatively impenetrable material such as rock or very dense sand and gravel. They derive most of their carrying capacity from the resistance of the stratum at the toe of the pile.
- **Friction piles**: these obtain a greater part of their carrying capacity by skin friction or adhesion. This tends to occur when piles do not reach an impenetrable stratum but are driven for some distance into a penetrable soil. Their carrying

capacity is derived partly from end bearing and partly from skin friction between the embedded surface of the soil and the surrounding soil.

- **Settlement reducing piles**: these are usually incorporated beneath the central part of a raft foundation in order to reduce differential settlement to an acceptable level. Such piles act to reinforce the soil beneath the raft and help to prevent dishing of the raft in the centre.
- **Tension piles**: structures such as tall chimneys, transmission towers and jetties can be subject to large overturning moments, so piles are often used to resist the resulting uplift forces at the foundations. In such cases the resulting forces are transmitted to the soil along the embedded length of the pile. The resisting force can be increased in the case of bored piles by under-reaming. In the design of tension piles the effect of radial contraction of the pile must be taken into account, as this can cause about a 10–20% reduction in shaft resistance.
- Laterally loaded piles: almost all piled foundations are subjected to at least some degree of horizontal loading. The magnitude of the loads in relation to the applied vertical axial loading will generally be small and no additional design calculations will normally be necessary. However, in the case of wharves and jetties carrying the impact forces of berthing ships, piled foundations to bridge piers, trestles to overhead cranes, tall chimneys and retaining walls, the horizontal component is relatively large and may prove critical in design. Traditionally piles have been installed at an angle to the vertical in such cases, providing sufficient horizontal resistance by virtue of the component of axial capacity of the pile which acts horizontally. However, the capacity of a vertical pile to resist loads applied normally to the axis, although significantly smaller than the axial capacity of that pile, may be sufficient to avoid the need for such 'raking' or 'battered' piles which are more expensive to install. When designing piles to take lateral forces it is therefore important to take this into account.
- **Piles in fills**: piles that pass through layers of moderately to poorly compacted fill will be affected by negative skin friction, which produces a downward drag along the pile shaft and therefore an additional load on the pile. This occurs as the fill consolidates under its own weight.

6.5.2 Pile behaviour

Pile behaviour during construction/installation portrays a significant threat to foundation design which needs to be considered with a view to providing adequate solutions prior to construction. Such behaviours may include the following.

6.5.2.1 Geological behaviour

It is natural to assume that soil layers are horizontal in nature, including the underlying layer where foundations are situated. Poulos (2005) stated that the effects of construction processes on the piles are also frequently ignored or simplified. Such ideal conditions are rarely, if ever, encountered in real life. Both natural geological circumstances and the processes involved in construction will generally lead to the above assumptions being invalid, at least to some degree.

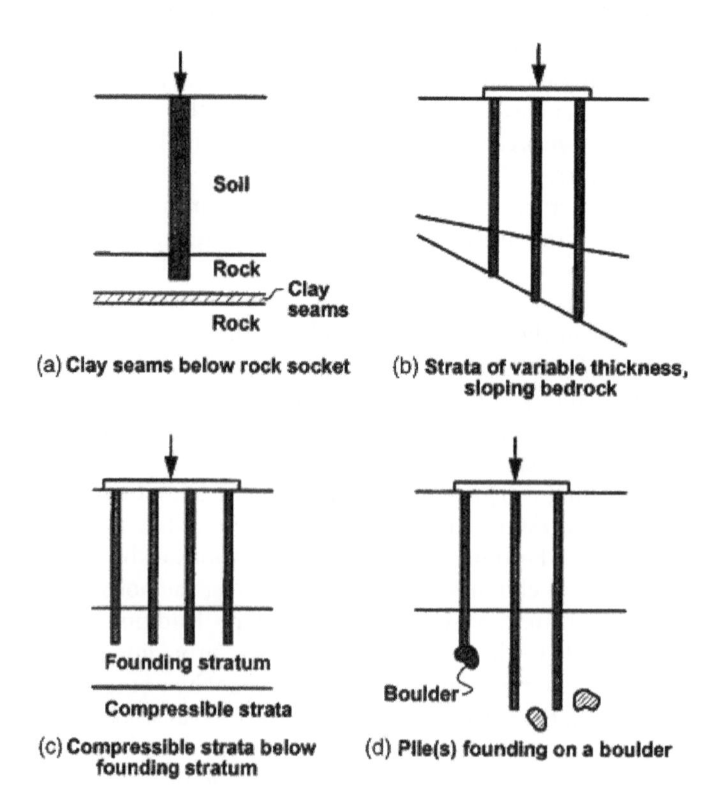

Figure 6.22 Imperfections due to geological behaviour (*after* Poulos, 2005).

In many problems, there may exist 'imperfections' that arise for natural geological or construction-related reasons, and which are generally unanticipated deviations from the expected circumstances. While it may be possible to handle these imperfections by appropriate analysis, the main difficulties are that the designer/analyst may not recognize the existence of the imperfections during the design process, or else they may only become manifest beyond the design process. Subsequent analysis may then become either a remedial or a forensic exercise Poulos (2005).

The imperfections may include (among many others) layers which are not horizontal or continuous, boulders within a soil layer, sloping bedrock, intrusions of rock over limited areas of the site, cavities in limestone rock, or the presence of softer layers below what might be regarded as suitable founding strata for the piles. Figure 6.22 illustrates some of these situations.

6.5.2.2 Inadequate ground investigation

These imperfections are generally related to those that arise from natural sources, but are exacerbated because the site is not properly characterized. Inadequacies are usually related to an insufficient number or depth of boreholes or probes to identify stratigraphic variations across the site, or inadequate testing to quantify the

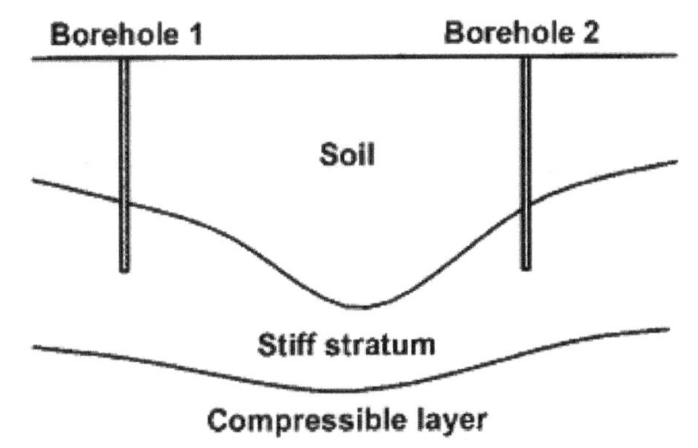

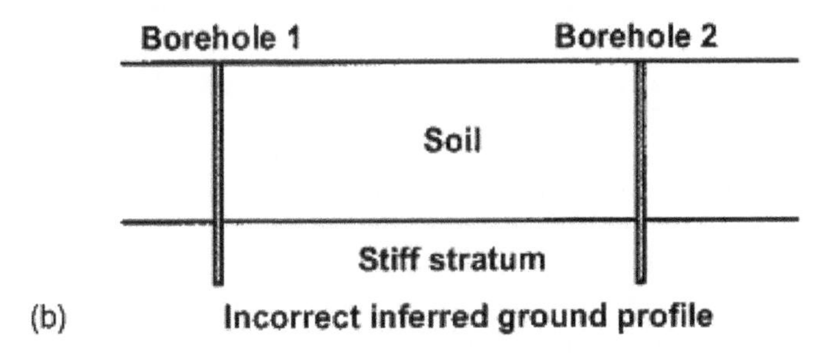

Figure 6.23 Imperfections due to poor ground investigation (*after* Poulos, 2005).

relevant geotechnical parameters. Figure 6.23 illustrates a typical example of such imperfections.

6.5.2.3 Construction behaviour

These imperfections, as highlighted by Poulos (2005), arise from processes related to the construction of the piles, either from inadequate construction control or from the inevitable consequences of construction activities. They may include:

- A soft toe on bored piles due to inadequate base cleaning (avoidable);
- Defects within the shaft of bored piles (avoidable);
- Inadequate founding conditions (avoidable);
- Ground movements developed due to drilling during the construction process (generally unavoidable);
- Excavation and dewatering effects, especially with remedial piling projects (generally unavoidable, but controllable); and
- Excessive driving of preformed piles (avoidable).

In general, construction-related imperfections in piles can be broadly classified into two main categories: structural defects and geotechnical defects (Poulos, 2005).

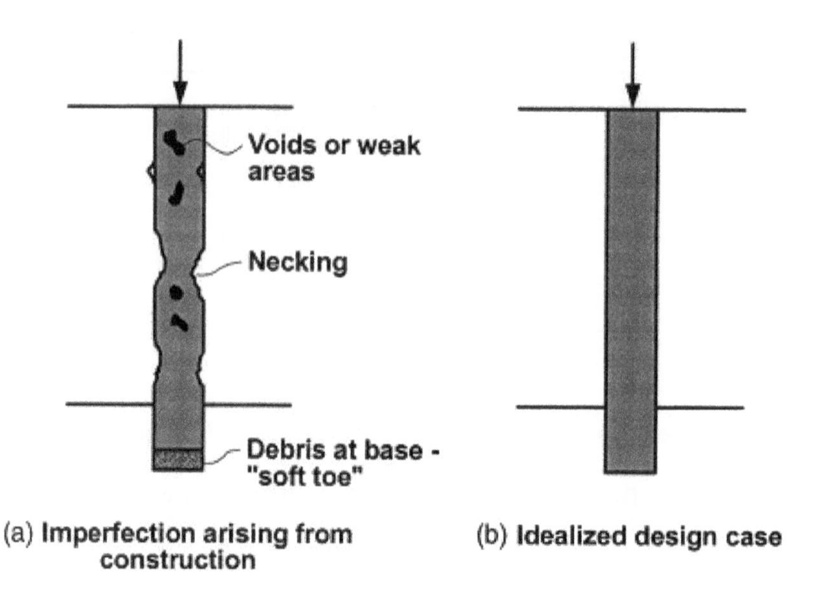

Figure 6.24 A pile with structural and geological defects (*after* Poulos, 2005).

Structural defects can result in the size, strength and/or stiffness of the pile being less than assumed in design. Examples of structural defects include the following: 'necking' of the shaft of bored piles, leading to a reduced cross-sectional area along part of the pile; poor quality control during the construction of bored piles, leading to some parts of the shaft having lower strength than assumed in design; tensile cracking of large diameter bored piles under the influence of thermal strains; damage during the driving of timber, precast concrete or steel piles, leading to reduced strength and stiffness of parts of the pile, especially near the top or tip of the pile; and bending of slender driven piles (Poulos, 2005). Geotechnical defects usually arise from either a misassessment of the *in situ* conditions during design, or else from construction-related problems, and may include reduced shaft friction and end bearing resistance arising from localized softer or weaker geotechnical conditions in the vicinity of one or more of the piles in the group; reduced skin friction; and end bearing resistance arising from construction operations such as the use of bentonite without due caution, and a 'soft base' arising from inadequate cleaning of the base of bored piles (Poulos, 2005).

Figure 6.24 illustrates an example of the usual idealization of a pile which may have both structural and geotechnical defects.

6.5.3 Piled raft foundation

A piled raft foundation may be described as a composite structure made up of piles connected to the raft from underneath, exercising a variety of responsibilities jointly and independently in providing safe transfer of bearing pressure to the underlying subsurface or sub-base structure. The raft and pile method (Figure 6.25) is an economical ground improvement technique that has technical advantages derived from both raft and pile foundations. The raft carries the embankment loading by distributing it partly

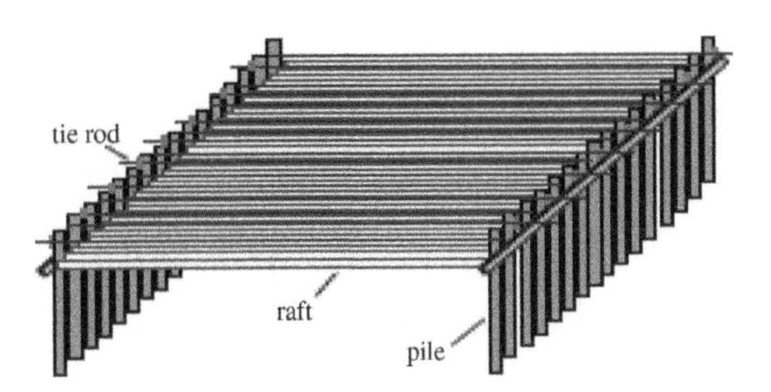

Figure 6.25 Raft and pile method (isometric view of raft (made of reapers), piles, side reapers and tie members) (*after* Poungchompu, 2009).

to the ground surface and partly to piles. The piles would then transfer the load to a deeper and stronger soil layer.

Piled supports of structures are a fundamental mean of construction over all soft soils. By carrying the structural forces to a competent layer, the problem of settlement of the structure can be largely avoided. However, one major problem with this method of construction is the relative movement between the supported and unsupported sections of the structure/embankment. In the past, the concept of fully piling an earth embankment with the provision of an extensive pile cap to support the earth fill was viewed as uneconomic. Some recent developments, however, have led to a review of the use of pile supports as a means of fully or partially supporting lengths of earth embankments on soft compressible soils. These methods may be used to provide transition zones between the completely piled structures and settling earth embankments or to support earth embankments with only limited settlements.

Where settlement is to be kept to minimum, such as adjacent to a road culvert or overhead bridge, the piles are usually driven to set. A transition zone is where the piles are driven to length (not set) in order to grade the settlements as uniformly as possible. The lengths of each row of floating piles are steadily reduced over a distance to achieve the specified differential settlement tolerance. A typical section is shown in Figure 6.26.

Large differential settlements between the piled and un-piled section of an embankment have caused a number of failures to road structures built using this technique (Huat *et al.*, 1994).

An example in which this method was employed specifically for soft organic deposits is in a section of the Bakun access road project in Sarawak Malaysia, where the soft deposits are rather extensive, 20 m or so in thickness. 200 × 200 mm RC piles at a spacing of 1.5 m to 2.0 m were used, capped with a 375 mm thick continuous RC slab (Figure 6.27).

Alternatives to a continuous RC slab above the piles are to rely on the principle of arching to distribute the weight of the earth fill to the piles with individual caps (Huat *et al.*, 1994), and the possibility of reinforcing the embankment base with geosynthetic materials such as geogrids. This technique was used for the construction of the Tungku Link road, Brunei, over peaty soil (Younger *et al.*, 1997). Individual piles with pile caps of 2.0 m diameter were used, above which was placed a geogrid mattress in a 200 mm crushed stone layer.

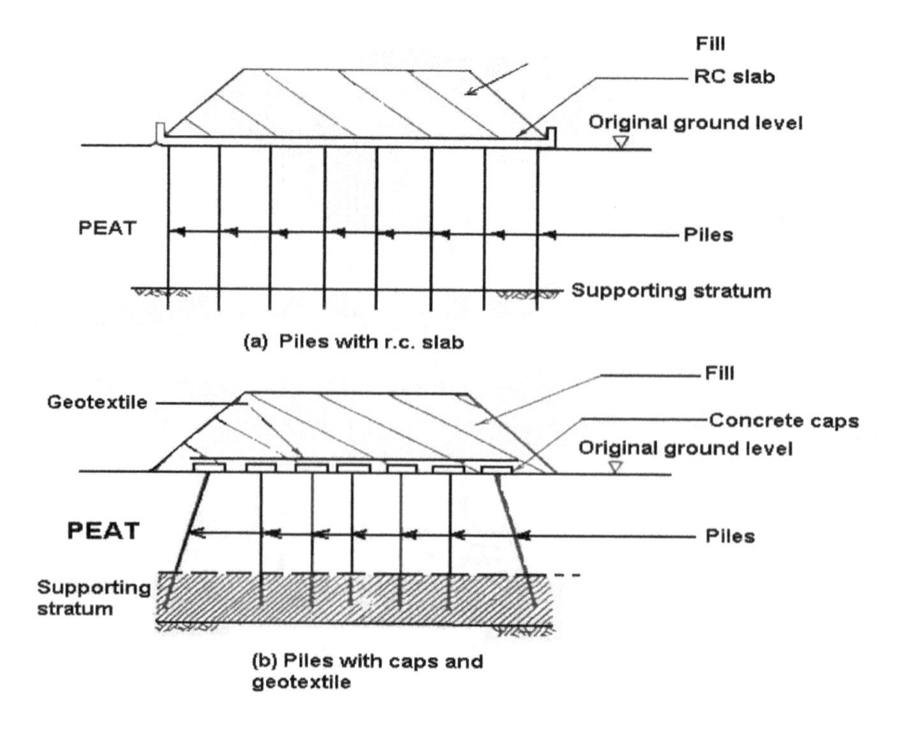

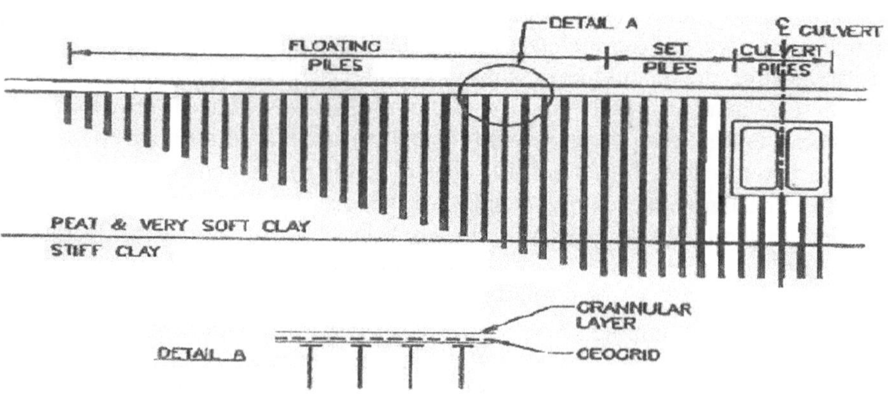

Figure 6.26 Pile embankment and transition zone.

A solution such as pile support would be too expensive for normal earth structures, except for very high embankments such as bridge approaches. The later option (use of individual pile caps rather than a continuous slab) minimizes the size of the pile cap required and improves the economics. Even so these systems are still very costly. A cost estimate by Greenacre (1996) was that geotextile-reinforced embankments were five times more expensive than normal embankments, and piled embankments were yet five times more expensive.

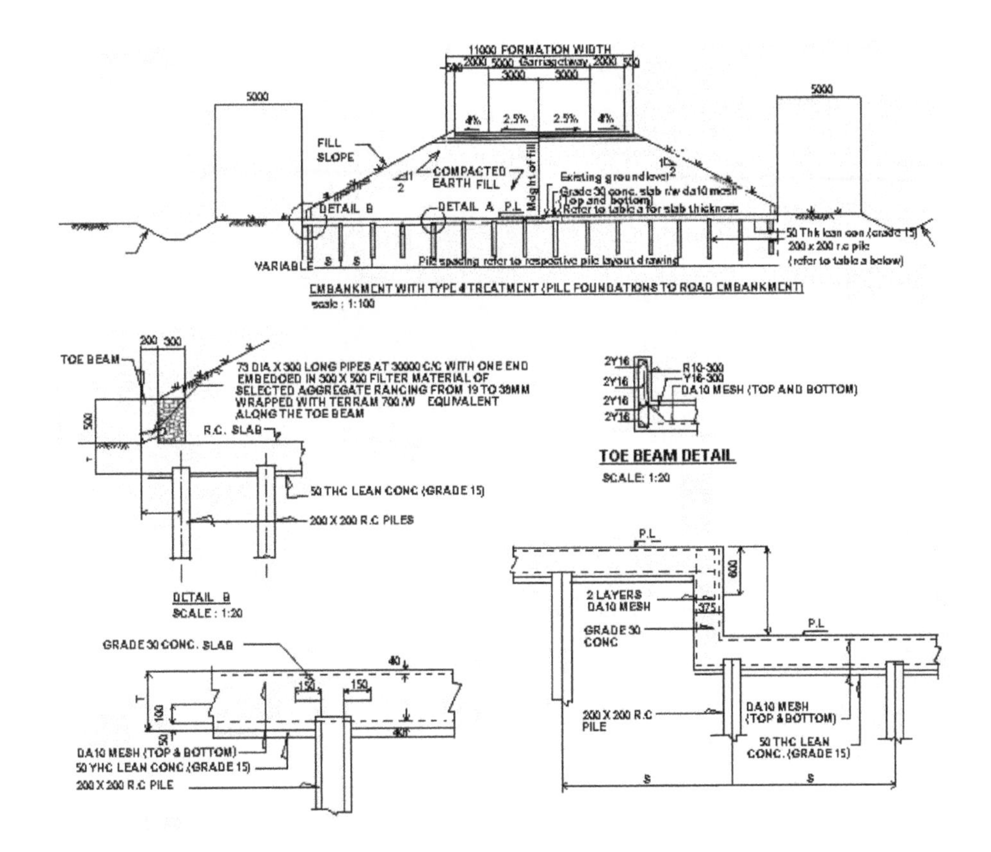

Figure 6.27 Cross-section of the Bakun access road pile embankment.

6.5.4 Pile mat-JHS system

Another version of the piled support system is the used of mini piles known as the pile mat-JHS system. This system of mini concrete piles and mats is actually similar to the pile-raft system. JHS system is in general a kind of pile-raft system consisting of square plates resting on mini concrete piles with a pin connection. Each pile-plate component is connected to the other components with connectors made of polyethylene. There are 3 to 4 connectors at each side of the plate. Each plate has dimension of 90 cm by 90 cm with an average thickness of 8 cm. Details of the system are shown in Figure 6.28.

Theoretically the whole system is capable of withstanding the embankment load and reducing the settlement to a certain amount. The longer the piles, the larger will be the reduced settlement of the system. The one interesting problem is the interaction of the plates and the soil under it and of the piles and soil around them. Because of the flexibility of the plate to plate connection, part of load is transferred to the soil below it and part goes to pile-soil interaction. Analysis of the JHS system is still in progress and is not yet available. However, a field trial of performance using this system at a site in Central Kalimantan reported by Rahadian *et al.* (2003) and Djajaputra and Shouman (2003) does not seems to be satisfactory.

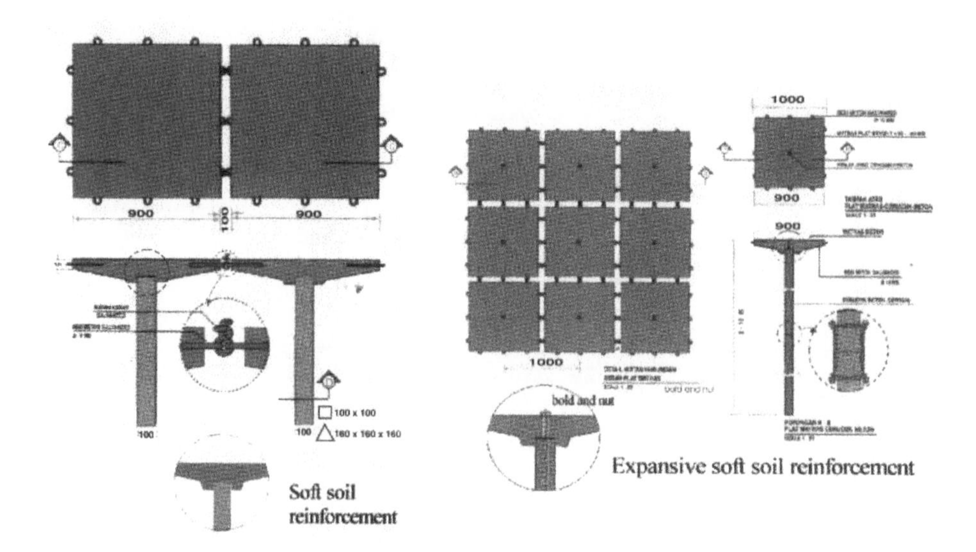

Figure 6.28 JHS system and its components.

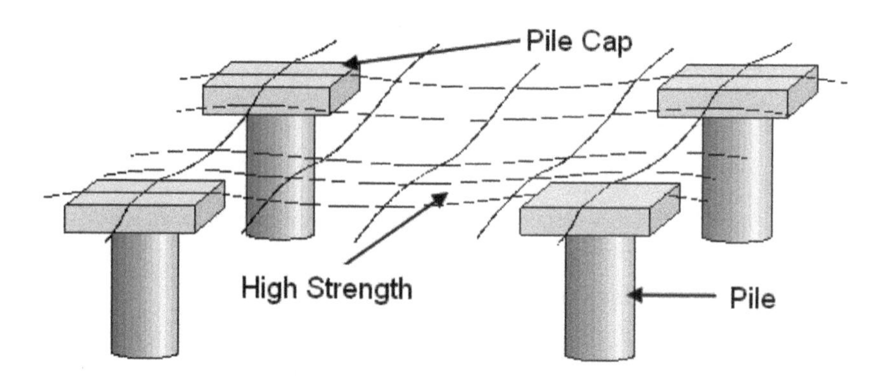

Figure 6.29 Schematic view of the AuGeo pile system.

6.5.5 AuGeo pile system

A variant of the pile support system is the AuGeo pile system. This system is theoretically cheaper than the conventional pile system (Abdullah *et al.*, 2003). The AuGeo piled embankment system consists of lightweight piles with an enlarged individual pile cap and pile-foot or pile base. A gravel mattress wrapped in two layers of high-strength geogrids is placed on top of the pile caps to transfer the load from the embankment to the pile caps (Figure 6.29). The piles are pushed into the soil in a close square grid using modified vertical drain stitcher equipment at high speed. In this manner the loads are theoretically transferred directly to the competent layers and the soft compressible layers are not loaded. In this way excessive settlement is avoided and the construction time can be limited.

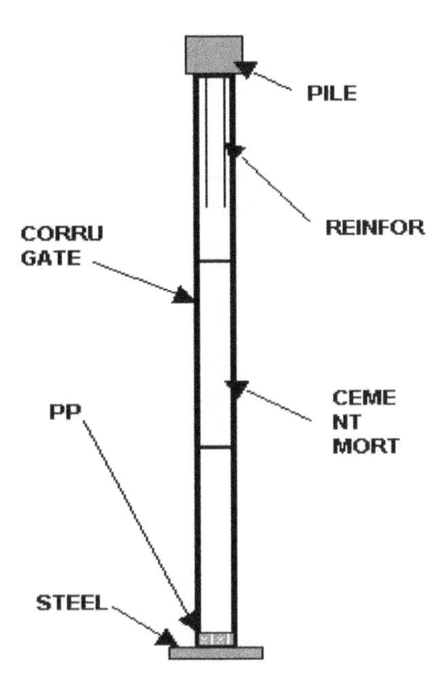

Figure 6.30 AuGeo pile.

The pile consists of the base, cement mortar-filled plastic casing, sand blanket, pile cap granular mattress and geogrids layer (Figure 6.30).

A typical embankment treatment with the AuGeo pile system for a 2 metre high embankment is presented schematically in Figure 6.31.

6.5.6 Friction/floating piles

Orr and McEnaney (1994) describe the use of timber friction poles for the construction of a walkway across the Clara bog in Ireland.

Mortar-sand-rammed-column (MSRC) is a variant of the floating pile foundation that combines the concept of a vertical drain (Figure 6.32).The vertical drain component is the sand surrounding the mortar, and the whole system is a floating pile foundation. Theoretically the ramming process increases the matrix soil lateral earth pressure in the vicinity of the columns and between the columns, thus making the soil stiffer. In time, as excess pore water pressure dissipates to the nearest column, the compressibility as well as shear strength of the underlying soil will improve.

These columns can be constructed in a five-step process shown in Figure 6.33. Holes of 0.5 m to 1.0 m diameter are drilled to a depth that typically varies from approximately 2 m to 10 m below the ground surface. A temporary casing should be employed to separate the mortar and compacted sand. Ramming the sand layer with a high-energy tamper to form undulating very dense sand stabilizes the bottom layer of sand. The process should be carried out layer by layer to avoid difficulties in withdrawing the casing. The subsequent step is preloading, and the duration of the waiting period

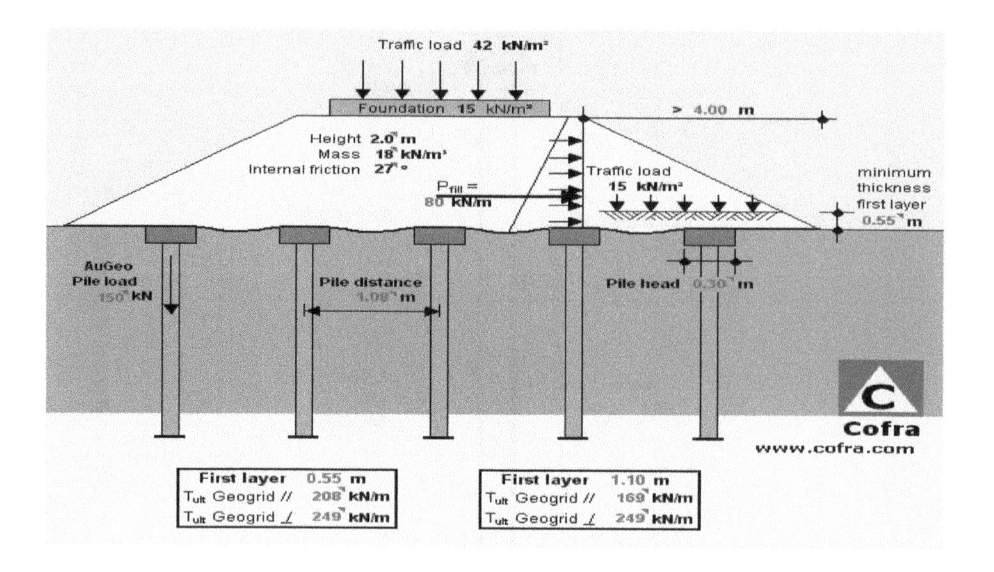

Figure 6.31 Typical section of the AuGeo pile system (*after* Abdullah *et al.*, 2003).

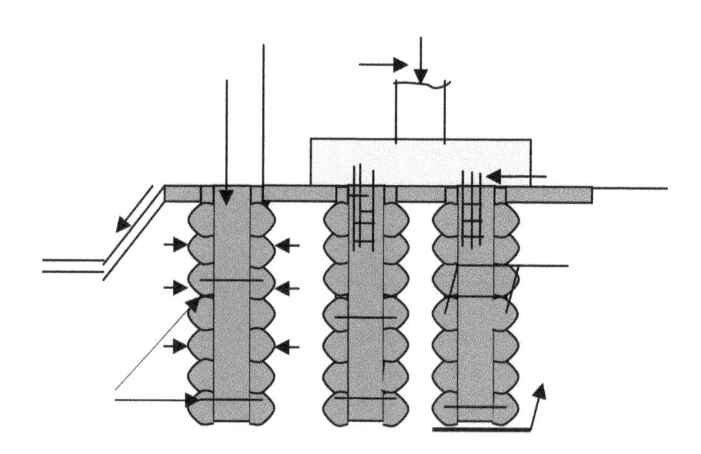

Figure 6.32 Concept of MRSC (*after* Suleaman, 2003).

is based on a calculation of soil parameters and prescribed performance. The preloading further prestresses and prestrains the columns and the surrounding soil matrix to increase the stiffness and strength of this foundation system (Sulaeman, 2003).

Preload piers. or geopiers, is another example of a floating foundation technique for stabilizing soft soil, including peat (Lien *et al.*, 2002; Biringen and Edil, 2003). These piers are inserted in the ground and expanded to about three times their original cross-sectional area. As a result of this expansion the surrounding soil is radially stressed, which initiates consolidation and general stiffening of the surrounding soil. If

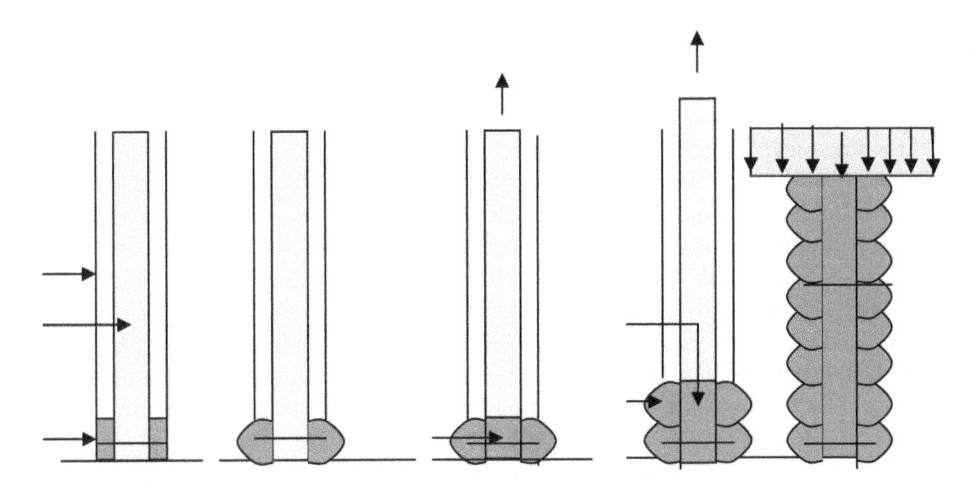

Figure 6.33 MRSC construction (*after* Suleaman, 2003).

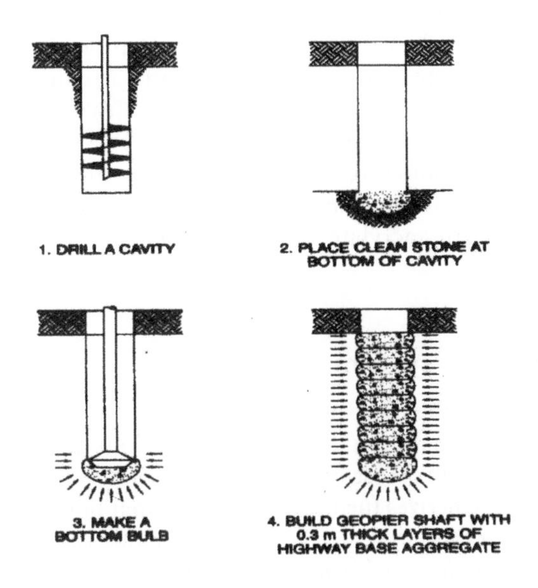

Figure 6.34 Geopier construction (*after* Lien *et al.*, 2002).

needed, vertical wick drains inserted between the preloaded piers can accelerate consolidation. The end result is a pier-soil composite with enhanced stiffness and strength to support the foundation loads within a tolerable degree of settlement.

This process can be viewed as a cavity expansion problem. Figure 6.34 shows the construction of geopiers.

6.6 CHEMICAL STABILIZATION

6.6.1 Chemical and cementation grouts

Chemical stabilization is an effective method to improve the soil properties by mixing additives to soils. Usually the additives are cement, lime, fly ash and bituminous material. These additives enhance the properties of soil. Generally, two major reactions for chemical stabilization are cation exchange reaction and cementation (Mitchell, 1993). Common chemical agents for the cementation process are Portland cement, lime, fly ash, sodium silicate polyacrylamides and bituminous emulsion (Kazemian*et al.*, 2010).

Many chemical grouts are based on the combination of sodium silicate and a reagent to form a gel. The Joosten process used in coarse granular soils uses calcium chloride as a reagent. Other reagents are organic esters, sodium aluminates and bicarbonates. The reagent and the proportion can be chosen to control the gel time, the initial viscosity and the order of strength of the grouted soil (Kazemian *et al.*, 2010).

Chemical grouts are injected into voids as a solution, in contrast to cementitious grouts, which are a suspension of particles in a fluid medium. The difference between chemical grout and cementitious grout is that chemical grout can be used to fill the finer voids of soil particles up to 10 to 15 µm in diameter. In other word, it has better penetration ability than cementitious grout (US Army Corps of Engineers, 1995).

Chemical grout can be classified as either single-step or two-step processes. In a one-step process, all the ingredients are premixed prior to injection, and the system is designed so that the reaction takes place *in situ*. In a two-step process, the initial chemical is injected into the soil mass and then followed by a second chemical material to react with the first *in situ* and stabilize the mass. There are several types of chemical grout, each of which has different characteristics and different applications. The most common are sodium silicate, acrylate, lignin, urethane and resin grouts (Shroff and Shah, 1999).

6.6.2 Sodium silicate system

Sodium silicate grouts are the most popular grouts because of their safety and environmental compatibility. It has been developed into various grout systems, such as the silicate chloride amide system. Most systems are based on reacting a silicate solution to form a colloid which polymerizes further to form a gel that binds the soil particles. The silicate solution concentration that may be used in grouting is in the range 10–70% by volume, depending on the material being grouted and the desired result. For a system using amide as the reactant, the amide concentration may vary from less than 1 to greater than 20% by volume. In practice, the amide concentration ranges from 2% to 10% (US Army Corps of Engineers, 1995).

The initial minimum viscosity of a grout that can produce a gel has a SiO_2:Na_2O ratio of 3.6 with a pH value of 8.5 to 9.2 for a given dilution within an ideal framework of gel time. The rate of reaction and strength of gel are directly proportional to the concentration of silicate and catalysts in the grout at constant temperature respectively (Shroff and Shah, 1999). Sodium silicate is noncorrosive to metals. Reactants such as amide and their water solutions will attack copper and brass, but they are noncorrosive to aluminates and stainless steel. Chloride solutions are not corrosive to iron and steel in the sense that acids are; however, if steel in a chloride solution is exposed to air, rusting will occur at the junction of the liquid and air. Bicarbonate is noncorrosive (US Army Corps of Engineers, 1995, Kazemian *et al.*, 2010).

6.6.3 Silicate chloride amide system

The silicate chloride amide system is a widely use silicate grout system containing sodium silicate as a gel forming material. The silicate aluminates-amide system has been used for strength improvement and water cut-off. Its behaviour is similar to the silicate chloride amide system but is better for shutting off seepage or the flow of water. The cost is slightly higher, and this system can be used in acidic soils. Amide will act as a reactant and the calcium chloride and sodium aluminates will be used as the accelerator. These reagents bring an almost instant setting time and produce a very low penetrability-type gel that is unsuitable for permeation treatments (Rawlings *et al.*, 2000; Kazemian *et al.*, 2010).

The function of the accelerator is to control the gel time and impart strength to the gel. The effect of the accelerator is important at temperatures below 37°C and increases in importance as the temperature decreases. Excessive amounts of accelerator may result in undesirable flocculation or formation of local hardening. This causes variations in both the gel and setting times that would tend to plug injection equipment or restrict penetration, resulting in a poorly grouted area. Therefore, a retarder should be added in the mixture to delay the setting time and formation of gel (US Army Corps of Engineers, 1995).

Other chemical grouts such as, acrylamide, N-methylolacrylamide, polyurethane, epoxy resins, aminoplasts, phenoplasts and lignosulfonates have been used for stabilization of sandy soils and clays (Nonveiller, 1989; Kazemian *et al.*, 2010; Magill and Berry, 2006; Vinson, 1970; Rawlings *et al.*, 2000; Karol, 2003).

6.7 CHOOSING THE GROUT

In order to choose a grout type, several properties of grout should be considered, such as rheology, setting time, toxicity, strength of grout and grouted soil, stability or permanence of the grout and grouted soil, and the penetrability and water tightness of the grouted soil (Rawlings, 2000). Moreover, the spreading of grout plays an important role in the development of grouting technology. In the actual field, the grouting method requires extensive consideration of the grout hole equipment, distance between boreholes, length of injection passes, number of grouting phases, grouting pressure and pumping rate (Shroff and Shah, 1999). Guidance on selecting a particular grout can be found in Table 6.3.

6.8 LIGHTWEIGHT FILL

An alternative is to utilize a very light material such as polystyrene blocks to cope with extreme soils. A typical density of polystyrene is about 20 kg m^{-3}, which may possibly increase to 100 kg m^{-3} as the material absorbs water from the ground. This

Table 6.3 Ranking based on toxicity, viscosity and strength (*after* Shroff and Shah, 1999; Kazemian *et al.*, 2010).

Grouts	Toxicity	Viscosity	Strength
Silicate			
Joosten process	Low	High	High
Siroc	Medium	Medium	Medium-High
Silicate-Bicarbonate	Low	Medium	Low
Lignosulphates			
Terra Firma	High	Medium	Low
Blox-All	High	Medium	Low
Phenoplasts			
Terramier	Medium	Medium	Low
Geoseal	Medium	Medium	Low
Aminoplasts			
Herculox	Medium	Medium	High
Cyanaloc	Medium	Medium	High
Acrylamides			
AV-100	High	Low	Low
Rocagel BT	High	Low	Low
Nitti-SS	High	Low	Low
Polyacrylamides			
Injectite 80	Low	High	Low
Acrylate			
AC-400	Low	Low	Low
Polyurethane			
CR-250	High	High	High

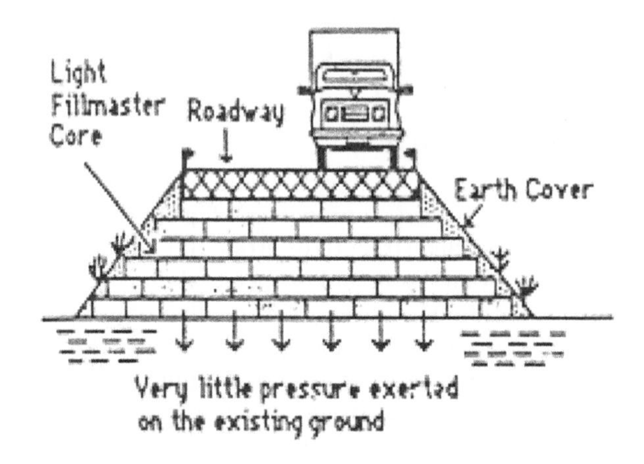

Figure 6.35 Lightweight embankment.

is still about 20 times lighter than conventional fill such as clay and sand. Figure 6.35 illustrates the principle of using polystyrene blocks as an alternative to heavy earth core structures built over soft ground. Because of their light weight, very little pressure is actually exerted on the existing ground, hence minimizing the stability and settlement problem.

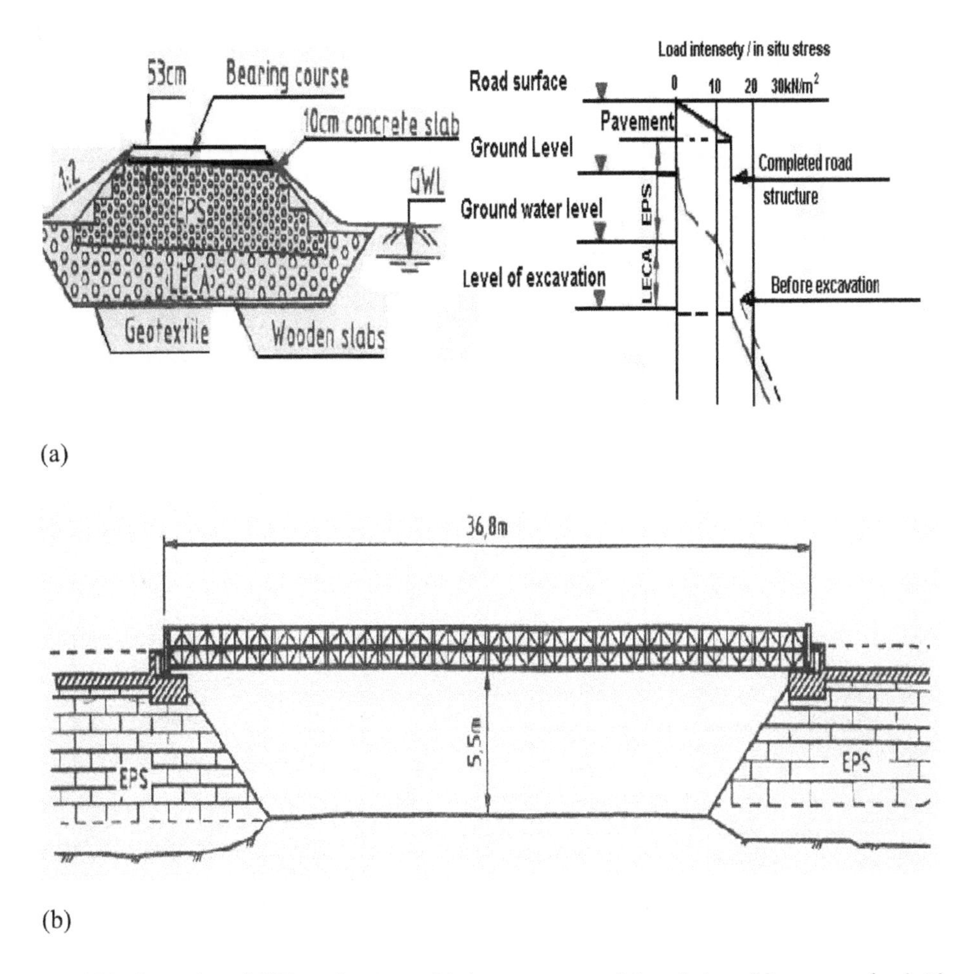

Figure 6.36 Examples of EPS applications. (a) As compensated foundation; (b) support for bridge abutment; (c) arrangement of lightweight polystyrene blocks at culvert transition; and (d) section view of culvert transition (*after* Gan and Tan, 2003).

By excavation and replacement with the polystyrene blocks, it is theoretically possible to completely float the embankment, thus imposing zero net stress to the underlying ground. This technique is also known as 'weight credit construction'. An interesting aspect of this construction is the need to have a stable water table, as any changes will alter the state of buoyancy and potentially caused movement in the systems. The blocks need also to be protected from fire, usually by mean of earth cover and top concrete slabs. A general review of the literature on the use of EPS is provided by Frydenlund and Aaboe (1997), Horvath (1995) and Gan and Tan (2003). Figure 6.36 illustrates some of their application examples.

Lauritzsen and Lee (2002) suggested that it is possible to use EPS as a foundation for building two-storey houses and garden directly on peat, in addition to road.

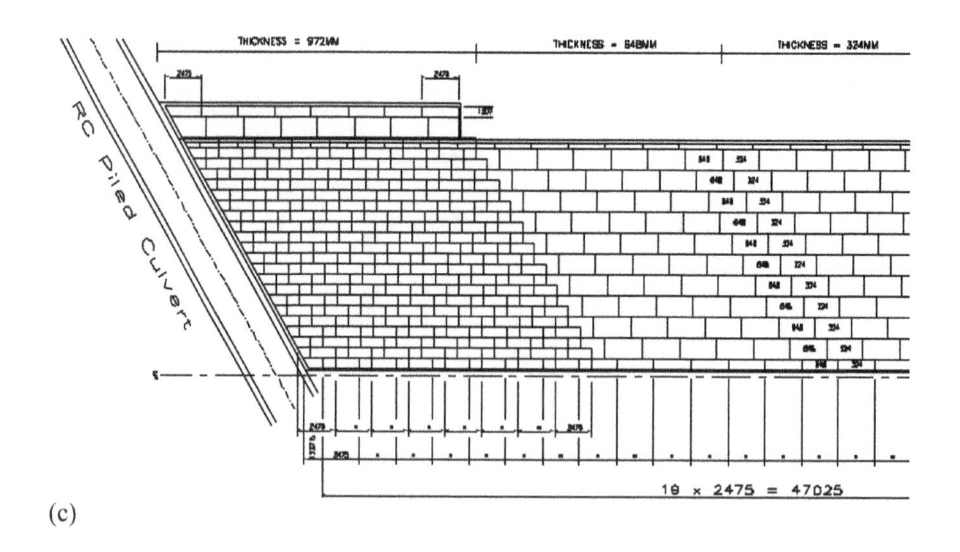

(c)

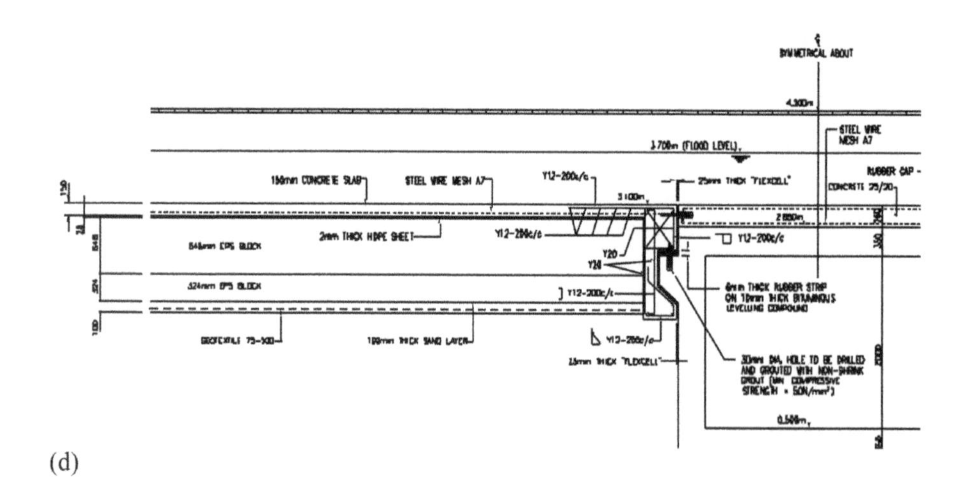

(d)

Figure 6.36 (continued).

Other materials that could be considered for the weight credit construction technique include shredded waste tyres, wood fibres or wood chips, sawdust or bales of peat.

The use of shredded waste tyres or tyre chips (Figure 6.37) proved to be advantageous. Its low unit weight served to limit settlement on compressible ground and applications using this material for road projects have been successful (Jones *et al.*, 1986).

Embankments constructed of wood fibres have also been found to perform well over 20 years, and more than 50 years can be expected as service life for wood fibre fills

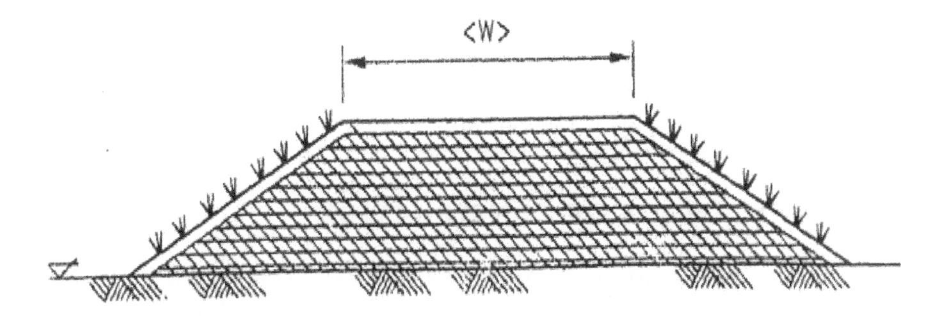

Figure 6.37 The use of scrap tyres as lightweight fill.

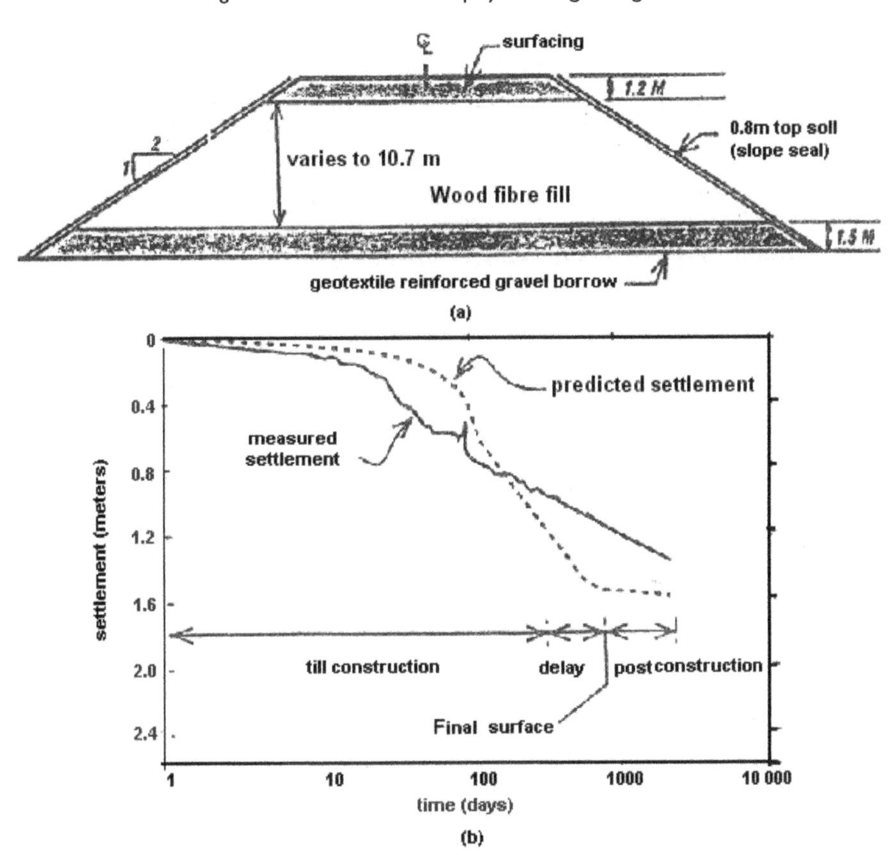

Figure 6.38 (a) Cross-section of wood fibre filled embankment; (b) settlement performance (*after* Kilian and Ferry, 1993).

(Kilian and Ferry, 1993). When using wood fibre, it is recommended that the height of the wood fibre should not exceed 5 m and ventilation to the wood fibre should be reduced to prevent spontaneous combustion. Figure 6.38 shows a wood fibre filled embankment.

In Japan, foamed cement paste (FCP) (material prepared by mixing foam with cement paste) is widely used as a lightweight geomaterial for road or embankment construction (Yasuhara *et al.*, 2003). FCP has a larger wet density (0.5–0.8 $Mg\,m^{-3}$) compared with EPS, but it has the advantage in stability against chemicals and its strength can be controlled by its composition.

6.9 OTHER METHODS OF CONSTRUCTION

The above methods of construction are not exhaustive. There are other methods that could be used, such as geocells, vibrated concrete columns, sand drains and sand/stone columns, vacuum preloading, vibrated concrete column, thermal precompression, friction piles and deep stabilization.

6.9.1 Geocells

These are essentially a series of strips of stiff polyethylene plastic about 200 mm wide spot welded together such that when they are stretched out they form a 'honeycomb' type of arrangement. The honeycomb is laid on the soft ground and filled with granular backfill. The underlying principle is simple: granular material is stronger and stiffer when it is confined. During construction, it is also common to lay a light geotextile beneath the geocells to act as a separator membrane.

6.9.2 Thermal precompression

A novel method to improve peat is the thermal compression technique. According to Fox and Edil (1994), peat compresses more rapidly under load when heated. Two trial embankments were constructed, one (A) unheated, the other (B) heated by circulating water at a temperature of 65°C flowing through heating wells installed in the peat soil (Figure 6.39). Heating causes a drastic increase in strain rate and a corresponding rapid decrease in the peat void ratio. Once the soil is cooled, it has a substantially reduced void ratio and creeps very slowly. In addition, the soil develops a significant quasi-preconsolidation effect and has very low compressibility with respect to further loading.

6.9.3 Gap method

In Holland the so-called gap method has been experimented with to reduce horizontal deformation during widening of highway embankments. The construction of the new embankment is performed from the outside inwards (Figure 6.40), rather than in layers as in conventional techniques. When consolidation of the outside sections has proceeded far enough the middle section, the 'gap' is filled, with an increase in arching and reduction in horizontal deformation.

6.9.4 Reinforced overlay

The use of bituminous overlay reinforced with geotextile or geogrids is an option in order to keep the road open and minimize maintenance.

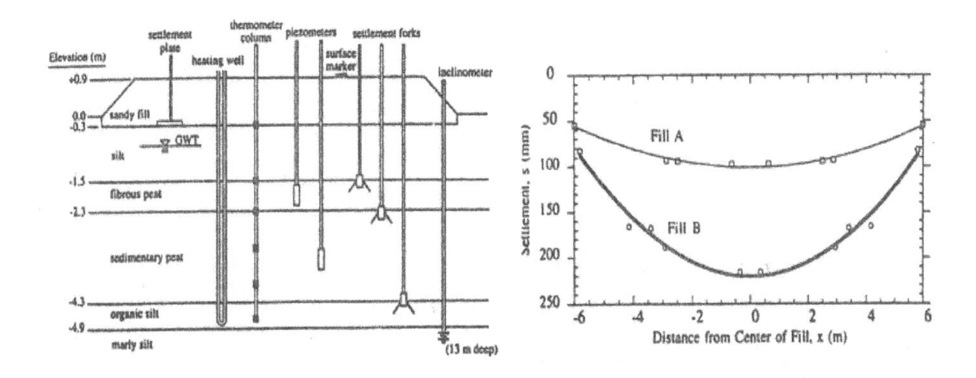

Figure 6.39 Cross-section and performance of heated and control fill (*after* Fox and Edil, 1994).

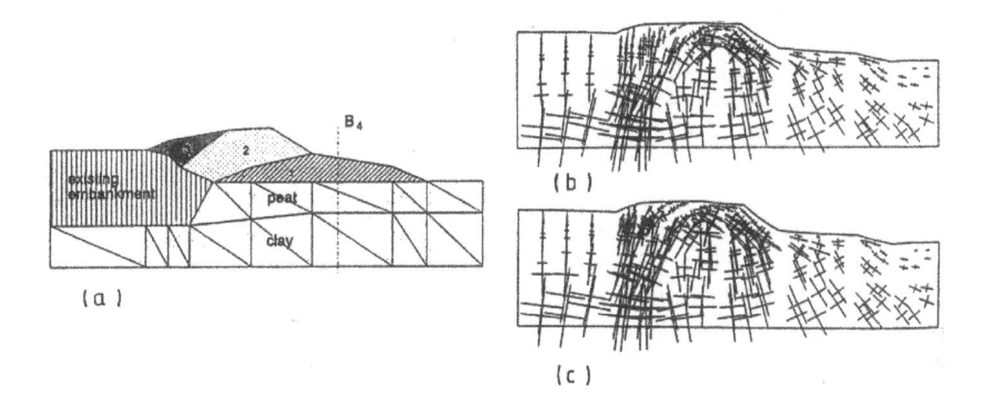

Figure 6.40 Principle of the gap method.

6.10 TRIAL EMBANKMENTS

Rahadian *et al.* (2003) describe the performance of five full-scale trial embankments built by the Institute of Road Engineering, Indonesia, at Berengbengkel, Central Kalimantan, in order to study the performance of embankment construction on peat deposits for the purpose of developing a design guide. They are:

1. Normal embankment, which was built over 3 m of peat deposit.
2. Surcharge embankment, which was built over 2.5 m of peat deposit.
3. Timber piled embankment which was built over 6 m of peat deposit
4. JHS (mini concrete) piled embankment which was built over 11 m of peat deposit
5. EPS embankment, which was built over more than 11 m of peat deposit.

The Berengbengkel peat is categorized as fibrous, hemic (H2–H5), low ash and highly acidic peat. The thickness of the deposit varies from 2 m to 15 m. Pollen analysis

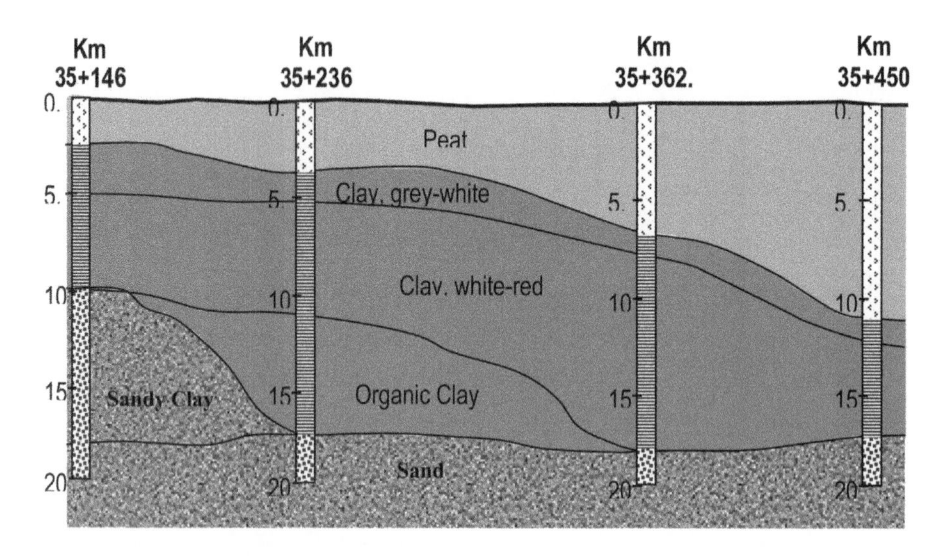

Figure 6.41 The soil stratigraphy at the Berengbengkel, Central Kalimantan, trial embankments (*after* Djajaputra and Shouman, 2003).

and carbon dating measurement identified that the age of the peat is approximately 4,000 years. The natural water content of the peat range from 467–1224%, with organic content 41–99%, fibre content 6–56%, unit weight of 9–10.5 kN m^{-3}, pH of 2–3, undrained shear strength 6–17 kPa, and coefficient of vertical permeability, k, of 3.5×10^{-7}–8.4×10^{-8} m s^{-1}. Figure 6.41 shows the soil stratigraphy of the trial site, Figure 6.42 shows cross-sections of the field trial and Figure 6.43 shows the settlement performance of two of the trials: surcharge embankment and EPS embankment.

From this trial, Rahadian *et al.* (2003) concluded that the surcharging method works rather well, with settlement reaching a steady state condition after 200 days, but the other four trials (i.e. normal embankment, timber and JHS (mini concrete), and EPS embankment) were not successful. Both piled embankments (wooden and JHS (mini concrete) piles), which are friction rather than end bearing piles, still showed signs of settlement after 650 days, with total settlements of 1.2 m and 2 m respectively (Figure 6.44). They attributed the possible cause of this to the very low friction between peat and pile material (wood and concrete). The longer and heavier concrete pile JHS system actually has more self weight than the shorter and lighter timber pile system. The EPS embankment also showed a large settlement (about 1.6 m).

However, Rahadian *et al.* (2003) admitted that direct comparison of the EPS and pile embankments is not possible, as they have different geometries and rates of construction. Furthermore, the embankments are built on different peat thicknesses. In any case, they conclude that the surcharging technique is effective for a thin layer of peat (less than 4 m). Should piles be used, they should be driven to the hard stratum as end bearing in order to minimize settlement, except of course for the transition zone.

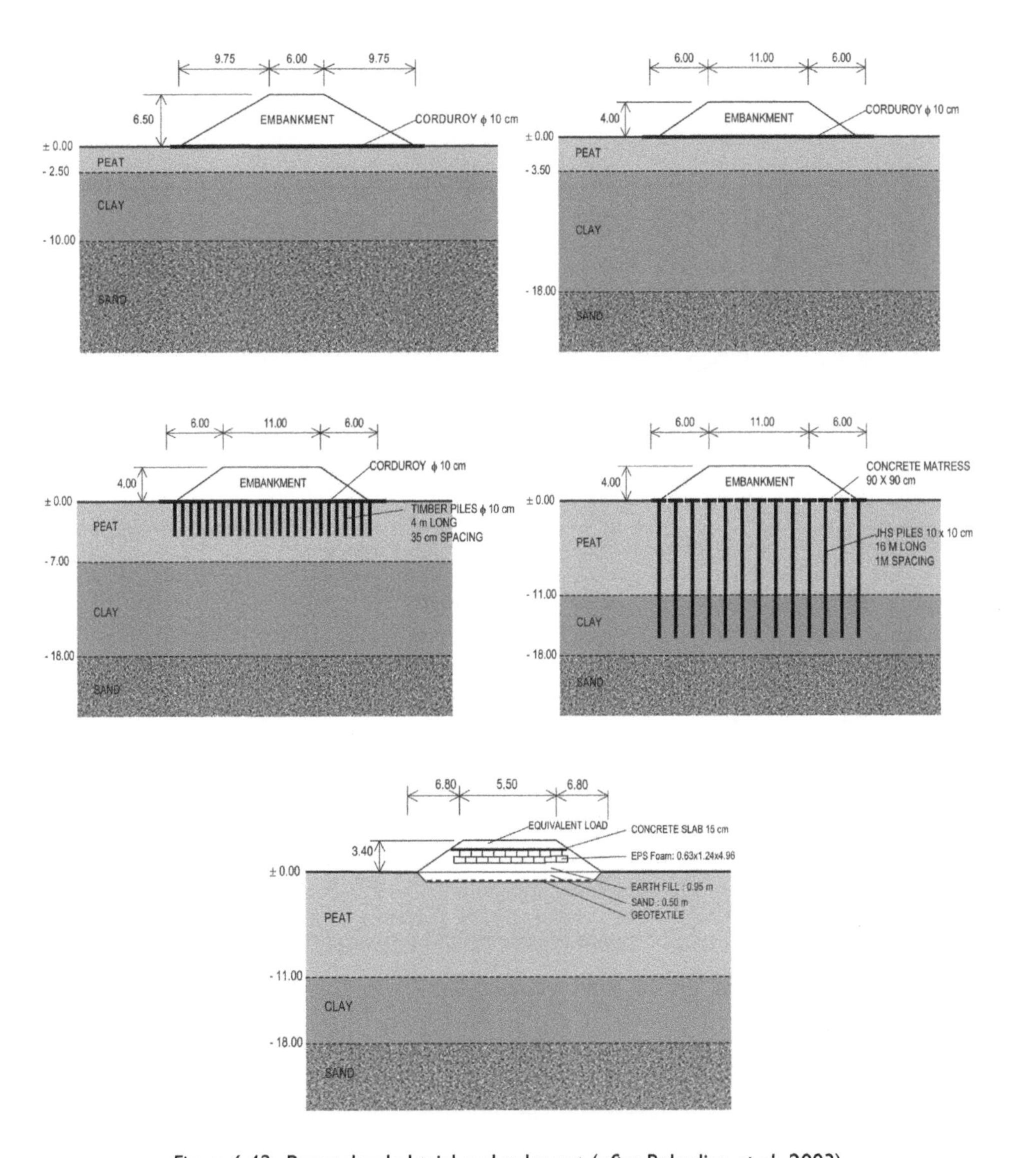

Figure 6.42 Berengbenkel trial embankment (*after* Rahadian *et al.*, 2003).

6.11 CHEMICAL AND BIOLOGICAL CHANGES

An important characteristic of peat is the potential chemical and biological changes with time. Further humification of the organic constituents would alter the mechanical properties such as compressibility, strength and hydraulic conductivity. Lowering of ground water may cause shrinking and oxidation of peat, leading to humification with a consequent increase in permeability and compressibility. Oxidation also leads to gas formation, which may contribute to excess pore pressures (Vonk, 1993). The significance of these effects for the long-term performance of structures placed on peat

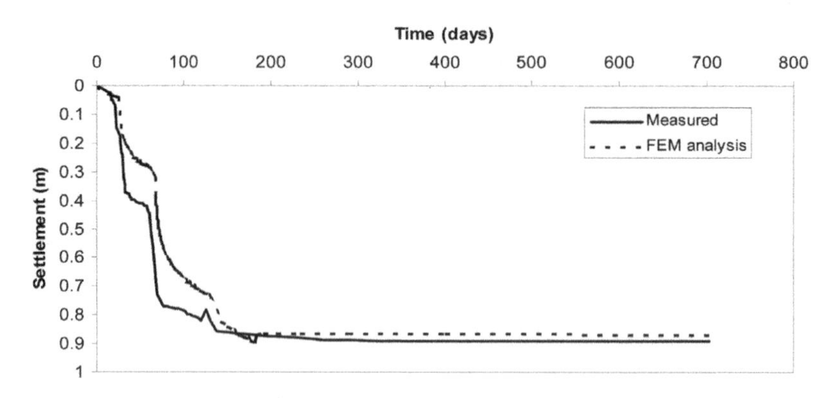

(a) Surcharge embankment

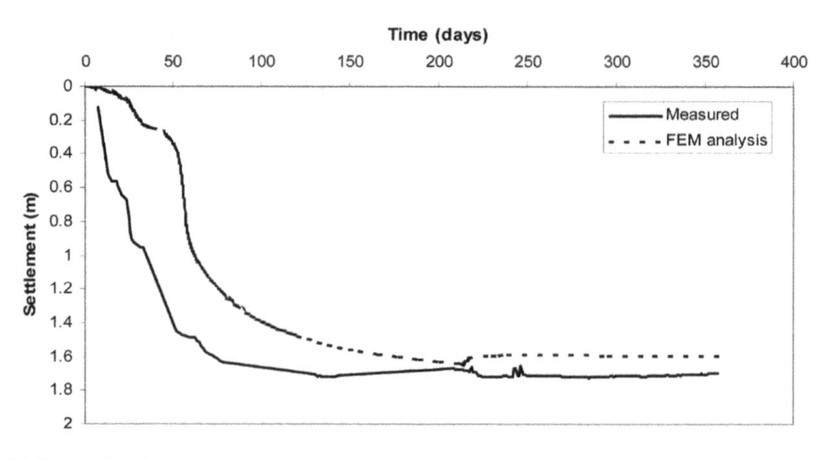

(b) EPS embankment

Figure 6.43 Settlement performance of the field trials.

is often not taken into account rigorously. However, in the absence of any significance change in the submergence of these deposits, the long-term chemical and biological degradation of the peats may not be significant.

6.12 EFFECT OF DRAINAGE

A study done by Mutalib *et al.* (1991) showed the development of a peat profile from sapric-hemic-fibric morphology over an estimated period of 20 years due to subsidence of 0.8 m, as shown in Figure 6.45. The soil's bearing capacity is reported to below. It is also unclear, however, if drainage over a shorter period of time and a lower magnitude of subsidence would result in the similar morphology.

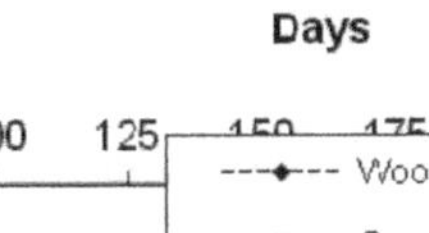

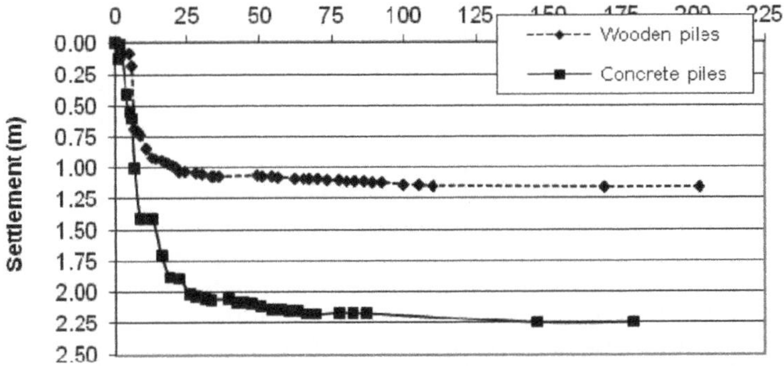

Figure 6.44 Result of settlement observation of the trial (mini) piled embankment.

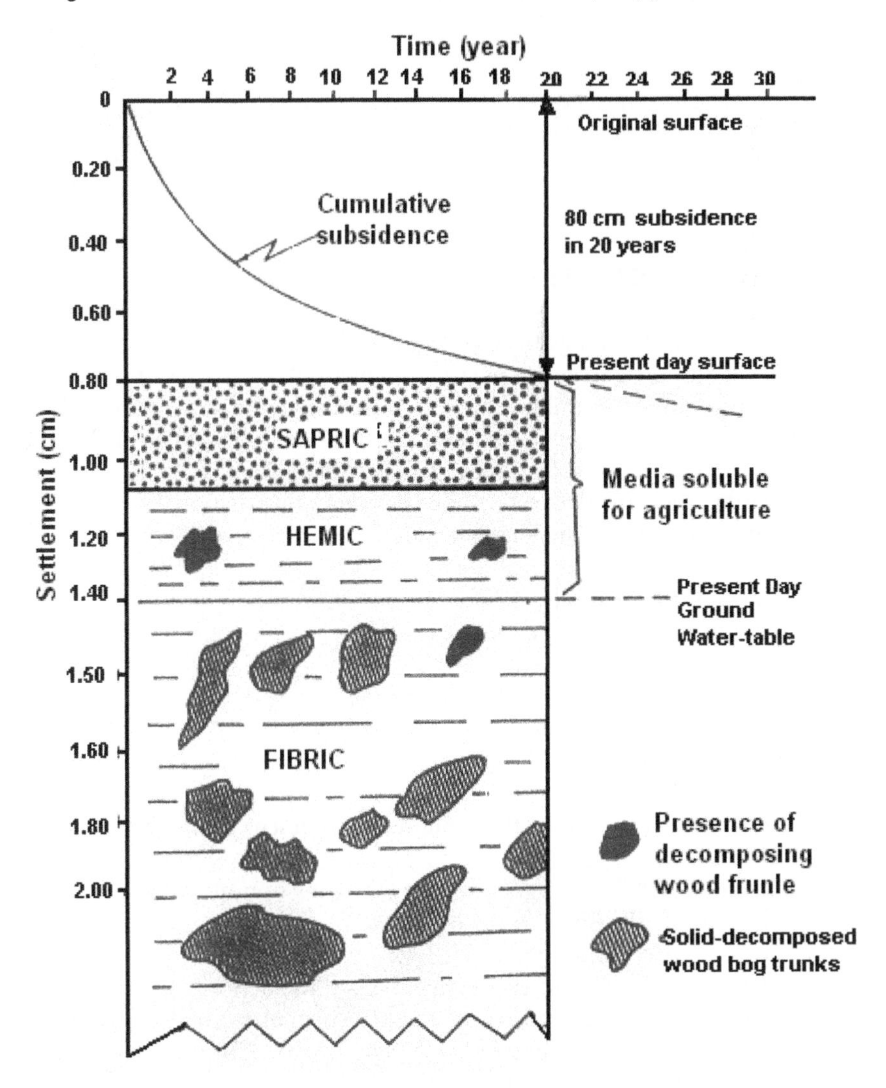

Figure 6.45 Effect of drainage.

Table 6.4 General ground improvement methods for soft ground (*after* Sin, 2003)

Type of treatment	*Technical input*	*Advantages/ disadvantages*
Piled embankment	Instead of consolidating the soft compressible layer, it acts as a structure and transfers all of the embankment load to the tip of the piles. Nominal post-construction settlement.	Expensive/minimum post-construction settlement
Prefabricated vertical drain (PVD)	Suitable and economical method to consolidate the soft compressible layer. Nevertheless, progress of embankment construction is slow due to the time taken to increase the undrained shear strength being much longer.	Cheap/long construction period
Stone column (SC)	Close spacing of SC expedites consolidation of soft compressible layers. Compaction of stone will be a difficulty due to the low undrained shear strength ($<15\,kN\,m^{-2}$) and lack of confinement strength.	Expensive, potential for column bulging effect
Total replacement (R/R)	With proper compaction of backfilling material, post-construction settlement can be kept to a minimum. Disposal of unsuitable material will be a great concern. Replacement ratio is 100%.	Construction problem due to high ground water, disposal site due to environmental restraint

6.13 CHOICE OF CONSTRUCTION METHODS

As mentioned before, buildings on peat are often constructed on piles. But for line structures like roads, the choice of construction method is a matter of finding optimal solutions between economic and technical factors, available construction time and the target performance standards. Undoubtedly construction in this difficult terrain is not easy, but with better understanding it can be more manageable.

In evaluating the effectiveness of the various construction techniques, many variables need to be considered. These include:

- Effectiveness of the methods
- Availability of materials and equipments
- Cost of materials
- Construction time
- Post construction maintenance
- Environmental effects

Table 6.4 presents a review of the ground improvement techniques applicable for general soft soils.

McManus *et al.* (1997) recommended the following options when constructing on peat, depending on the thickness of the deposits:

- Removal and replacement when the thickness of the deposit is ≤ 3 m.
- When the thickness of the deposit is more than 3 m but less than 10 m, the following options may be considered: preloading, stage construction with vertical and sand drains, lightweight fills and surface mattresses.

Table 6.5 (a) Analysis of the effectiveness of the various foundation techniques

Technique	*Special equipment*	*Pile spacing (m)*	*Pile diameter (m)*	*Pile depth (m)*	*Strength increase (%)*	*Compressibility decrease (%)*
Limit piles	*	1–2	0.3–0.5	3	40–70	25–50
Lime/cement columns	Required *	1–2	0.5–3.5	30	1,000–5,000	40–60
Compacted sand piles	Required *	1.5–2.2 max. 2 in peat	0.5–1.5	15	400–500 in clay 30–70 in peat	30–50
Stone columns	Required *	1.2–3.0	0.6–1.0	5	400–500	80

(b) Application of methods to soil types (*after* McManus *et al.*, 1997)

Deep improvement technique	*Soft clays*	*Peat*	*Organic soils with high water content*
Lime piles	✓		
Lime columns	✓	✓	
Lime/fly ash column		✓	✓
Lime/gypsum column	✓		✓
Cement column	✓	✓	✓
Compacted sand pile	✓	✓	✓
Stone column	✓		

* Modified standard bored piles equipment

- For deposit of greater than 10 m, one of the deep stabilization techniques, including the pile support method, can be considered.

Analysis of the effectiveness of the various foundation techniques and application of methods to soil types is included in Table 6.5.

Chapter 7

Recent advances in the geotechnics of organic soils and peat

7.1 INTRODUCTION

In recent years, in terms of economic development, concerns about peat and its difficulties from the geotechnical standpoint have led to the development of many new techniques for improving it. Peat is soil with a high compressibility and low shear strength. This implies that the engineers have to build structures, like roads and highways, railways, ports and airports, land reclamations, storage tanks, chemical plants, bridges, tunnels and residential buildings in less favourable areas (Myślińska, 2003). A variety of methods (which have been explained in Chapter 6) are widely used for stabilizing soft soils. Various new ground improvement techniques have been suggested, and recently there have been new advances in organic soils and peat stabilization, such as electrokinetics, new deep mixing method, biocementing and reinforced stone columns.

7.2 ELECTROKINETICS

Electrokinetics is defined as the physicochemical transport of charge, action of charged particles and effects of applied electric potentials on the formation of and fluid transport in porous media (Acar and Alshawabkeh, 1993; Acar *et al.*, 1995).

Isomorphous substitution and broken continuity of a clay structure presents a net negative charge at the surface of clay particles. The clay attracts positively charged ions from the pore water to balance the negative charge. The charge distribution at the colloid surface is shown in Figure 7.1. This distribution shows an increase in anion concentration and a decrease in cation concentration with distance from the clay surface. Dipolar water molecules can be electrically attracted toward the surface of clay when the clay or colloid is placed in water. The negative surface charge of the clay or colloid and the electrically attracted water are together termed the diffuse double layer (Hunter, 1993). The nature and thickness of the double layer affects the clay's characteristics. In general, the tendency (for particles) for flocculation decreases with increasing thickness of the double layer.

The diffuse layer can be compressed in solutions of high concentration, mainly because diffusion away from the surface is lower. The properties of the cations also have an effect, the higher their charge, the more strongly they are attracted. Sodium

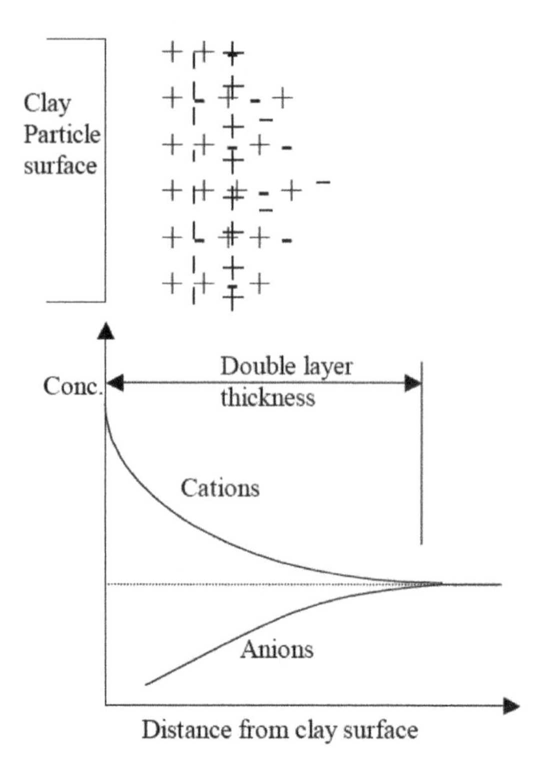

Figure 7.1 Schematic diagram of charge distribution at the colloid surface (*after* Mitchell and Soga, 2005).

ions, because of their single charge, are weakly held, and in solution each has a large shell of water molecules. On the other hand, calcium ions have double charges and they are more strongly held, and thus the diffuse layer of calcium-saturated clay is more compact than sodium-saturated clay.

The presence of the diffuse double layer causes several electrokinetic phenomena in soil, which may result from either the movement of different phases in respect of each other including transport of charge, or the movement of different phases relative to each other due to the electric field (Acar and Alshawabkeh, 1993).

The electrokinetic phenomena include electroosmosis, streaming potential, electrophoresis and sedimentation potential. Electroosmosis is defined as fluid movement as a result of an applied electric potential gradient (Figure 7.2(a)). Streaming potential is the reverse of electroosmosis (Figure 7.2(b)). Electrophoresis is the movement of solids suspended in a liquid due to the application of an electric potential gradient (Figure 7.2(c)). Sedimentation potential is an electric potential generated by the movement of particles suspended in a liquid (Figure 7.2(d)).

Electroosmosis and electromigration play a key role in soil stabilization (Acar and Alshawabkeh, 1993; Barker *et al.*, 2004).

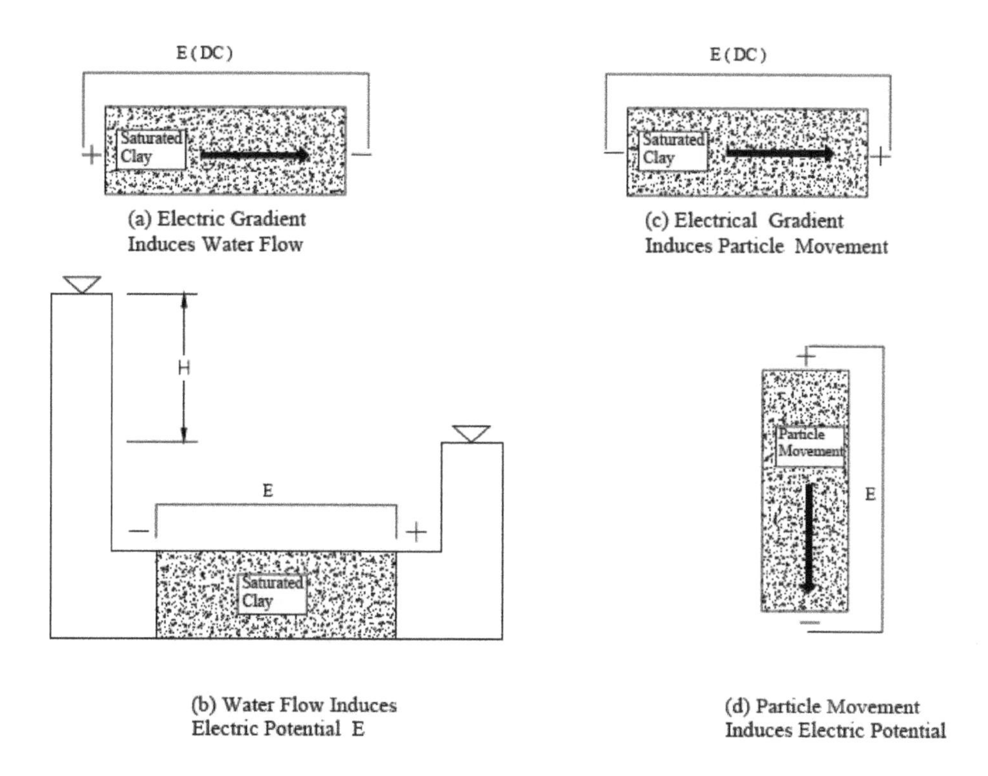

Figure 7.2 Electrokinetic phenomena in soil (*after* Mitchell and Soga, 2005).

7.2.1 Electroosmotics

Electroosmosis is the movement of a fluid with respect to a solid wall as a result of an applied electric potential gradient (Cashman and Preene, 2002). In other words, electroosmosis involves water transport through a continuous soil particle network, where the movement is primarily generated in the diffuse double layer or moisture film, where the cations dominate. When the direct electrical gradient is applied to a clay–water system, the surface or particle is fixed, but the mobile diffused layer moves and the solution is carried with it. According to Hausmann (1990), a greater soil particle surface results in higher moisture film transfers. The electrical potential applied and viscosity can also affect this phenomenon (Hausmann, 1990). The main mechanism in electroosmosis is the migration of ions, meaning that the cations migrate to the cathode and the anions move towards the anode (Tajudin *et al.*, 2008) (Figure 7.3).

Reuss (1809) discovered that water flow could be persuaded through a capillary by an external electrical gradient (Das, 2008). There are several theories that describe electroosmosis, including Helmholtz–Smoluchowski theory, Schmid theory, the Spiegler friction model, Buckingham π theory and ion hydration theory (Gray and Mitchell, 1967).

Helmholtz–Smoluchowski theory is widely used to describe electrokinetic processes (Helmholtz, 1879; Smoluchowski, 1921). The theory assumes that the pore

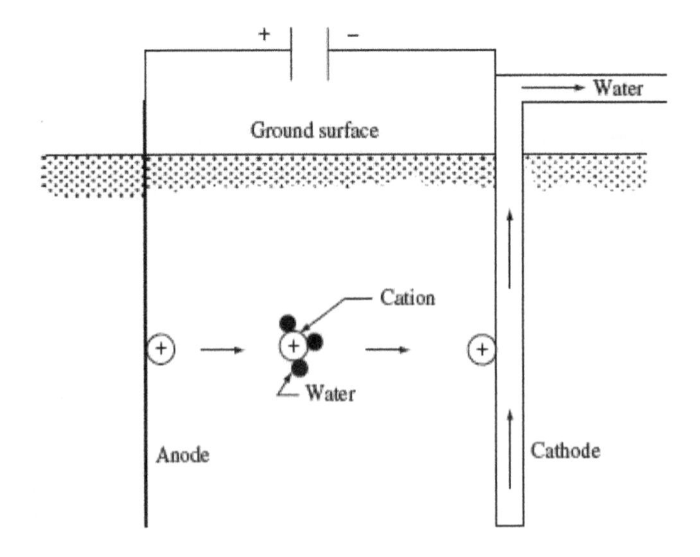

Figure 7.3 Principles of electroosmosis (*after* Das, 2008).

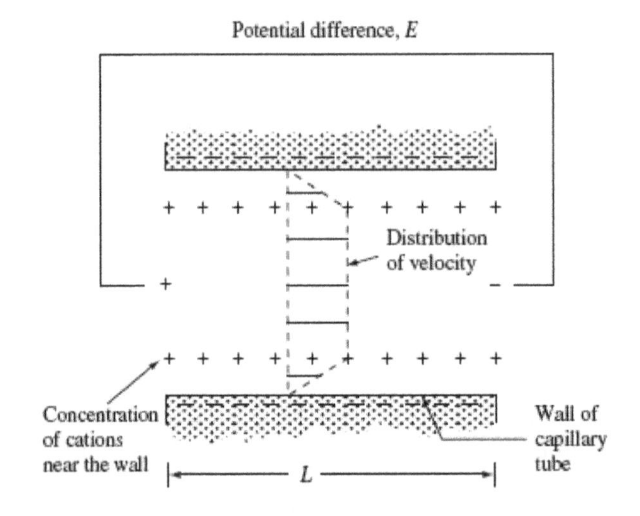

Figure 7.4 Helmholtz–Smoluchowski theory for electroosmosis (*after* Das, 2008).

radii are relatively large in comparison with the thickness of the diffuse double layer and the mobile ions are concentrated near the soil–water interface (Figure 7.4). Based on Hemholtz–Smoluchowski, the zeta potential (ζ) and the charge distribution in the fluid adjacent to the soil surface play important roles in determining the electroosmotic flow (see Chapter 3).

The rate of electroosmotic flow is controlled by the coefficient of electroosmotic permeability of the soil, k_e, which is a measure of the fluid flux per unit area of the soil per unit electric gradient. The value of k_e is a function of ζ, the viscosity of the pore fluid, the soil's porosity and the soil's electrical permittivity. The coefficient of electroosmotic permeability is given by Equation (7.1):

$$q = \frac{\varepsilon\zeta}{V_t} n \frac{E}{L} A \tag{7.1}$$

where:
ζ = zeta potential
V_t = viscosity of the pore fluid
n = soil porosity
ε = soil electrical permittivity
A = gross cross-sectional area perpendicular to water flow
L = length
q = flow rate

Hydraulic conductivity, k_{h}, is significantly affected by the pore size and distribution in the medium, but k_{e}, based on Helmholtz–Smoluchowski theory, depends mainly on ζ.

The value of k_{e} is assumed to be constant during the electrokinetic process as long as there is no change in the concentration of ions or pH of the pore fluid (Gray and Mitchell, 1967). Based on Hemholtz–Smoluchowski theory, ζ and the charge distribution in the fluid adjacent to the soil surface play important roles in determining the electroosmotic flow.

Das (2008) reported that Schmid (1951) proposed a theory in contrast to the Helmholtz–Smoluchowski theory. It was assumed that the capillary tubes formed by the pores between clay particles are small in diameter and results in the excess cations would be uniformly distributed across the pore cross-sectional area (Figure 7.5). Based on this theory:

$$q = n \frac{r^2 A_o F}{8 V_t} \frac{E}{L} A \tag{7.2}$$

where:
r = pore radius
A_{o} = volume charge density
F = Faraday constant
n = porosity
A = gross cross-sectional area perpendicular to water flow
L = length
V_t = Viscosity
q = flow rate

However, the most widely used electroosmotic flow equation for the soil system is proposed by Casagrande (1949):

$$q = k_e i_e A \tag{7.3}$$

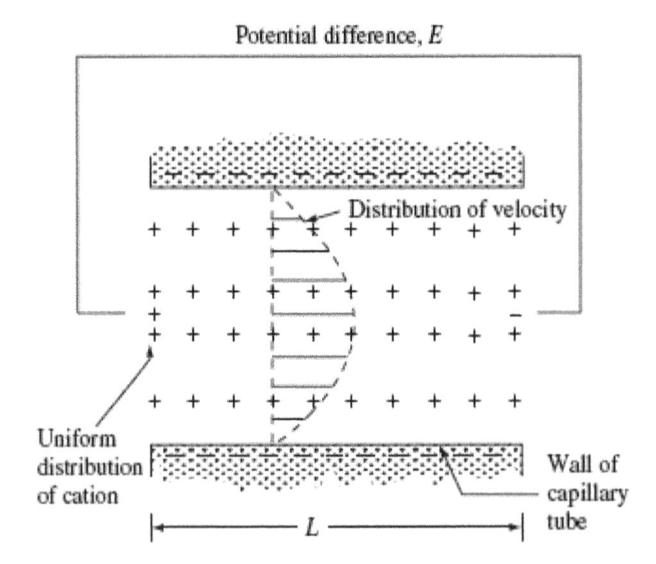

Figure 7.5 Schmid theory of electroosmosis (*after* Das, 2008).

Table 7.1 k_e of various clay soils (*after* Mitchell and Soga, 2005).

Soil type	k_e ($cm^2\ s^{-1} V^{-1}$)
London clay	5.8×10^{-5}
Boston blue clay	5.1×10^{-5}
Kaolin	5.7×10^{-5}
Clay silt	5.0×10^{-5}
Na-montmorillonite	2.0×10^{-5} to 12×10^{-5}

where:
A = gross cross-sectional area perpendicular to water flow
i_e = applied electrical gradient
k_e = coefficient of electro-osmotic permeability
q = flow rate

Table 7.1 shows some typical values of k_e for several soils (Alshawabkeh *et al.*, 2004; Mitchell and Soga, 2005; Das, 2008).

Application of direct current through electrodes causes electrolysis reactions at the electrodes (Acar *et al.*, 1990; Cherepy and Wildenschild, 2003). Oxidation of water at the anode generates an acid front and reduction at the cathode generates a base front. Electrolysis reactions are described by Equations (7.4) and (7.5).

$$2H_2O - 4e^- \rightarrow O_2 \uparrow + 4H^+ \quad \text{(anode)} \tag{7.4}$$

$$4H_2O - 4e^- \rightarrow 2H_2 \uparrow + 4OH^- \quad \text{(cathode)} \tag{7.5}$$

Within the first few days of electrokinetic processing, electrolysis reactions decrease the pH at the vicinity of the anode and increase the pH in the vicinity of the cathode. These changes depend on the total current applied (Acar *et al.*, 1990; Acar and Alshawabkeh, 1993).

The acid generated at the anode advances through the soil towards the cathode. This is due to ionic migration and electroosmosis. The base generated at the cathode initially advances towards the anode. The base advance is because of diffusion and ionic migration. However, the counter flow because of electroosmosis makes the back-diffusion and migration of the base front slower. The advance of the base front is slower than the advance of the acid front because of (1) the counteracting electoosmotic flow and (2) the ionic mobility of H^+ is higher than OH^- (Acar *et al.*, 1990; Alshawabkeh and Acar, 1992; Probstein and Hicks, 1993). Geotechnical reactions in the soil pores significantly impact electrokinetic phenomena and can enhance or make slower the electrokinetic process. Geomechnical reactions, including precipitation, dissolution, sorption and complexation reactions, are highly dependent on the pH conditions (Acar *et al.*, 1990; Alshawabkeh and Acar, 1992).

7.2.2 Electroosmosis in organic soils and peat

Peats generally have a very high water content, which can be in excess of 1500%, compared with mineral soils (sand, silt, and clay) whose values in the field may range from 3% to 100% (Huat, 2004; Tan, 2008). Peats tend to have high water content due to their organic content. Fibrous peat also tends to have higher water content than humified peat. The bulk densities of peat are in the range 0.8–1.2 $\mathrm{Mg\,m^{-3}}$ compared with the bulk densities of mineral soils which are in the range 1.8–2 $\mathrm{Mg\,m^{-3}}$ (Huat, 2004). This is due to the lower specific gravity (1.2–1.4) of the solids found and the higher water-holding capacity in the peat (Huat, 2004). A saturated mass is needed in electrokinetic phenomena and the peat environment is a good medium.

In electroosmotic dewatering, the frictional drag is produced by the movement of hydrated ions (Yeung and Datla, 1995). The quantity of these ions depends on the soil cation exchange capacity (CEC). The CEC range of humus is from 100 to 300 $\mathrm{cmol^+\,kg^{-1}}$, which is highest among colloids. The mineral fractions of tropical regions are dominated by kaolinite, aluminium oxides and iron oxides. The CEC range of kaolinite and Al, Fe oxides is 5–10 and 2–6 $\mathrm{cmol^+\,kg^{-1}}$ of soil, respectively. Therefore the humus is the key component in creating the potential for water momentum in electroosmotic phenomena (Asadi *et al.*, 2011b).

The charge on humus, kaolinite and Fe and Al oxides is affected by pH (Stevenson, 1994). Dissociation of H^+ is under the control of pH. In all Al and Fe oxides, as well as some silicate clays, exposed OH^- groups in moderate to acid conditions experience protonation. This occurs as an H^+ attaches to the OH^-. Since humus has charge, the occurrence of a water flow is expected in electroosmotic phenomena. However, the charge behaviour of peat needs to be investigated (Asadi *et al.*, 2010).

Electroosmosis can affect the pore fluid and distribution of exchangeable ions in soil and consequently can change the intensity of the forces holding the water films between soil and water (Fang and Daniels, 2006). Since the compressibility of a soil indicates the intensity of those forces, the electroosmotic environment can affect the compressibility behaviour of a soil. It is noteworthy that the water is attracted to a soil due to the electrical potential of the diffuse double layer and the space of the bulk phase. In peat, the bulk phase due to the presence of fibres in fibric peat is higher than in amorphous peat (Huat, 2004). Therefore the electroosmotic behaviour of peat is under control of its structure. On the other hand, the structure of a peat is related to its degree of humification. Therefore the electroosmotic behaviour of a peat is expected to be influenced by the degree of humification (Asadi *et al.*, 2009f; Asadi 2010).

In tropical regions, the cations may be removed from the topsoil due to heavy rainfall (Huat *et al.*, 2006). On the other hand, in coastal areas, because of the sea water, the presence of cations could be different, and as a result, there might be different behaviour of electroosmotic flow with depth and area (Asadi *et al.*, 2009f; Asadi 2010; 2011b).

Peats have noticeable qualities that make a suitable environment for utilization of electrokinetic techniques, i.e. (1) the saturated mass is a good environment, (2) the high charge and high specific surface area of the humus increase the presence of cations for water momentum in electroosmotic phenomena, (3) the surface charge can cause electroosmosis to occur (Asadi, 2010), (4) the decomposition processes affect the electroosmotic behaviour, and (5) the resistivity can affect the electroosmotic behaviour (Asadi *et al.* 2009c; Asadi et al, 2011f).

Asadi *et al.* (2011c) have done comprehensive investigations of the electroosmotic behaviour of organic and peaty soils using electroosmotic cells (Figure 7.6). The cumulative outflow volume profile over the testing period showed a continuous flow from anode to cathode. The cumulative outflow volume of H3, H5 and H7 peat over the 10-day periods of the experiments were 97, 279 and 450 mL, respectively. The rates of flow were observed to diminish with time (Figure 7.7). The cumulative outflow volume of H7 was higher than those of H3 and H5, meaning the cumulative outflow volume of humified peat is higher than that of unhumified peat (Asadi *et al.*, 2011c).

The average coefficients of electroosmotic permeability (k_e) of the H3, H5 and H7 peats were 4.91×10^{-6}, 1.12×10^{-5} and 1.57×10^{-5} cm^2 V^{-1} s^{-1} respectively. k_e is an indicator of the hydraulic velocity under unit electrical gradient. The maximum k_e of the H3 peat was less than that of H5 and H7. The k_e of the peats increased after 2 to

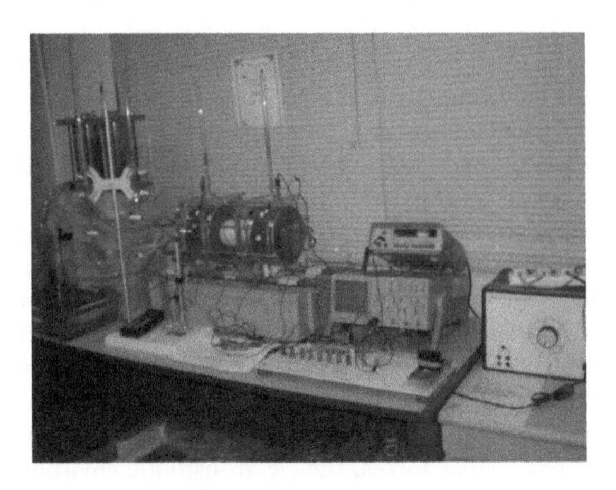

Figure 7.6 Electroosmosis apparatus (*after* Asadi et *al.*, 2011c).

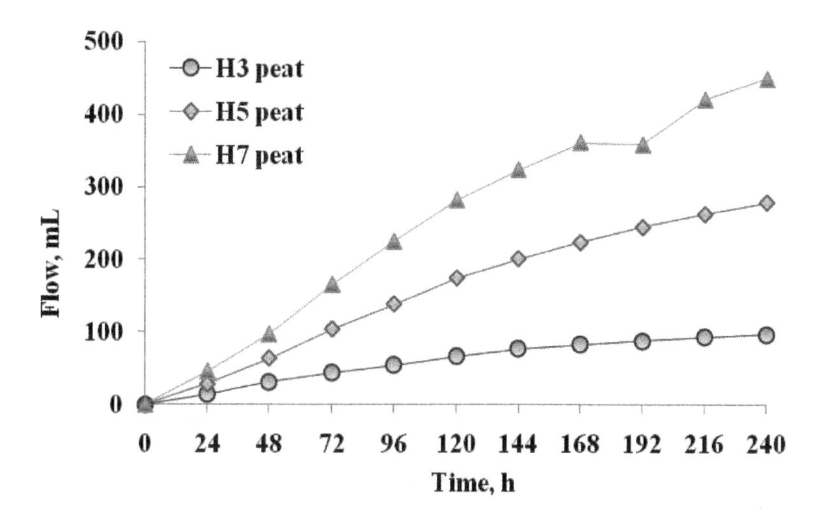

Figure 7.7 Cumulative outflow volume with time (*after* Asadi *et al.*, 2011c).

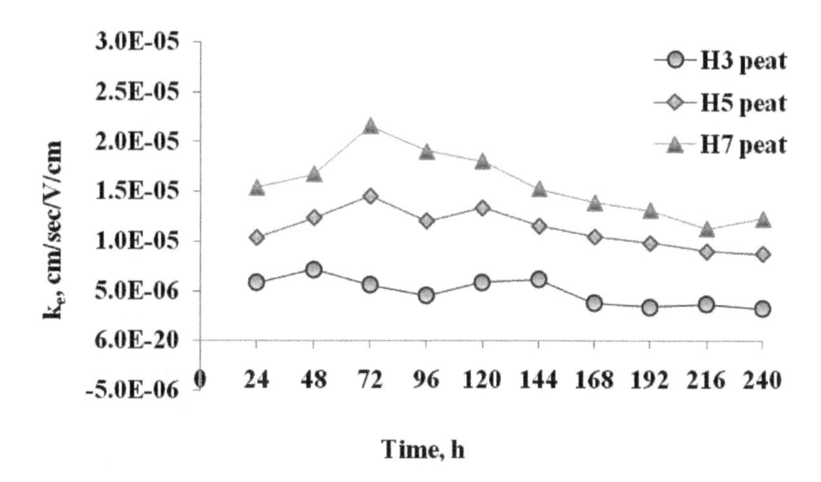

Figure 7.8 Variation of coefficient of electroosmotic permeability with time (*after* Asadi *et al.*, 2011c).

3 days and were observed to diminish to a minimum rate after 240 hours (Figure 7.8). Since the ζ of the H7 peat was higher than that of the H3 peat, the amorphous peat showed a higher k_e (Asadi *et al.* 2011c).

The average electroosmotic peat water transport efficiencies (k_i) of the H3, H5 and H7 peats were 2.92×10^{-2}, 5.32×10^{-2}, and 6.36×10^{-2} mL A^{-1} s^{-1}, respectively. k_i was calculated from the flow rate and current of each peat specimen. k_i (net flow per sec/(average voltage gradient × cross-sectional area)) is an indicator of the efficiency of the electroosmotic dewatering. The maximum k_i of the H7 peat was higher than that of H3 and H5. The k_i of the peats increased after 2 to 3 days and were observed to diminish to a minimum rate after 240 hours (Asadi, 2010) (Figure 7.9).

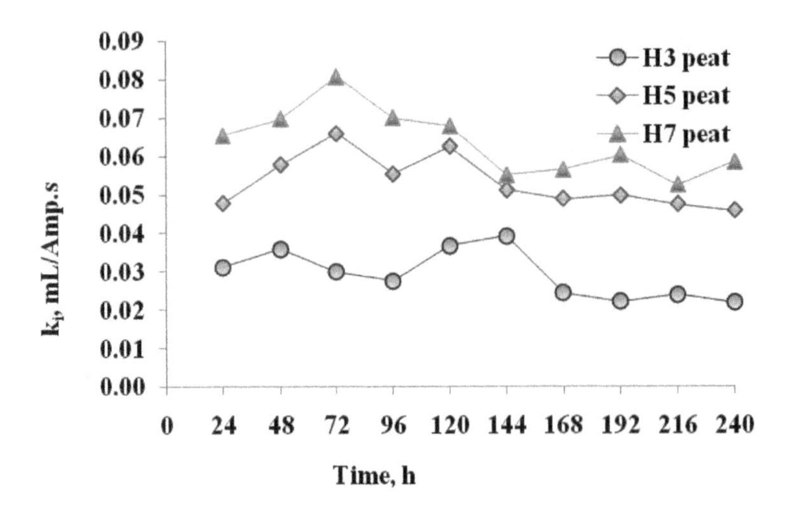

Figure 7.9 Variation of coefficient of electrooosmotic water transport efficiency with time (*after* Asadi *et al.*, 2011c).

The flow through a soil varies considerably depending on many factors, such as the minerals present, particle size distribution, porosity, soil fabric and the nature of the pore fluid. Despite the fact that H3 peat had less fine particles than those of H5 and H7 peat, and was therefore, more permeable, the H3 peat had a lower coefficient of electroosmotic permeability. A good understanding of humus as the most chemically active fraction of the peat colloids could make clear the underlying reasons for the significant differences. The humified peat had a more active fraction than unhumified peat, resulting in a higher coefficient of electroosmotic permeability (Asadi *et al.*, 2011e). The zeta potential measurements also proved that the surface charge of the humified peat was higher than the surface charge of the unhumified peat (see Chapter 3). The H7 peat had a larger surface area per unit mass (i.e. smaller particles) and had a higher proportion of the sample passing the #100 sieve. Therefore the quantity of the humus portion in H7 peat was higher than in H3 and H5 peat (Asadi, 2010).

Since the humus is dynamic and very active in charge (Stevenson, 1994), the H7 peat had a higher electroosmotic conductivity. Therefore the very highly decomposed peat had significant differences in electroosmotic properties compared with undecomposed peats. The study confirmed that the coefficients of electroosmotic permeability of the peats were dependent on the CEC, specific surface area and the degree of peat humification (Asadi *et al.*, 2011c).

7.3 ELECTROKINETIC CELL

Asadi (2010) proposed two types of setup for electrokinetic investigations on organic and peaty soils. Figure 7.10 shows a schematic electroosmotic cell (Type I) which most

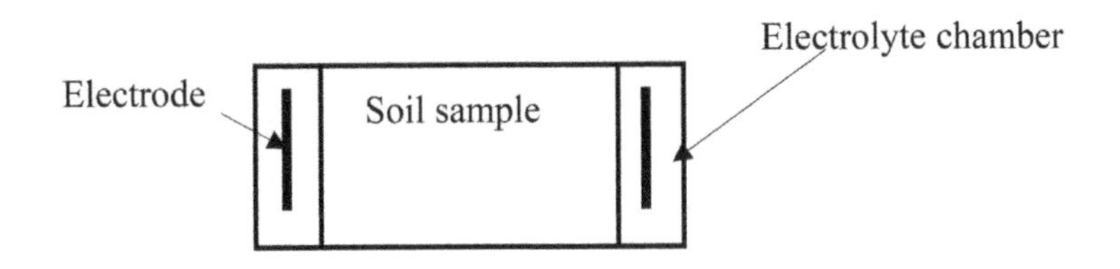

Figure 7.10 Schematic diagram of electroosmotic cell (Type I).

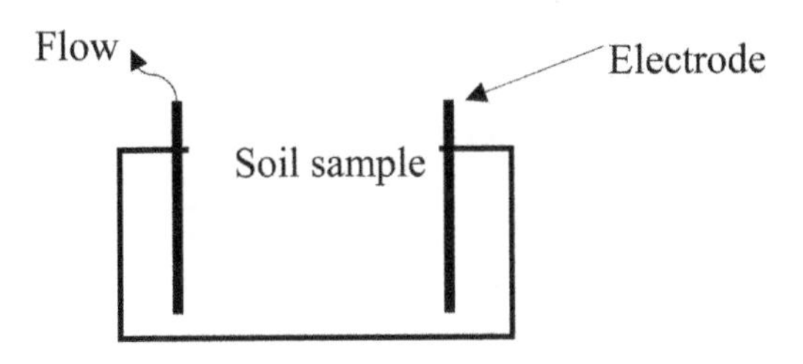

Figure 7.11 Schematic diagram of electroosmotic cell (Type II).

the researchers have used for their experiments. Some limitations and advantages of this cell are as follows:

1. It is hard to transfer the undisturbed soil samples into this cell.
2. The potential gradient is relatively uniform along the cell.
3. Natural soil samples are usually prepared vertically, while electroosmotic flow is horizontal.

Figure 7.11 shows a schematic for another setup (Type II) with the following limitations and advantages:

1. It is similar to field electroosmotic flow.
2. The potential gradient is less uniform relative to Type I
3. The analytical and experimental results may be less accurate in comparison with the previous cell.

A series of electroosmotic consolidation experiments were carried out on an organic soil in Malaysia (Kaniraj and Huong, 2008)). Commercially available EVD was used to induce electroosmosis and drainage of pore water. They concluded that electroosmosis using EVDs was effective in the electrokinetic treatment of organic soil (Kaniraj et al, 2011; Kaniraj and Yee, 2011) (Figure 7.12).

Kaniraj and Yee (2011) reported that electroosmotic consolidation improved the undrained shear strength of organic soil (Figure 7.13).

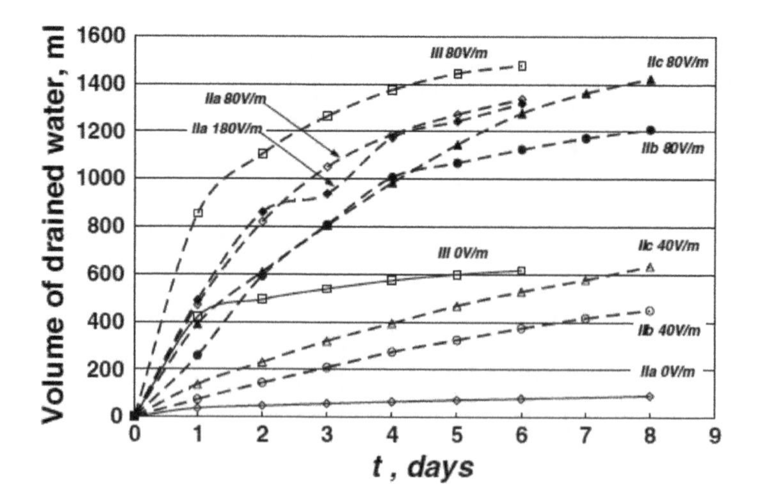

Figure 7.12 Variation of total volume of water drained with electroosmotics (*after* Kaniraj *et al.*, 2011).

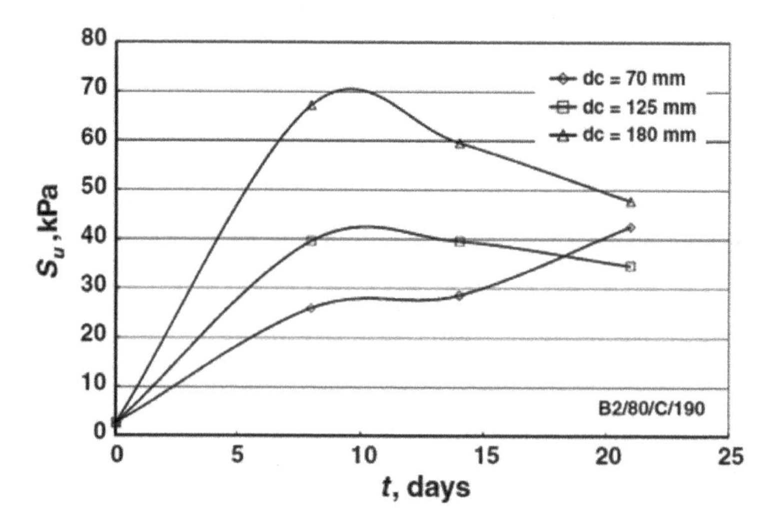

Figure 7.13 Variation in the undrained strength of organic soils (*after* Kaniraj and Yee, 2011).

7.4 ELECTROKINETIC STABILIZATION OF ORGANIC SOILS AND PEAT

In peats with lower organic content, clay–organic complexes and the formation of stable aggregates are key components of peat stabilization. Several bonding forces may operate between humus and clay minerals, including cation bridging, water bridging, anion exchange, ligand exchange and van der Waals forces (Stevenson, 1994). Since organic anions are normally repelled from negatively charged clay surfaces, adsorption of humus by clay minerals occurs only when polyvalent cations are present on the exchange complex. The main polyvalent cations responsible for the binding of humus to soil clays are Ca^{2+}, Fe^{3+} and Al^{3+}. The divalent Ca^{2+} ion does not form strong coordination complexes with organic molecules and would be effective only to the extent that a bridge linkage could be formed (Stevenson, 1994). In contrast, Fe^{3+} and Al^{3+} form coordination complexes with organic compounds and strong bonding of humus with clay could be possible through this mechanism.

Since the CEC of peat is high, and since the cations Na^{+}, K^{+}, Ca^{2+}, Mg^{2+} and silica are removed from top soil due to leaching, there is a probable higher presence of the trivalent ions (e.g. Fe^{3+} and Al^{3+}) at the surface in peat in tropical regions. Therefore, using sodium silicate systems, electromigration and injection of trivalent cations could be a winning combination in electro-strengthening of peat at the cathode area. Cationic surfactants are expected to flocculate peat through a combination of charge neutralization and bridging (Asadi *et al.*, 2011a–g).

7.5 BIOCEMENTING STABILIZATION

Biocementing is a new method to improve soils based on microbiologically induced precipitation of carbonate calcium. In particular, methods are being developed using microorganisms which are able to increase the strength and stiffness of granular soils by inducing the precipitation of calcium carbonate (Whiffin 2004; De Jong *et al.*, 2006; Whiffin *et al.*, 2007; Ivanov and Chu, 2008). Most studies on biological grouting (biogrouting) report the use of microorganisms containing the enzyme urease and, in particular, the bacterium *Sporosarcina pasteurii* (DSM 33, renamed from *Bacillus pasteurii*) (Whiffin 2004; De Jong *et al.* 2006, Whiffin *et al.*, 2007). This process induces carbonate calcium precipitation, increasing the strength of soils.

$$CO(NH_2)_2 + 2H_2O \rightarrow 2NH_4^+ + CO_3^{2-} \tag{7.6}$$

$$Ca^{2+} + CO_3^{2-} \rightarrow CaCO_3\downarrow \tag{7.7}$$

The enzyme hydrolyzes urea to CO_2 and ammonia, resulting in an increase in the pH and carbonate concentration. Microbial activity-induced $CaCO_3$ precipitation on concrete has also been indicated (Ramachandran *et al.*, 2001). *S. pasteurii* is a rod-shaped, gram-positive soil bacterium that secretes numerous enzymes to degrade a variety of substrates, enabling the bacterium to survive in a continuously changing environment (Westers *et al.*, 2004). In addition, the bacterial net cell surface charge is negative and absorbs cations (for example Ca^{2+}) from the environment to deposit on the cell surface (Achal *et al.*, 2010). Because the bacterial cell surface has many negative charges, if Ca^{2+} is added first without urea there will be enough time for Ca^{2+} to be attached to the bacterial surface and the bacteria activity will be greatly influenced and retarded.

The bacteria grow at 37°C and are facultative alkaliphiles which grow optimally at pH 9.2 in relatively high amounts of NH^{4+} (Wiley and Stokes, 1962, 1963) or urea (Larson and Kallio, 1954; Bornside and Kallio, 1956). The bacteria should have high negative zeta-potential, ureolytic ability, be alkalophilic (optimum growth rate occurs at pH around 9, and no growth at all around pH 6.5 (Dick *et al.*, 2006; de Muynck *et al.*, 2007) to increase adhesion), and produce urease in the presence of high concentrations of ammonium (Kaltwasser *et al.*, 1972; Friedrich and Magasanik, 1977) to promote $CaCO_3$ precipitation and ureolysis (Nemati and Voordouw, 2003).

In bacterium cells, calcium concentrations are high in extracellular compared to intracellular (as a result of alkaline pH regimes). The combination of extracellular alkaline pH and calcium ions presents an unavoidable stressful environment for bacteria: because of $Ca^{2+}/2H^{+}$ electrochemical gradients caused by the process, the passive calcium rush will lead to intracellular calcium build-up and excessive proton expulsion (Hammes and Verstraete, 2002). At the cellular level, this could be detrimental due to (1) the disruption of intracellular calcium-regulated signal processes, (2) the alkalization of intracellular pH and (3) the depletion of the proton pool required for numerous other physiological processes (Norris *et al.*, 1996). The microbiological $CaCO_3$ precipitation began at pH 8 and was completed at pH 9.0, consolidating 98% of the initial concentration of Ca^{2+}. Calcium carbonate precipitation appeared to be correlated with the growth of *S. pasteurii* and was completed within 16 h following inoculation (Stocks-Fischer *et al.*, 1999).

Calcite is produced by *S. pasteurii* that have been studied in standard nutrient broth (NB) and Corn steep liquor (CSL). 100 ml of NB and CSL media were added to 2% urea and 25 mM $CaCl_2$ mixed with 1 mM of overnight grown Bp M-3. The bacteria were grown at 37°C with continuous aeration at 120 rpm (Achal *et al.*, 2010). The urease activity was determined for bacterial isolates in NB media containing filter-sterilized 2% urea and 25 mM $CaCl_2$ by measuring the amount of ammonia released from urea. One unit of urease is defined as the amount of enzyme hydrolyzing one micro mole urea per minute (Achal and Pan, 2011).

7.6 BIOGROUTING AND ITS CHALLENGES

Biogrouting is intended as a ground improvement technique for sandy soils. The crystals of calcium carbonate precipitate in the presence of dissolved calcium, which form bridges between the sand grains and hence increase the stiffness and strength up to 12 MPa (Van Paassen *et al.*, 2010). Biogrouting significantly reduce the permeability of the strengthened soil, which hinders ground water flow and limits long-distance injection, making large scale treatment unfeasible. Biological techniques (biogrouting) can provide the solution (Whiffin, 2004; De Jong *et al.*, 2006; Ivanov and Chu, 2008).

By injecting specific groups of microorganisms into the soil, in combination with substrates, precipitation of inorganic minerals is induced at the desired location. These minerals connect the existing sand grains, thereby increasing the strength of the material. The product has similar properties to natural sandstone and it remains permeable, thereby enabling large-scale applications (Van Paassen *et al.*, 2010). Microbial grouting decreased the permeability after two injections by about 98%. Enzymatic formation of $CaCO_3$ *in situ* is an effective method for reducing the permeability of porous media (Nemati and Voordouw, 2003). The attachment of bacteria depends on many factors, including grain size distribution, mineralogy, the properties of the pore fluid and the properties of the bacteria themselves (Scholl *et al.*, 1990; Torkzaban *et al.*, 2008).

Transport of bacteria is limited in fine grained soils. As bacteria have a typical size of 0.5 to 3 μm, they cannot be moved through silt or clay soils; nor can they induce carbonate precipitation (Mitchell and Santamarina, 2005). Also, in fine sands or coarser materials bioclogging could occur when bacteria are adsorbed or strained by the solid grains, which could result in limited treatment distance for ground reinforcement purposes (Van Paassen *et al.*, 2010).

There is an important problem with biogrouting: the limited dispersion of bacteria injected into soils. Bacteria often stick fast to solid surfaces. In the absence of a strong hydrogeological gradient the organisms remain localized at the origin of injection, resulting in the fouling of wells and inadequate dispersion of bacteria. The direct transportation of bacteria from injection wells to other zones would be advantageous to augmentation approaches used for *in situ* remediation (De Flaun and Condee, 1997). It causes a heterogeneous diffusion of $CaCO_3$ away from the injection points.

The lack of $CaCO_3$ close to the injection points could be the result of a higher flow velocity, causing more bacterial flush out and hence lower activity and less $CaCO_3$. Another explanation for the lack of $CaCO_3$ around the injection points considers the kinetics of $CaCO_3$ precipitation and transport of crystals. Initially the crystals are still small or not even present if the solution is not yet sufficiently oversaturated that nucleation has taken place, which is likely in quartz sand (Lioliou *et al.*, 2007), that they are still easily transported through the pores. Once the flow velocity drops or crystals become bigger, they are more easily trapped in the narrow pores (Van Paassen *et al.*, 2010).

The control and predictability of the *in situ* distribution of bacterial activity and reagents and the resulting distribution of $CaCO_3$ and related engineering properties in the subsurface are not yet sufficient, and form the greatest challenge for further optimization, especially if biogrouting is applied in an open system (Van Paassen *et al.*, 2009). Further research should demonstrate what mechanisms are responsible for the observed heterogeneity in the deposition of carbonate and the consequent geotechnical parameters and what the implications of this heterogeneity are for the designed purpose (Van Paassen *et al.*, 2010).

7.7 ELECTRO-BIOGROUTING IN ORGANIC SOILS AND PEAT

Electro-biogrouting is a new method to stabilize the organic soils. There are noticeable qualities in organic soils that make a suitable environment for utilization of electro-biogrouting techniques, i.e. (1) the saturated mass is a good environment, (2) the high charge and high specific surface area of the humus increase the presence of cations for water momentum in an electroosmotic phenomenon, (3) the surface charge can cause electroosmosis to occur, and (4) the decomposition processes affect the electroosmotic behaviour. The bacteria are rod-shaped with many negative charges on their surface, and they can move across the soil under an electric gradient. The urease is negatively charged and can diffuse with electric potential. Indeed, the urease produced can be mixed with ammonia and transported in organic soils under an electric gradient. The calcium chloride solution is then added in an electric injection process. This method can induce carbonate precipitation ($CaCO_3$). It can operate in fine organic clay and peat. However, the surface charge of bacteria in the electrokinetic environment, the effect of pH on the bacteria, and the transport rates of ammonia, urease and calcium chloride in organic soils are all challenges that need to be studied.

7.8 CONVENTIONAL ADDITIVES AND/OR FIBRE REINFORCEMENT IN ORGANIC SOILS AND PEAT

Conventional additives are cement, blast furnace slag, silica fume, lime, fly ash and bituminous materials. These additives enhance the properties of soil. Cementation and pozzolanic reactions have been investigated in detail by Taylor (1997) and Hwan Lee and Lee (2002). When water comes in contact with cement, three phenomena take place: (1) cement reacts with water (called hydration), (2) pozzolanic reactions between $Ca(OH)_2$ from burnt cement and pozzolanic minerals in the soil, and (3) ion exchange between calcium ions (from cement and additives) with ions present in the clay, which leads to an improvement in the strength of the treated soil (Kazemian, 2011).

7.8.1 Ground granulated blast furnace slag (BFS)

Ground granulated blast furnace slag is the granular pozzolanic material formed when molten iron blast furnace slag is rapidly chilled (quenched) by immersion in water. It is a granular product with very limited crystal formation, is highly cementitious in nature and, ground to cement fineness, hydrates like Portland cement with a specific surface around $450\,m^2\,g^{-1}$, specific gravity of 2.9 and bulk density of $1150\,kg\,m^{-3}$ (Fernandez and Puertaz 1997; YTL cement 2008). Axelsson *et al.* (2002) reported that ground blast slag used as stabilizer has latent hydraulic properties. This means that, like pozzolanic materials, the slag can form strength-enhancing products with calcium hydroxide ($Ca(OH)_2$).

Blast furnace slag, being a latent hydraulic material, needs an activator such as calcium hydroxide to react. Its reactivity also depends on the CaO/SiO_2 ratio. As this ratio is increased, the more hydraulic the material will be. The CaO/SiO_2 ratio for blast furnace slag is about 1, while it is up to 3 for ordinary Portland cement (OPC). Hydraulic materials such as OPC react spontaneously with water. Since the reaction produced by blast furnace slag is slower than that of OPC, it results in slower strength gain and lower heat evaporation than with cement. However, the long-term strength of the slag admixture can be higher (Axelsson *et al.* 2002; Janz and Johansson 2002).

The effects of BFS as an additive on stabilized fibrous peat with OPC have been examined by Kalantari *et al.* (2011) using unconfined compressive stress (UCS) tests. Various trial mixtures of OPC treated with and without BFS were tested for their UCS values. Undisturbed samples and samples made of only 5, 15 and 30% OPC were tested as control measure samples. The stabilized samples were mixed initially with the fibrous peats (FPt) with natural moisture content $w = 200\%$ and air cured for three months.

The results of various trial mixtures for UCS stabilized samples are shown in Figure 7.14.

Four types of mixtures chosen for compaction test were fibrous peat with;

(a) 3.75% OPC and 1.25% BFS
(b) 11.25% OPC and 3.75% BFS
(c) 22.5% OPC and 7.5% BFS
(d) 22.5% OPC and 22.5% BFS

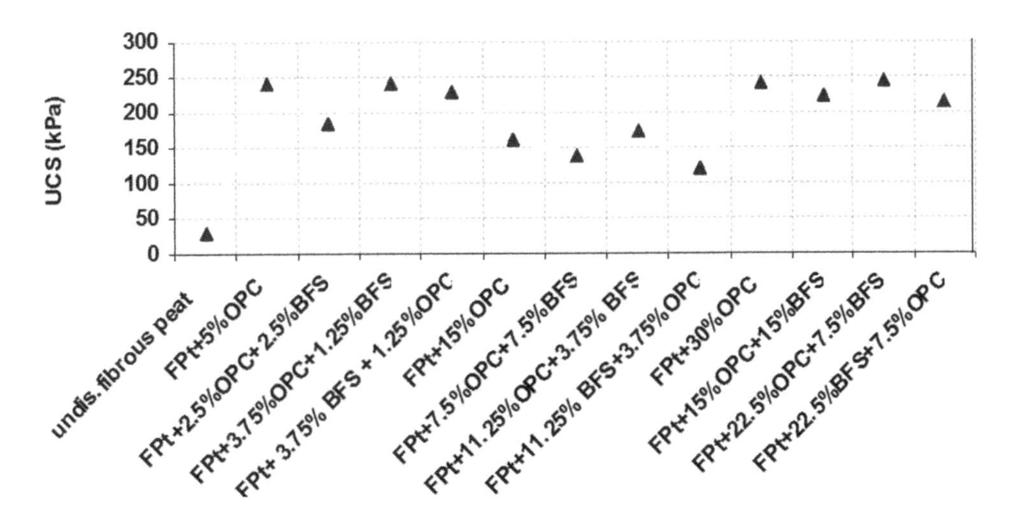

Figure 7.14 Unconfined compressive strength for various types of stabilized fibrous samples with and without blast furnace slag.

The results in Figure 7.14 indicate that, as the blast furnace slag contents of the stabilized samples are replaced by ordinary Portland cement, the UCS values decrease.

7.8.2 Pulverized fuel ash/fly ash (FA)

Pulverized fuel ash, also known as fly ash, is the fine residue produced when hard or bituminous coal is burnt in power station. Because of variations in coals from different sources, as well as differences in the design of coal-fired boilers, not all fly ash has the same properties.

Black coal gives a fly ash with lower CaO content than brown coal, but different black coals and brown coals also may vary widely. Thus a change in the in the coal burnt by a power station can result in a fly ash with entirely different properties.

In order for a pozzolanic material such as fly ash to react with water at all an external source of lime is necessary, e.g. in the form of Portland cement. When cement reacts with water it forms calcium hydroxide $Ca(OH)_2$, which in turn may form either a strength-enhancing CSH gel of the same type as Portland cement but with a lower CaO/SiO_2 ratio, or calcium aluminate silicate hydrate (CASH), which is closely similar to CSH but contains aluminium (Janz and Johansson, 2002).

Fly ash to be used in combination with Portland cement has been classified by ASTM C 618 into two classes, F and C, based on the chemical composition of the fly ash. The most influential factors dividing these two classes of fly ash are the amounts of calcium, silica (SiO_2), alumina (Al_2O_3) and iron content (Fe_2O_3): see Table 7.2.

Past research has shown that the addition of fly ash to cement or lime when mixed with mineral soils can improve the performance of the final product (Sukumar *et al.* 2008; Douglas 2004).

The effect of fly ash as an additive on stabilized fibrous peat with ordinary Portland cement has been examined by Kalantari and Huat (2009a) through UCS tests. Various

Table 7.2 Parameters for classifying fly ash (*after* Samsuri, 1997).

Properties	*Fly ash class*	
	F	*C*
$SiO_2 + Al_2O_3 + Fe_2O_3$ (min %)	70	50
SO_3 (max %)	5.0	5.0
Moisture content (max %)	3.0	3.0
Loss of ignition (max %)	6.0	6.0

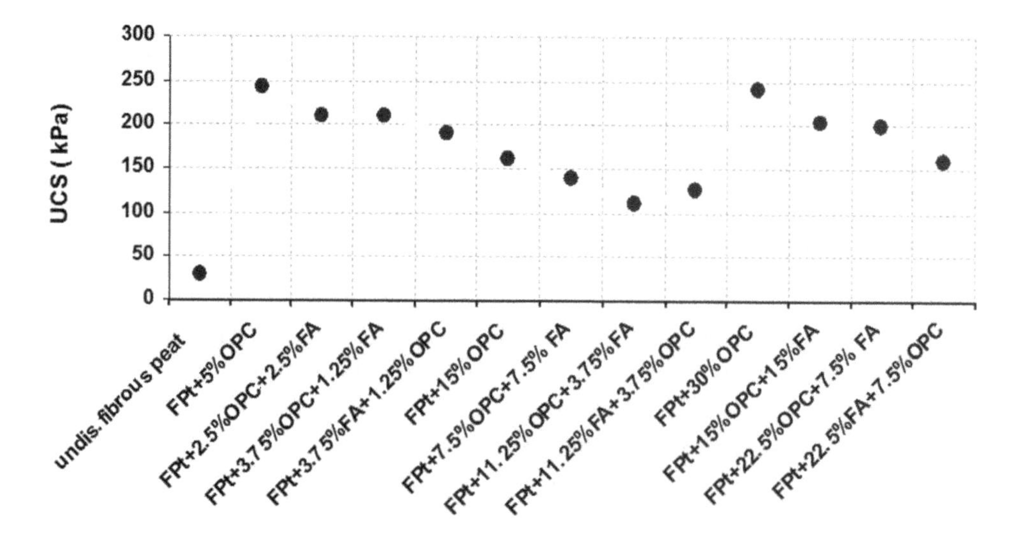

Figure 7.15 Unconfined compressive strength for various types of stabilized fibrous samples with and without fly ash.

trial mixtures of OPC treated with and without fly ash were tested for their UCS values. Undisturbed, as well as 5, 15 and 30% OPC samples, with no additives, were tested as control measure samples. The stabilized samples were mixed initially with the fibrous peat (FPt) with natural moisture content $w = 200\%$ and air cured for three months. The results are shown in Figure 7.15.

Four types of mixtures chosen for compaction test were fibrous peat with:

(a) 2.5% OPC and 2.5% FA
(b) 7.5% OPC and 7.5% FA
(c) 15% OPC and 15% FA
(d) 22.5% OPC and 22.5% FA

7.8.3 Silica fume/micro silica (SFU)

Silica fume or micro silica is an extremely fine product of high amorphous silica content arising from the condensation of rising vapour given off in the manufacture of

Table 7.3 Physical properties of silica fume (*after* Toutanji *et al.*, 1999).

Properties	*Values*
Loss on ignition (%)	1.44
Specific gravity	2.25
Particle size (average) (micron)	0.1

ferrosilicon and metallic silicon in high-temperature electric arc furnaces. Silica fume is extremely fine and dusty, with a typical particle size equal to 1.016×10^{-4} mm and surface area of around 19,000 m^2 kg^{-1}. Its pozzolanic activity with cement is estimated at 120–200%. Cement and silica fume mix show a higher strength. Silica fume is said to be a very effective pozzolanic as well as siliceous material, which in itself possesses little or no cementitious property, but which in finely divided form and in the presence of moisture will react chemically with calcium hydroxide to form compounds possessing cementitious properties (Agarwal, 2006). Silica fume as a pozzolanic additive gives cement increased strength, density and durability (Fleri and Whetstone, 2007).

There is limited available literature on cement, silica fume and peat, although silica fume is used as an additive to produce high-performance concrete. This is more homogeneous than normal strength concrete, and if made with small aggregates it can be compared to a strong rock. One of the benefits of using silica fume in Portland cement-based composites is its performance as filler in capillary pores and the cement paste–aggregate interface (Toutanji *et al.*, 1999). Detwiler and Metha (1989) observed that cement reacts with water to form calcium silicate hydrate and calcium hydroxide. Silica fume reacts with the calcium hydroxide in the presence of water to form calcium silicate hydrate. The increased calcium silicate hydrate gel and reduction in capillary pores in the cement paste are the main factors in its increased strength and impermeability. Silica fume and fly ash undergoes the same type of pozzolanic reaction. The strength gain is more rapid than with fly ash but slower than that of cement. The pozzolanic reaction is highly temperature dependent, so that the short-term strength is lower at low temperatures. However, long-term strength is increased by replacing part of the cement with silica fume (Janz and Johansson, 2002).

Some of the physical properties of silica fume are shown in Table 7.3.

The usual dosage of silica fume used along with cement in concrete mixtures is 5–10% by weight of cement (Bunke, 1988). Hooton (1993) has observed that addition of 15 and 20% of silica fume decreases the tensile strength of 91 day-old concrete by 15 and 20% respectively. Toutanji *et al.* (1999) showed that the partial replacement of cement by 8% of silica fume resulted in an increase in tensile strength of cement mortar (cement and sand). The replacement of cement by a higher dosage of silica fume (25%) resulted in a decrease in the tensile strength of mortar.

Kalantari *et al.* (2010) have attempted to stabilize or improve peat by using cement as binders and silica fume (Figure 7.16) as additive. This study was carried out by adding 5–50% of cement to peat and silica and measuring the compressive strength (UCS) and California bearing ratio (CBR). Silica fume is a proven pozzolanic material and its pozzolanic activity is estimated at 120–200% that of cement. The cement and silica fume mix shows a much higher strength compared with cement when used alone.

Figure 7.16 Silica fume (*after* Kalantari *et al.*, 2010).

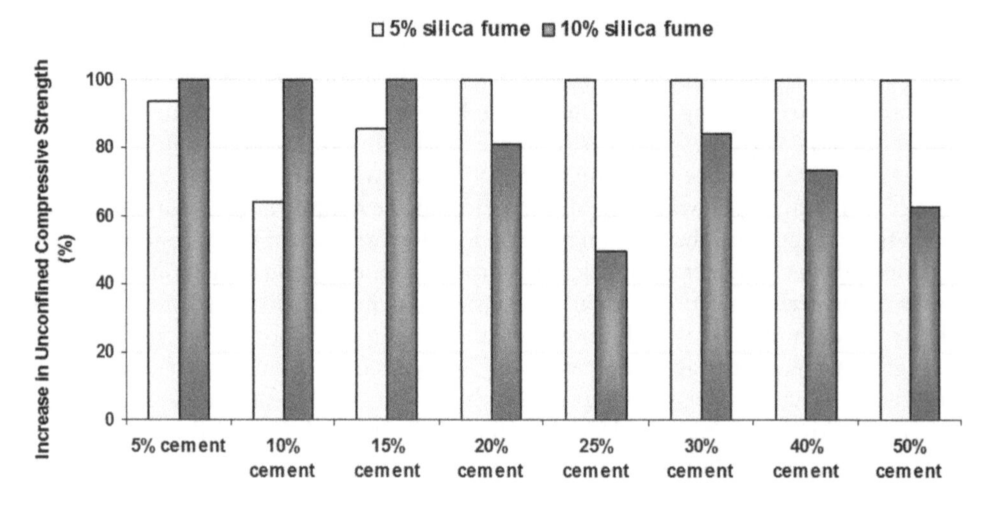

Figure 7.17 Percentage increase in UCS versus different percentages of cement and silica fume (*after* Kalantari *et al.*, 2010).

Figure 7.17 shows the increase in UCS values for different percentages of cement and silica fume. Similarly, it was observed that the CBR values of stabilized soil increased, and the percentage increase in CBR is shown in Figure 7.18.

Different types of UCS test samples, made of fibrous peat that had a moisture content of 187% and various amounts of OPC (5, 15 and 25%) and SFU (5 and 10%) were prepared, and then air cured for up to 90 days. UCS tests were then conducted for stabilized samples. Soaked UCS tests were also conducted on 90 days air cured samples by submerging them in water for seven days prior to being tested as soaked samples. The results are shown in Figure 7.19.

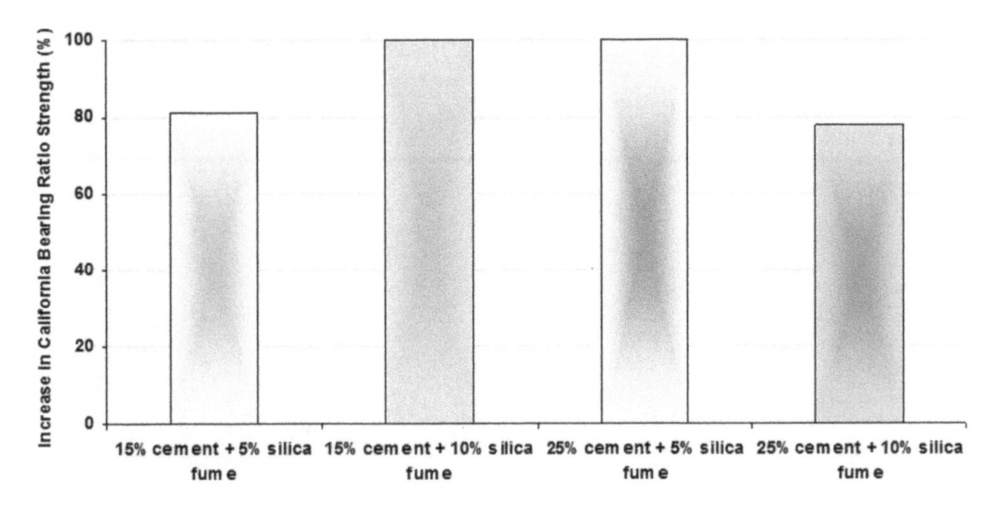

Figure 7.18 Percentage increase in CBR versus different percentages of cement and silica fume (*after* Kalantari *et al.*, 2010).

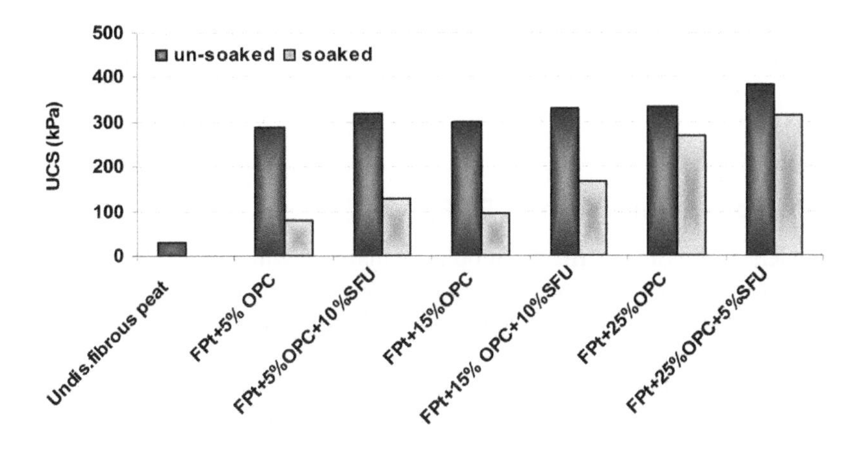

Figure 7.19 UCS values for different percentages of ordinary Portland cement and silica fume for unsoaked and soaked samples.

From the results of UCS test (Figure 7.19), it is observed that;

(a) Use of 10% silica fume in three months air cured, 15% OPC treated stabilized fibrous peat increases the unsoaked and soaked UCS values by 10% (from 300 to 330 kPa) and 72% (from 96 to 165 kPa), respectively, compared with when it is stabilized with only 15% OPC.

(b) UCS values of the samples increase with an increase in cement content.

(c) 5% silica fume along with 25% OPC has caused the 90 days UCS of unsoaked and soaked stabilized samples to increase by 15% (from 332 to 381 kPa) and 17% (from 268.5 to 315 kPa) respectively.

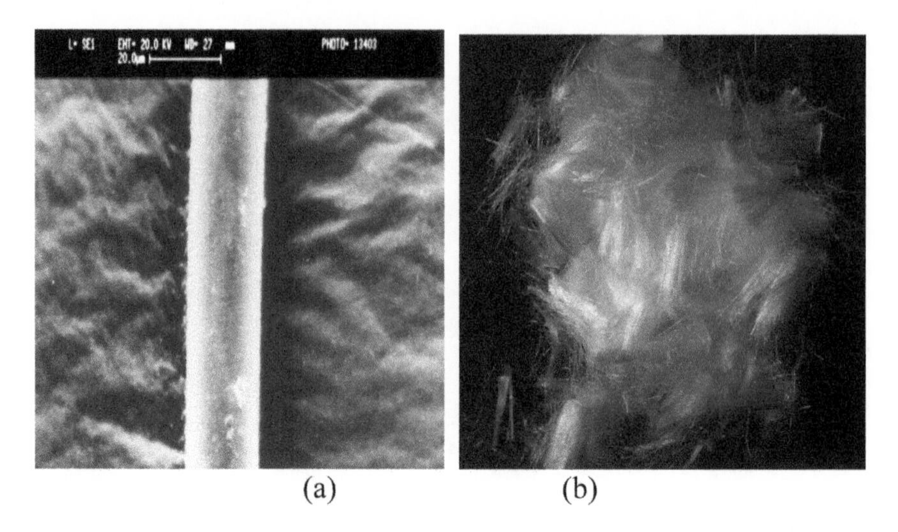

(a) (b)

Figure 7.20 Polypropylene fibres: (a) SEM image (*after Kaniraj* and Gayathri, 2003); (b) photograph showing the discrete short PP fibre (*after* Tang *et al.*, 2007).

(d) Dosage rates of 5, 15 and 25% of OPC along with 5 and 10% of silica fume increase the soaked UCS values of undisturbed fibrous peat by a factor as high as 10.

7.8.4 Polypropylene fibres (PPF)

Polypropylene (PP) is a versatile thermoplastic material which is produced by polymerizing monomer units of polypropylene molecules into very long polymer molecules or chains in the presence of a catalyst under carefully controlled heat and pressure. It has a good combination of properties, is cheaper than many other materials that belong to the family of polyolefins and it can be manufactured using various techniques. In general, polypropylene (Figure 7.20) is resistant to alcohols, organic acids, esters and ketones. Its general use in the construction industry is to reinforce concrete to control cracks as well as for increasing their load-bearing capacity and durability (Mullik *et al.*, 2006). Some properties of polypropylene fibres are listed in Table 2.14. Polypropylene fibres have also been used successfully as a non-chemical admixture to strengthen various types of soil, such as sands and clays, in conjunction with a binding agent such as Portland cement and or lime (Kaniraj and Gayathri, 2003; Tang *et al.*, 2007).

Studies have also been carried out by several researchers (Gray and Ohashi, 1983; Consoli *et al.*, 1998, 2002, 2009; Yetimoglu and Salbas 2003; Yetimoglu *et al.*, 2005; Ranjan *et al.*, 1996; Park and Tan, 2005; Tang *et al.*, 2007; Chauhan *et al.*, 2008) on the influence of fibre inclusion on the mechanical behaviour of cemented soil. In general, reports in the literature show that the randomly distributed fibres can be used to overcome the drawbacks of using cement alone, such as the high stiffness and brittle behaviour of the stabilized soil. In a series of laboratory unconfined compressive strength tests, the raw specimens attained a distinct axial failure stress at an

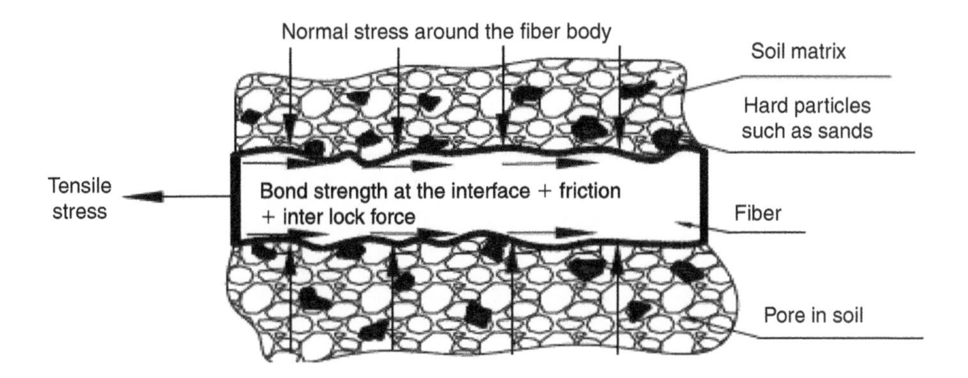

Figure 7.21 Sketch of mechanical behaviour at the interface between fibre surface and soil matrix (*after* Tang *et al.*, 2007).

Table 7.4 Polypropylene fibres specifications (*after* Sika Fibres 2005; Brown *et al.*, 2002).

Property	*Specification*
Colour	Natural
Specific gravity	0.91 g cm^{-3}
Fibre length	12 mm
Fibre diameter	18 micron (nominal)
Tensile strength	25–33 MPa
Elastic modulus	6,000–9,000 N mm^{-2}
Flexural modulus	1.2–1.5 GPa
Elongation at break	150–300%
Water absorption	None
Softening point	160°C

axial strain of about 1.5–2.5%, following which they collapsed. However, the fibre-reinforced specimens exhibited highly ductile behaviour (Kaniraj and Gayathri, 2003). Also, from the experiments on field test sections in which a sandy soil was stabilized with polypropylene fibres, Santoni and Webster (2001) concluded that the technique showed great potential for military airfield and road applications and that a 203 mm thick sand fibre layer was sufficient to support substantial amounts of military truck traffic.

According to Tang *et al.* (2007), in fibre-reinforced soils the fibre surface is attached by many soil particles which make a contribution to the bond strength and friction between the fibre and the soil matrix. The distributed discrete fibres act as a spatial three-dimensional network to interlock soil grains, help the grains to form a unitary coherent matrix and restrict their displacement. Consequently, the stretching resistance between soil particles and strength behaviour are improved. Because of the interfacial force, the fibres in the matrix are difficult to slide and they can bear tensile stress, as shown in Figure 7.21.

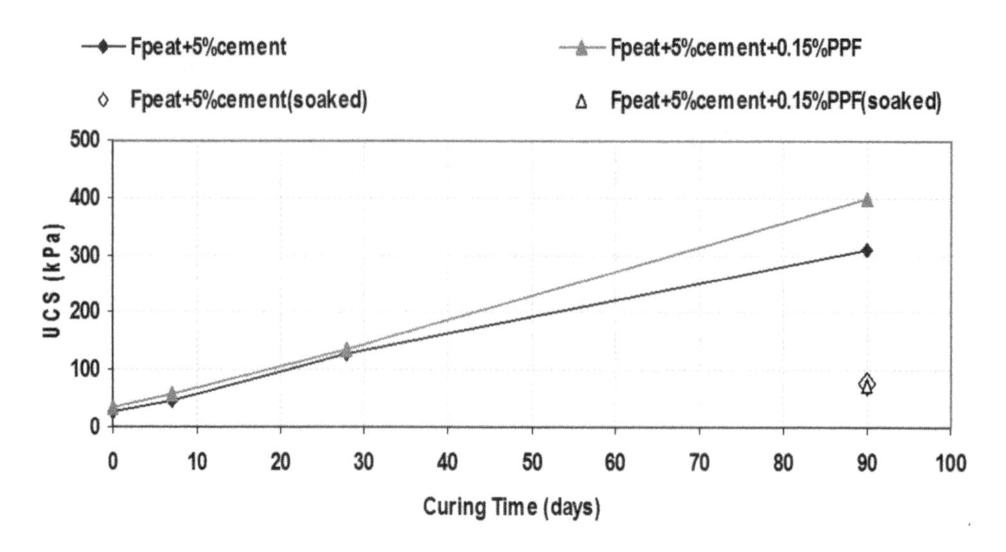

Figure 7.22 Unconfined compressive strength (UCS) values for fibrous peat and 5% OPC with and without polypropylene fibres (PPF) vs. curing time.

The minimum dosage rate of PPF recommended for concrete mixes starts at 0.6 $kg\,m^{-3}$ (Sika Fibres 2005). Also, Tang *et al.* (2007) tried 0.05, 0.15 and 0.25% (by soil weight) of polypropylene fibres (PPF) through UCS tests to find the optimum PPF amount to strengthen cement-treated clay soil.

Kalantari *et al.* (2011) examined polypropylene fibre (PPF) as an additive used in OPC-treated fibrous peat samples, and its effect was evaluated through UCS. Air curing was used to cure the OPC treated fibrous peat samples with PPF.

Different types of UCS test samples made of fibrous peat with a moisture content of 198% and various amounts of OPC (5, 15, 20, 30 and 50%) and 0.15% PPF were prepared, and then cured for up to 90 days. UCS tests were then conducted on stabilized samples immediately after mixing (0 day), 7, 28 and 90 days. Soaked UCS tests were also conducted on 90 days air cured samples by submerging them in water for seven days prior to being tested. The results are shown in Figures 7.22–7.26.

UCS tests were also conducted on plain (untreated) fibrous peat samples immediately after mixing (0 day) and after 90 days, in unsoaked (cured in air), and soaked (cured 90 days in air, and then soaked for seven days). The result is shown in Figure 7.27.

The results in Figures 7.22–7.27 show that:

(a) As the OPC amount is increased in stabilized samples with or without PPF, the immediate (0 day) and soaked UCS values increase as well.
(b) Addition of 0.15% polypropylene to the OPC-treated fibrous peat samples increases the UCS values of the air cured stabilized samples.
(c) UCS values increase from 28 days to 90 days of curing periods for samples containing 5% OPC by over 142% (from 128 to 310), and 196% (from 135 to 400), and for 15% by over 135% and 196%, and also for samples containing

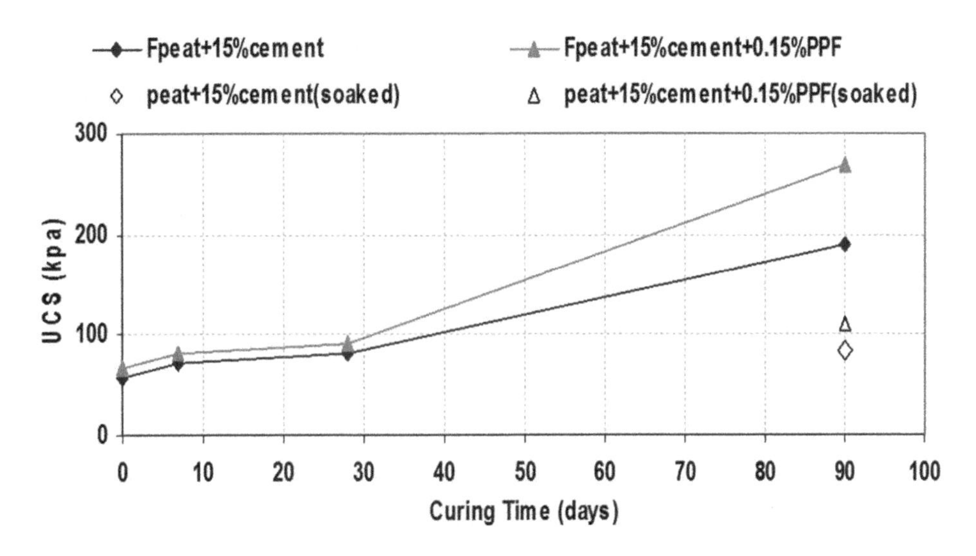

Figure 7.23 Unconfined compressive strength (UCS) values for fibrous peat and 15% OPC with and without polypropylene fibres (PPF) vs. curing time.

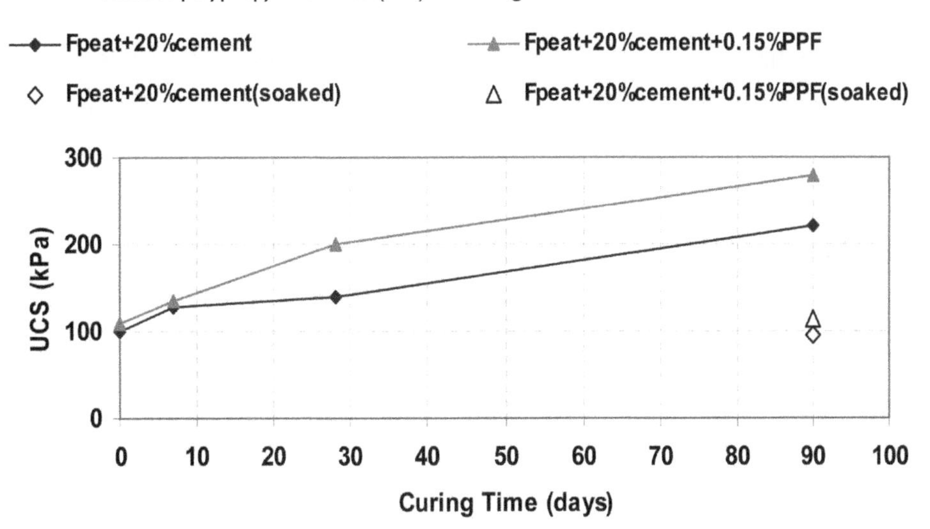

Figure 7.24 Unconfined compressive strength (UCS) values for fibrous peat and 20% OPC with and without polypropylene fibres (PPF) vs. curing time.

30% OPC by 5% (from 400 to 420 kPa) and 7% (300 to 320 kPa) with and without PPF respectively.

(d) The UCS values of 15% OPC treated stabilized fibrous peat cured for 90 days increase by a factor of 7 (from 28.5 to 190 kPa), and the inclusion of 0.15% polypropylene increases the UCS further by a factor of over 9 (from 28.5 to 280 kPa). Also, their soaked values increase by factors of 3 (from 28.5 to 85 kPa) and 4 (from 28.5 to 115 kPa) without and with fibres, respectively.

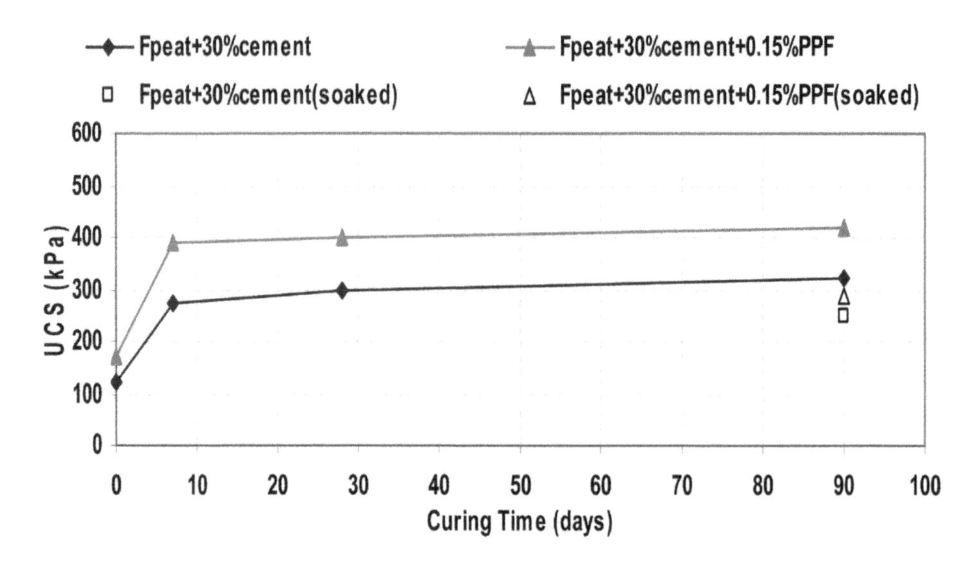

Figure 7.25 Unconfined compressive strength (UCS) values for fibrous peat and 30% OPC with and without polypropylene fibres (PPF) vs. curing time.

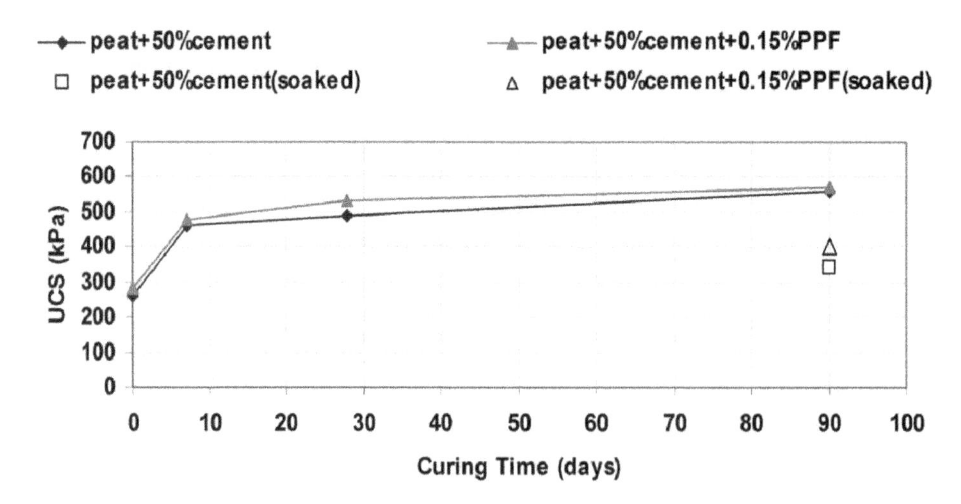

Figure 7.26 Unconfined compressive strength (UCS) values for fibrous peat and 50% OPC with and without polypropylene fibres (PPF) vs. curing time.

(e) In general, the soaking process of stabilized samples causes the UCS values of the respective unsoaked samples to decrease. As the cement amount in the samples increases, the magnitude of this decrease is less significant.

From the results obtained, it is also possible to conclude that stabilizing fibrous peat with OPC at its field moisture content increases the UCS value during curing. In general, for a lower dosage of cement content stabilized fibrous peat, the slope of the

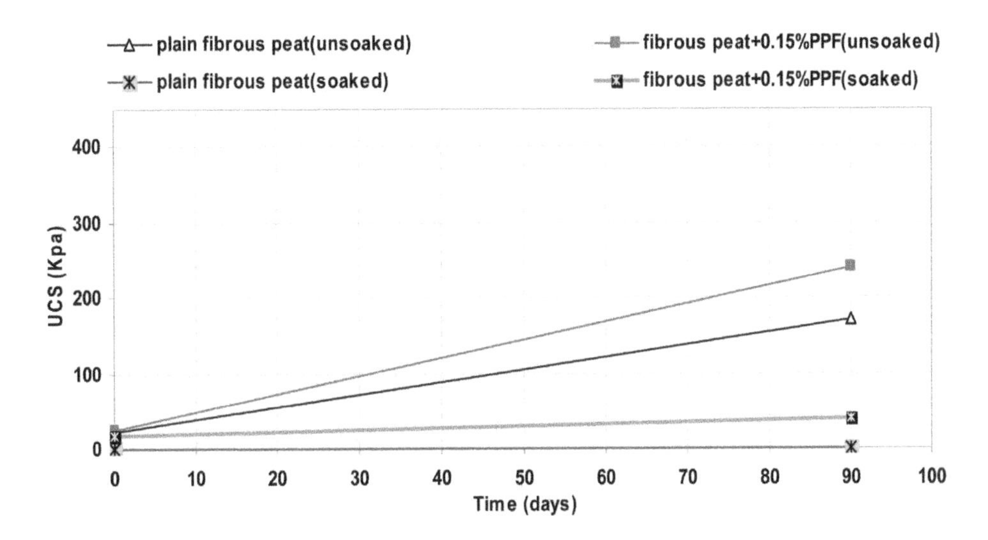

Figure 7.27 UCS values of various types of plain fibrous peat vs. curing time.

curve up to 28 days is less than the slope beyond 28 days, but for the higher dosage of cement content (30 and 50% OPC) samples that is reversed. The reason may be that as the cement content of the samples is increased to higher amounts (30%, and specially 50% OPC) more strength is gained through first 28 days, and the stabilized samples may behave similar to the behaviour of cement mortar mixes (cement, sand and water) which they gain most of their strength within first 28 days (Neville, 1999).

Also, the results obtained from UCS values for plain fibrous peat, with and without PPF, cured for 90 days and tested as unsoaked and soaked samples, shown in Figure 7.27 indicate that plain fibrous peat treated with and without PPF increases its UCS after 90 days of air curing by factors of 8 (from 28.5 to 240) and 6 (from 28.5 to 170 kPa), respectively. Soaking the identical air cured fibrous samples, on the other hand, causes the UCS value of fibrous peat with fibres to drop by factor of 6 (from 240 to 39 kPa), and the sample without fibres loses all of its strength (from 170 to 0 kPa) when soaked.

Figure 7.28 shows the reduction in moisture content for various types of treated peat samples through an air curing period from just after mixing up to 90 days. Comparing the results of treated peat samples having various amount of OPC in this figure indicates that stabilized samples containing 5% ordinary Portland cement continue to lose moisture content at a faster rate than other types of stabilized samples, and with a sharper slope during the curing periods. The final moisture content of samples with 5% OPC at the end of the 90 days curing period is close to that of stabilized samples treated with 30% OPC.

At the same time, if the 90 days cured UCS (unsoaked) values of samples with 5% OPC shown in Figure 7.22 are compared with other results shown in Figures 7.23–7.25 belonging to stabilized peat samples containing 15, 20 and 30% OPC respectively, it is possible to observe that the unsoaked UCS values of samples having only 5% OPC is relatively higher than those of samples having greater OPC content. This phenomenon

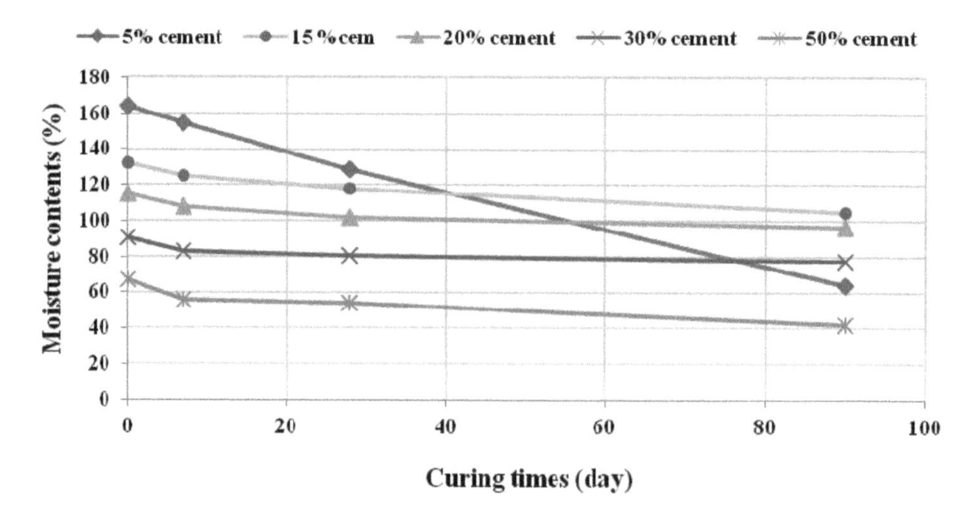

Figure 7.28 Moisture content reductions versus curing periods for various types of treated fibrous peat samples.

may be explained by the fact that as the treated fibrous samples lose their moisture content through the air curing period, they gain strength because of their wooden fibre content. According to Figure 7.28, this condition is more relevant to samples containing less OPC (5%). As these woody fibres, which make up most of the structural identity of fibrous peat, lose their moisture and become drier, the samples become more solid and thus gain more strength. As soon as the stabilized samples containing less OPC (samples with 5% (Figure 7.22) or 0% (Figure 7.27)) are soaked, their gain in strength decreases significantly, and their actual strengths are revealed.

CBR tests were conducted on various types of OPC treated fibrous peat for their unsoaked as well as soaked CBR values. The CBR samples first were air cured for three months, and then tested under unsoaked and soaked conditions. Figure 7.29 shows the results, indicating that:

(a) For stabilized fibrous peat with 15% OPC, cured for 90 days, the unsoaked and soaked CBR increase by factors of over 23 (from 0.8 to 19%) and 9 (from 0.8 to 7.2%), respectively
(b) For stabilized fibrous peat with 15% OPC, 0.15% polypropylene fibres, cured for 90 days, the unsoaked and soaked CBR increase by factors of over 28 (from 0.8 to 23) and over 18 (from 0.8 to 15).

Three months of air curing, as well as 15% OPC and 0.15% polypropylene fibres used to stabilize fibrous peat, will increase the general rating of *in situ* peat from very poor (CBR from 0 to 3%) to fair and good (CBR from 7 to above 20%) (Bowles, 1978).

Peat samples stabilized with OPC and polypropylene fibres show an increase in CBR values by as much as 39% (with 50% OPC). The OPC acts as a binding agent and is responsible for the increase in the mechanical strength of the samples. When

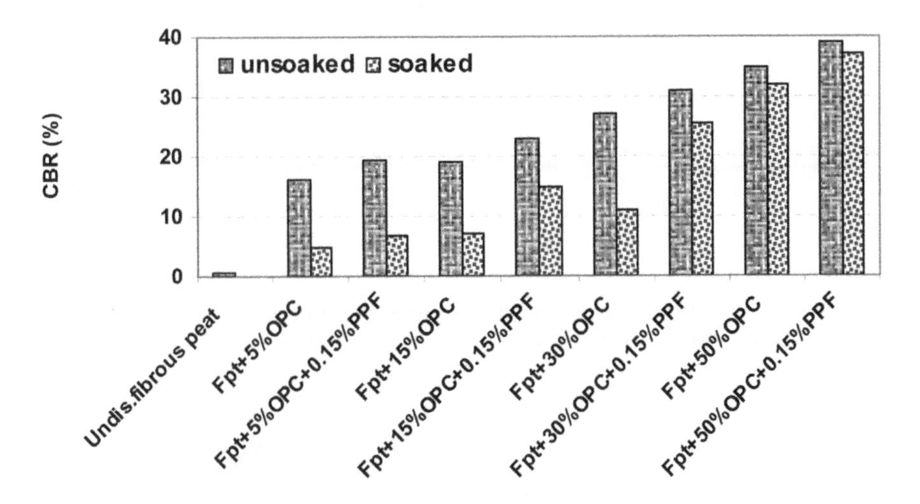

Figure 7.29 CBR (%) values for undisturbed peat and different percentages of OPC and polypropylene fibres for stabilized peat cured for 90 days.

cement and water are mixed together, the aluminates react with the water to form an aluminate-rich gel which reacts with sulfate in solution and the cement starts to hydrate, with the formation of calcium silicate hydrate and calcium hydroxide, and it gains strength. The polypropylene fibres act as reinforcements to the fibrous peat. It appears that it prevents the formation of cracks in the sample, and along with the cement, binds the peat particles together, leading to an increase in CBR values of the stabilized peat samples. The results agree well with the findings of researchers (Sivakumar *et al.*, 2008; Tang *et al.* 2007), who have also reported an increase in strength with the addition of cement and polypropylene fibres in clay.

There appear to be some micro-structural changes resulting from the addition of cement and polypropylene fibres or the interaction between cement and fibre reinforcement which is responsible for the increase in UCS, as well as CBR values. Using the air curing technique instead of normal moist curing also plays an important role in increasing the strength, as this method keeps lowering the moisture content, and as this reduces, stabilized samples gain more strength.

7.8.5 Steel fibres

Steel fibres are made of cold drawn steel wire with a low carbon (C) content or stainless steel wire. Steel fibres are manufactured in different types; hooked, undulating or flat, according to the construction project. These fibres are used in construction for concrete reinforcement. They are mostly used to reinforce of concrete, mortar and other composite materials. Steel fibre reinforcement products are designed to increase concrete material performance in a variety of applications. The technical advantages of steel fibres in concrete include: improve flexural toughness, increase fatigue, enhanced impact and abrasion resistance, additional load bearing capacity, controlled shrinkage behaviour and increased durability. Steel fibres can also be used as an addition

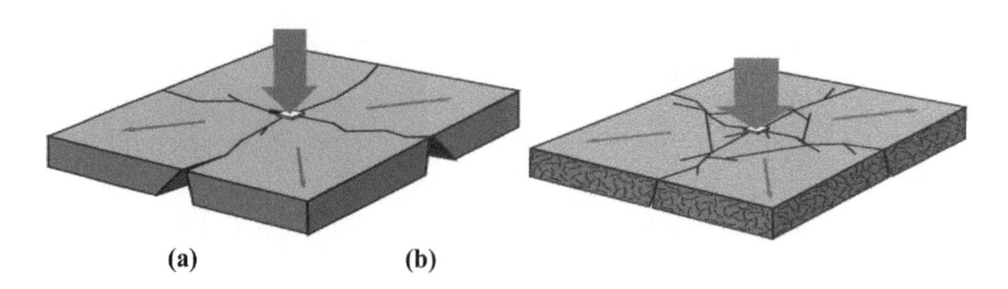

Figure 7.30 Schematic diagram of concrete block performance under load: (a) plain concrete and (b) steel fibre reinforced concrete (*after* Timuran Engineering, 2007).

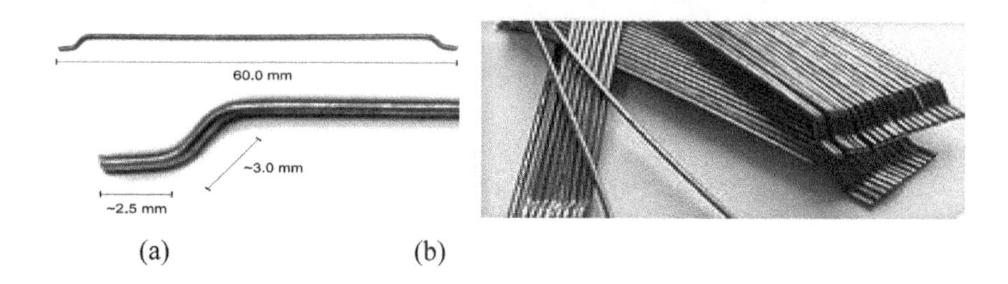

Figure 7.31 Hooked end steel fibres: (a) dimensions, (b) photograph (*after* Timuran Engineering, 2007).

Table 7.5 Hooked steel fibre specifications (*after* Timuran Engineering, 2007).

Property	*Specification*
Fibre length	60 mm
Equivalent diameter	0.75 mm
Aspect ratio (length/diameter)	80
Tensile strength	$1100 \pm 100\,\mathrm{N\,mm^{-2}}$
Piece per kg	4,600

or total replacement for conventional steel meshes (http://www.steelfibre.org). Figure 7.30 shows schematically the performances of two different concrete blocks upon loadings: plain concrete and fibre-reinforced concrete (FRC). Figure 7.31 shows the hooked end shape of steel fibres that are most often used to strengthen concrete mixes, and Table 7.5 shows the properties of steel fibres.

7.8.6 Cement and fibres

Chemical stabilization using cement has been used by many researchers to improve the performance of soft soils. Drawbacks in the behaviour of chemically stabilized soils

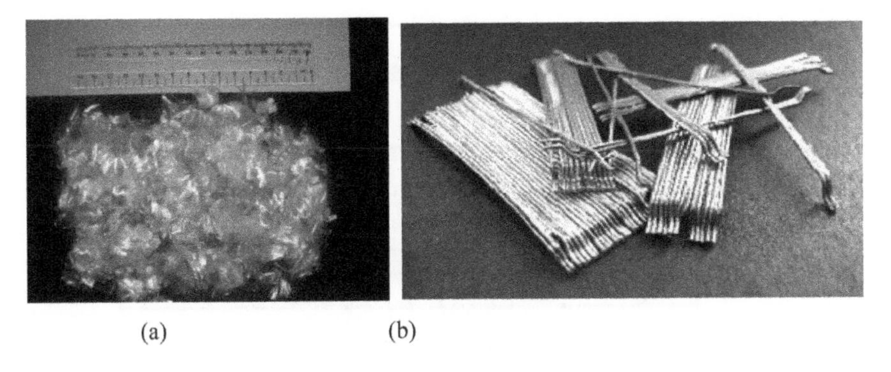

(a) (b)

Figure 7.32 Photographs of (a) polypropylene fibres; (b) steel fibres (*after* Kalantari *et al.*, 2011).

has been overcome with some success by incorporating randomly oriented fibres (reinforcement) within the soil which help by limiting the potential planes of weakness that develop with parallel oriented reinforcement. An increase in the compressive strength of the samples, due to the addition of fibres, has been reported by many researchers (Consoli *et al.*, 2009; Maher and Ho, 1993; Park and Tan, 2005; Tang *et al.*, 2007).

Kalantari *et al.* (2011) have attempted stabilizing peat using ordinary Portland cement (OPC) as a binding agent and polypropylene (Figure 7.32(a)) and steel fibres (Figure 7.32(b)) as chemically inert additives. CBR and UCS tests were carried out to evaluate the increase in strength of the stabilized samples.

The results of UCS tests on peat samples (showing an increase in UCS) with 5 and 15% OPC, different percentages of polypropylene and steel fibres, compacted at their optimum moisture content (OMC) and then air cured for 90 days, are presented in Figure 7.33.

Further, the results of the CBR test carried out on stabilized peat samples compacted at their OMC with 5% OPC, polypropylene fibres and steel fibres, and air cured for 1, 28 and 90 days are presented in Figure 7.34(a). Similarly, the results with 15% and 30% OPC and polypropylene fibres and steel fibres are presented in Figures 7.34(b) and (c) respectively.

Munro (2004) has also reported an improvement in the CBR when OPC is added to peat. An improvement in CBR of fly ash, upon addition of cement, has also been reported by Leelavathamma *et al.* (2005).

7.9 PEAT STABILIZATION BY REINFORCED COLUMNS

Huat (2004), Kalantari and Huat (2009b), Moayedi *et al.* (2009), Kazemian and Huat (2009b) and Edil (2003) summarized a number of construction options that can be applied to peat and organic soils, namely excavation–displacement or replacement; ground improvement and reinforcement to enhance soil strength and stiffness, such as by stage construction and preloading, stone columns, piles, thermal pre-compression and preload piers; or by reducing driving forces by lightweight fill; and chemical admixture such as cement and lime.

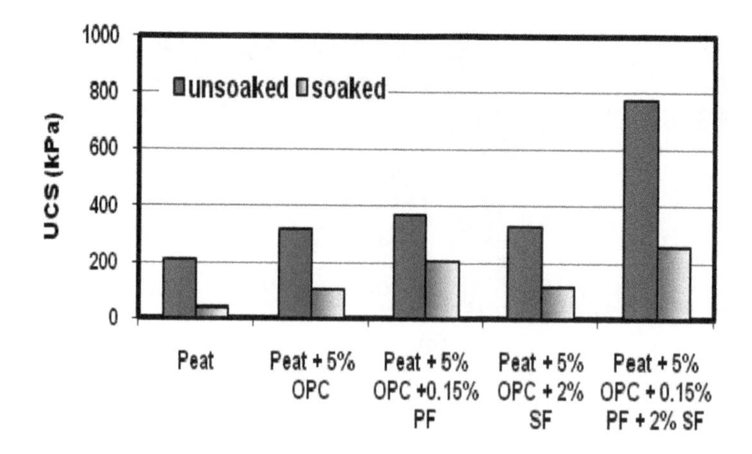

(a)

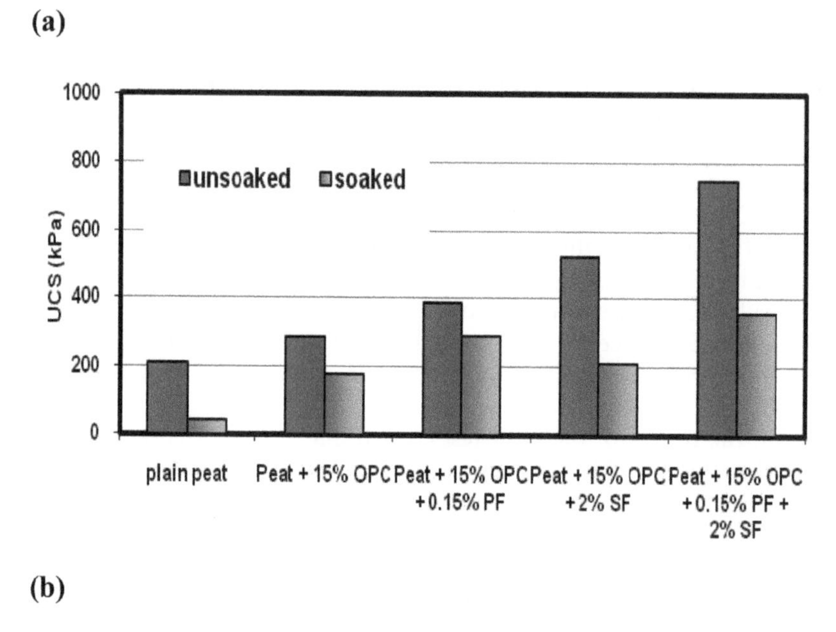

(b)

Figure 7.33 UCS (unsoaked and soaked) of peat stabilized with OPC, polypropylene (PF) and steel (SF) fibres and air cured for 90 days: (a) 5% OPC; (b) 15% OPC (*after* Kalantari *et al.*, 2011).

Ground improvement work by deep mixing methods using a binder is becoming the most popular method in construction industries. The deep soil stabilization technique is often an economically attractive alternative to the removal of deep peat or use as a deep foundation (Kazemian *et al.*, 2010). The essential feature of deep soil stabilization is that a column of stabilized material is formed by mixing the soil in place with a 'binder' and the interaction of the binder with the soft soil leads to a material that has better engineering properties than the original soil (Hebib and Farrel, 2003; Kazemian *et al.*, 2011).

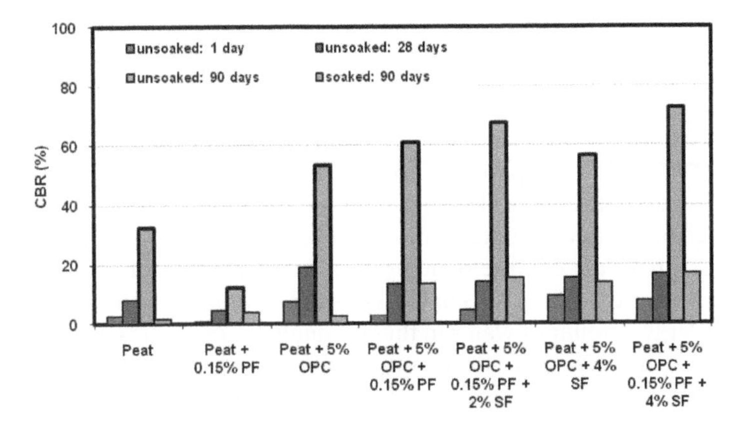

(a)

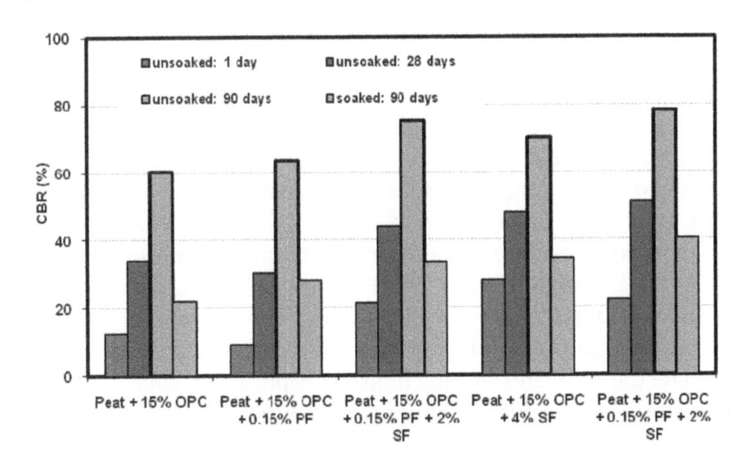

(b)

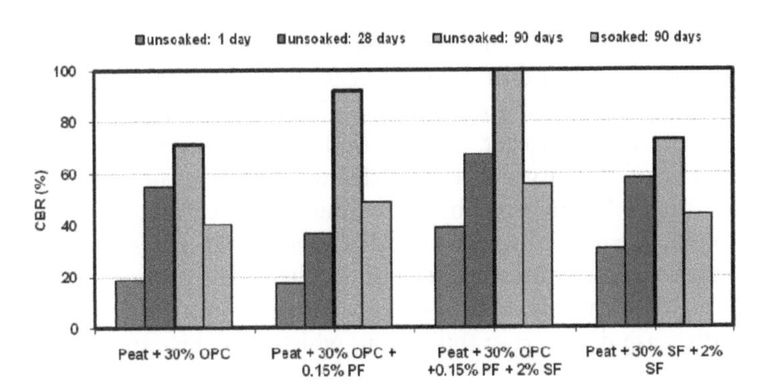

(c)

Figure 7.34 CBR (unsoaked and soaked) of peat stabilized with OPC and polypropylene (PF) and steel (SF) fibres: (a) 5% OPC; (b) 15% OPC; and (c) 30% OPC (*after* Kalantari *et al.*, 2011).

7.9.1 Cement-sodium silicate stabilized columns

As stated earlier, the organic materials in organic soil and peat contain substances such as humus and humic acids which act as retarding materials during hydration and other chemical reactions with cement. During stabilization with cement, the humic acids react with $Ca(OH)_2$ (from burnt cement) to form insoluble reaction products which precipitate out on the clay particles, thus inhibiting the strength gain via pozzolanic reactions. Secondly, the acids also cause the soil pH to drop and hinder cementation. Thirdly, since organic soils contain fewer solid particles to stabilize, and it is the solid particles that provide a definite structure, a greater quantity of stabilizer needs to be added. Furthermore, organic soils have a considerably higher water:soil ratio than clay, and hence a large amount of water in the soil implies larger voids, thus requiring more stabilizers, which is not cost-effective (Carlsten and Ekstrom, 1995; Moayedi *et al.*, 2011). As Müller-Vonmoos (1983) emphasized, in soils with high organic content, the quantity of binder needs to exceed a 'threshold'. As long as the quantity of binder is below the threshold the soil will remain unstabilized. These factors affect the reaction rate of the binders, resulting in a slower strength gain or higher cost of stabilizing organic soil than clay.

Sodium silicates have been developed into a variety of different grout systems and are widely using as a chemical grout. The sodium silicate systems consist of sodium silicate and a reactor/accelerator (i.e. calcium chloride) which can be compatible with cement to get strong bonding properties in a two-compound system. Two-compound systems have been used in grouting in the soil, below a water table or with high moisture content, and produce a high-strength permanent grout if not allowed to dry out (Clarke, 1984; Shroff and Shah, 1999). Sodium silicate is the common name for the compound sodium metasilicate, Na_2SiO_3, also known as water glass or liquid glass. Sodium silicates function as rapid-setting additives for cement or cement clay grouts and can alleviate the problems of cement grout application, such as separation of settle out due to excessive water bleed, equipment and manpower being idle while waiting for curing to take place, and surrounding water diluting the grout and preventing it from setting in flooded environments (Houlsby, 1990).

To overcome the difficulties in using cement alone for the stabilization of organic soils Kazemian *et al.* (2011c, 2012a) have used a cement–sodium silicate system (i.e. calcium chloride, cement and sodium silicate) to stabilize organic soils. It was observed by the authors that the trend of gain in shear strength of the organic soils changed with an increase in calcium chloride content. It increased until the net charge of the sample became zero. Thereafter, it decreased with an increase in calcium chloride as deflocculation of the larger size particles took place. The shear strength of the samples was observed to increase by increasing the concentration of sodium silicate. This behaviour is due to the hydration and pozzolanic reactions of cement with water and the rapid reaction between cement, sodium silicate and calcium chloride. However, upon increasing the concentration of sodium silicate, the strength of the mixture was observed to decrease due to re-stabilization, which prohibits an increase in the shear strength of the mixture.

Huat *et al.* (2011) and Kazemian *et al.* (2012b) carried out a series of tests on peat collected from Kampung Jawa, Selangor (Peninsular Malaysia), using a block sampling method. Various types of binder were used to stabilize peat and the shear strength of

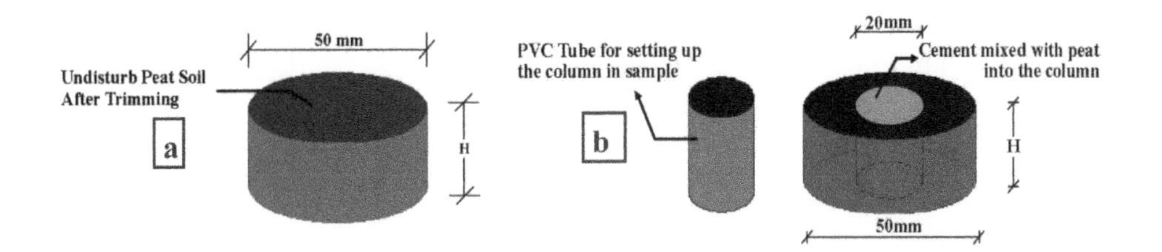

Figure 7.35 (a) A cylindrical test specimen from the original soil sample after trimming; (b) method used to set up the cement column in the specimen (*after* Kazemian *et al.*, 2012b).

the samples was determined using a triaxial compression test under unconsolidated undrained (UU) conditions to determine the shear strength of peat.

The procedure for setting up the specimen in the triaxial apparatus is shown in Figure 7.35. The specimen's height and diameter were 100 mm and 50 mm respectively.

To prepare a peat sample reinforced with cement–sodium silicate and kaolinite in order to investigate its undrained shear strength, a PVC tube was placed at the centre of the peat specimen and used to remove a portion of peat from the cell. The removed portion of peat was replaced with different ratios of ordinary Portland cement, calcium chloride as reactor, sodium silicate (Kazemian *et al.*, 2012b), water, cement, kaolinite and peat. The cement column's diameter was 20 mm.

Different compositions of chemical grouts (sodium silicate, calcium chloride, cement and kaolinite) were considered. The samples were then cured for 28 days in a soaking basin with natural peat water (collected from the peat site). A triaxial test (UU) on each sample was subsequently carried out.

The effect of sodium silicate (0, 1, 2.5 and 5.0%) on shear strength was investigated while other parameters remained constant (calcium chloride 2%, kaolinite 20% and cement 15%). Based on the results in Figure 7.24, there was an increasing trend in the shear strength with 1.0% and 2.5% sodium silicate compared to the conventional binder (i.e. 0% sodium silicate). However, higher shear strength was achieved with 2.5% sodium silicate. This indicates that 2.5% sodium silicate was an effective dosage to give a reasonable strength (Figure 7.36).

The mechanisms of peat soil stabilization under the effect of sodium silicate can be explained by following reactions (in some conditions):

$$\text{Cement} + \text{water} \longrightarrow \text{C-S-H gel} + Ca(OH)_2 \tag{7.8}$$

$$Na_2(SO_3) + Ca(OH)_2 \longrightarrow Ca(SO_3) + 2Na(OH) \tag{7.9}$$

The effect of ordinary Portland cement (0, 40 and 50%) was investigated while other parameters remained constant (Figure 7.37). From the results obtained (Figure 7.36), the strength increased as the percentage of cement increased. A 50% proportion of cement gave the highest value, but the effective dosage is yet to be calculated since the strength gain from 0–40% is better than from 40–50%. These findings agree well with Kazemian *et al.* (2010a), Kalantari *et al.* (2010, 2011) and Duraisamy *et al.* (2007a, 2009).

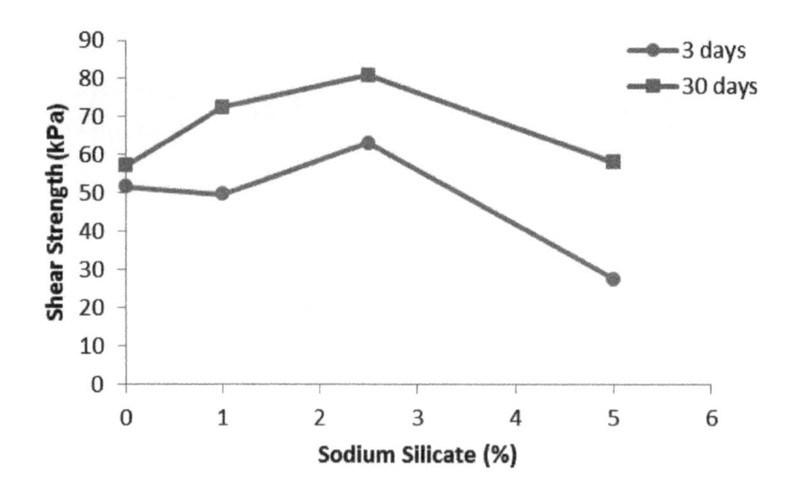

Figure 7.36 Effect of sodium silicate system grout on peat (*after* Huat *et al.*, 2011).

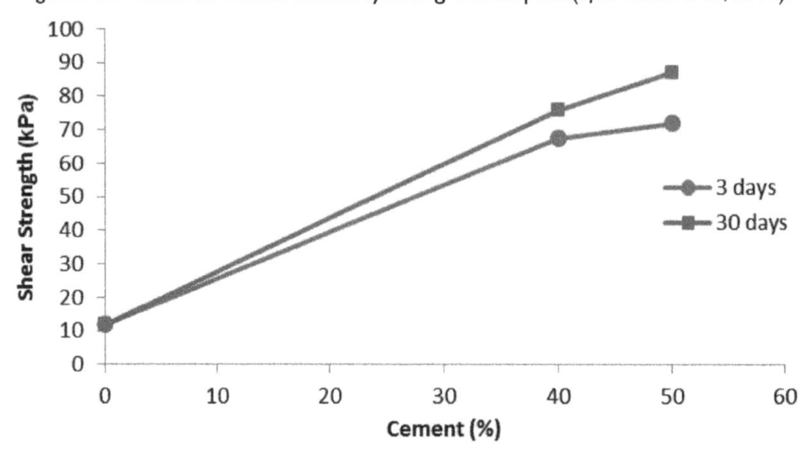

Figure 7.37 Effect of ordinary Portland cement on reinforced peat (*after* Huat *et al.*, 2011).

The mechanism of peat stabilization under the effect of cement can be explained by the following reactions in some conditions:

$$Ca(OH)_2 + SiO_2 \longrightarrow CaO\text{–}SiO_2\text{–}H_2O \tag{7.10}$$

$$Ca(OH)_2 + Al_2O_3 \longrightarrow CaO\text{–}Al_2O_3\text{–}H_2O \tag{7.11}$$

Compared to the conventional binders, the rate of hydration process of stabilized peat was accelerated by adding sodium silicate, ordinary Portland cement and kaolinite. The solid particles contributed to the hardening of stabilized peat by providing the cementation bonds to form between the contact points of the particles. The ordinary Portland cement provides a better strength gain compared with kaolinite and sodium silicate. Kaolinite acts as the pozzolanic material for cement and forms a connection

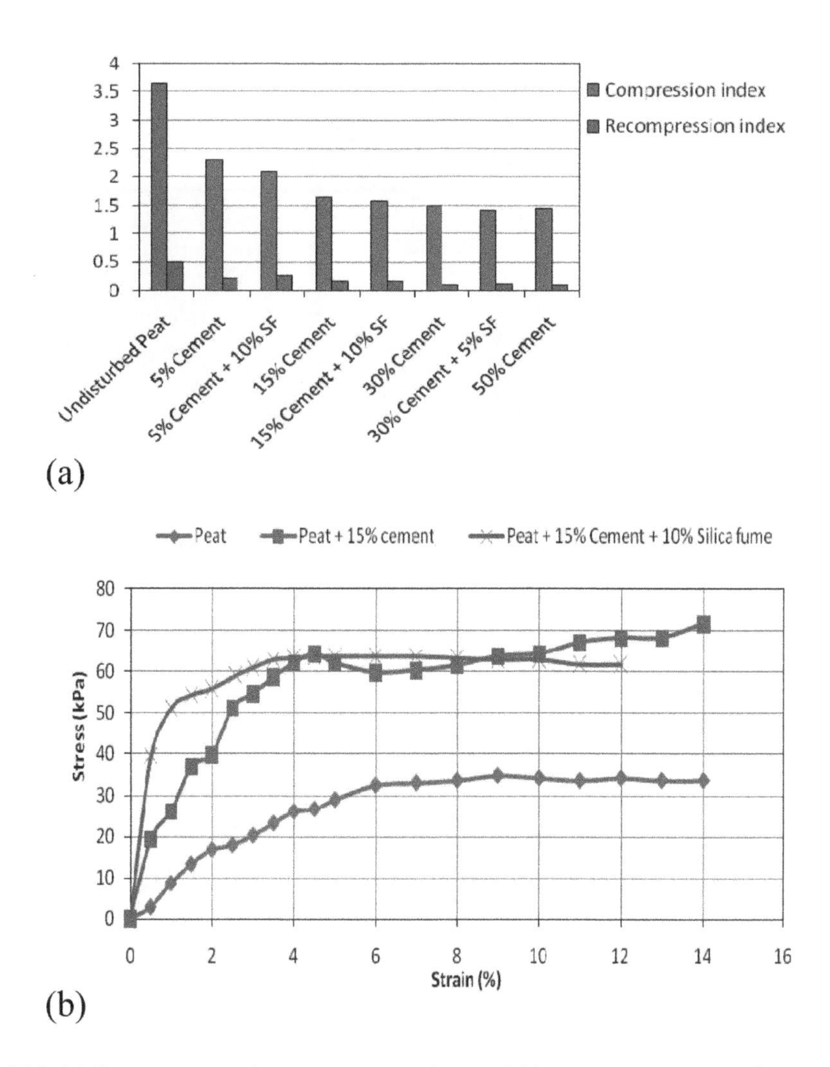

Figure 7.38 (a) Compression and recompression indices and (b) stress–strain curves of peat and peat with additives.

with peat. The addition of sodium silicate to peat at effective dosage yields a substantial increase in the undrained shear strength.

7.9.2 Cement and silica fume stabilized precast columns

Kalantari *et al.* (2012) have used precast cement stabilized peat columns, with and without the addition of silica fume to study the compressibility behaviour of peat. Plate load tests were also conducted to study the bearing capacity of precast stabilized peat columns.

The compression (C_c) and recompression (C_r) indices of undisturbed peat and peat stabilized with precast columns are shown in Figure 7.38(a) and the stress–strain

Figure 7.39 Plate load test: (a) schematic diagram of test tank; (b) precast stabilized column; (c) precast column after being installed in the test tank.

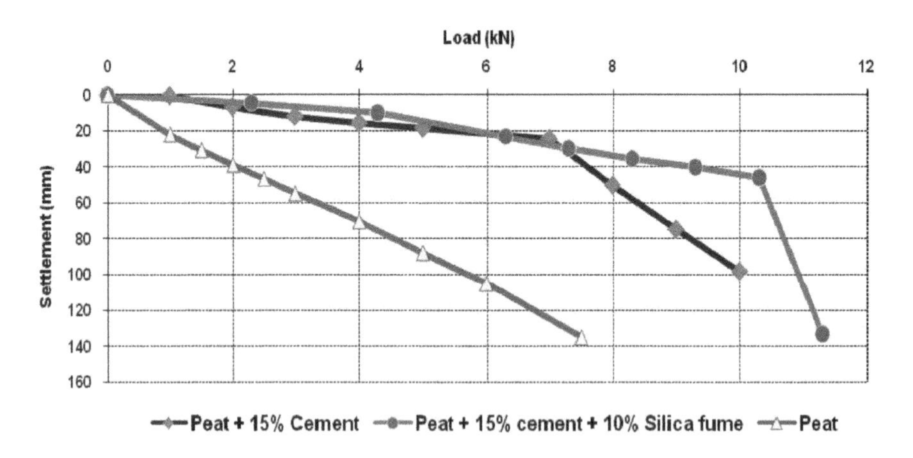

Figure 7.40 Load–displacement curves.

curves for peat and stabilized peat are shown in Figure 7.38(b). The addition of micro silica to the mixture of peat and cement for the columns also reduced the compression index.

In order to evaluate the bearing capacity of peat stabilized by precast columns, plate load tests (PLT) were carried out in a specially designed and fabricated circular steel test tank (Figure 7.39). The load displacement curves are shown in Figure 7.40.

The authors observed that in the case of peat only, punching failure was observed. It is obvious that the load-bearing capacity increased with the precast column, and at the same time there is also a further increase in the bearing capacity with the addition of silica fume. The use of 10% silica fume as an additive to the column that had 15%

cement caused an increase of about 35% in the bearing capacity of the column (failure load increased from 7 to 10.5 kN).

7.10 GEOGRID REINFORCED VIBROCOMPACTED STONE COLUMN

Ground improvement techniques such as compacted stone have been used increasingly to reinforce soft soils and increase the bearing capacity of the foundation soil (Al-Homud and Degen, 2006; Ambily and Gandhi, 2007; Chen *et al.*, 2008). This ground improvement technique has been successfully applied for the foundations of structures like liquid storage tanks, earth embankments and raft foundations, where a relatively large settlement can be tolerated by the structure. The stone columns develop their load-carrying capacity through bulging, and near-passive pressure conditions are developed in the surrounding soil. In weak deposits, the lateral support is significantly low and the column fails by bulging. In order to improve the performance of stone columns when treating weak deposits, it is imperative that the tendency of the columns to bulge should be resisted/prevented effectively. This will facilitate an increase of load transfer through the stone column and thus enhance the load-carrying capacity. Such a condition can be achieved by encasing the stone columns with geosynthetics over the full or partial height of the column (Alexiew *et al.*, 2005; Black *et al.*, 2007; Gniel and Bouazza, 2009; Raithel and Kempfert, 2000; Murugesan and Rajagopal, 2006, 2009, 2010; Raithel *et al.*, 2002; Yoo and Kim, 2009). The geosynthetic encasement will significantly increase the load-carrying capacity of stone columns due to the additional confinement by the geosynthetic. The geosynthetic encasement will also prevent lateral squeezing of stones when the stone column is installed in some extremely soft soils, leading to a minimal loss of stones.

The effectiveness of geogrid encasement on vibrocompacted stone columns was investigated by Prasad *et al.* (2012) through a parametric study carried out using the commercially available finite element package PLAXIS. The influence of parameters such as the stiffness of geogrid encasement, the depth of encasement from ground level, the diameter of the stone columns, spacing of the stone columns and the shear strength of the surrounding peat were analyzed.

In order to evaluate the improvement achieved due to the geogrid encasement, two cases were analyzed: stone columns without geogrid encasement (SC) and stone columns encased with geogrid (GC). In order to directly assess the influence of the confinement effects due to encasement, the analyses were performed by applying uniform pressure on the stone column portion alone. Analysis was also performed by applying a load on the entire area of the unit cell, and finally loading was applied to a group of columns having seven columns arranged in triangular pattern.

All the analyses, for column diameters, 0.6 m and 1.0 m and group of seven columns, were carried out by varying s/d from 2 to 4, geogrid stiffness from 50 to 5,000 kN m^{-1}, length of encasement from $1d$ to $4d$ from the top (where d is the diameter and s is the centre to centre spacing of the columns). The improved performance was evaluated based on the reduced settlement and lateral bulging of the stone column.

Figure 7.41 shows the typical deformed mesh, at a prescribed displacement, for the case of a single column loaded for SC and GC for $s/d = 3$ and $c = 6$ kPa. It is

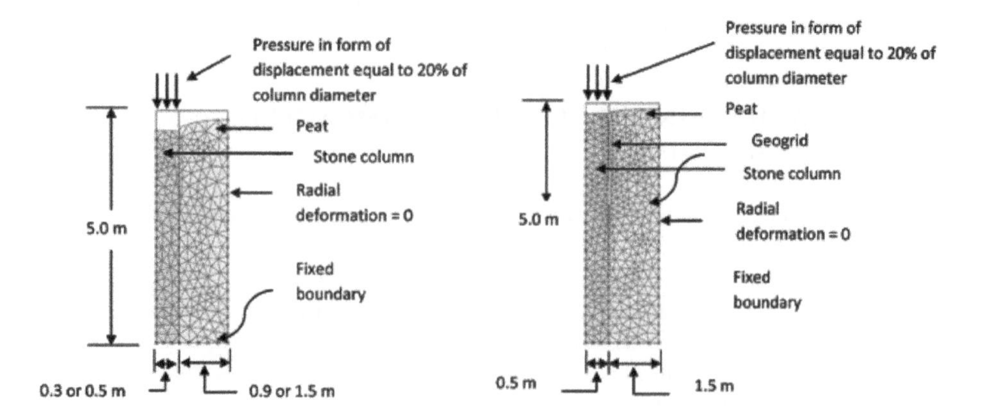

Figure 7.41 Deformed mesh for SC and GC, $s/d = 3$, $c = 6$ kPa, diameter = 1.0 m, geogrid up to $3d$: (a) deformed mesh for column area only loaded for SC; (b) deformed mesh for column area alone loaded for GC (*after* Prasad *et al.*, 2012).

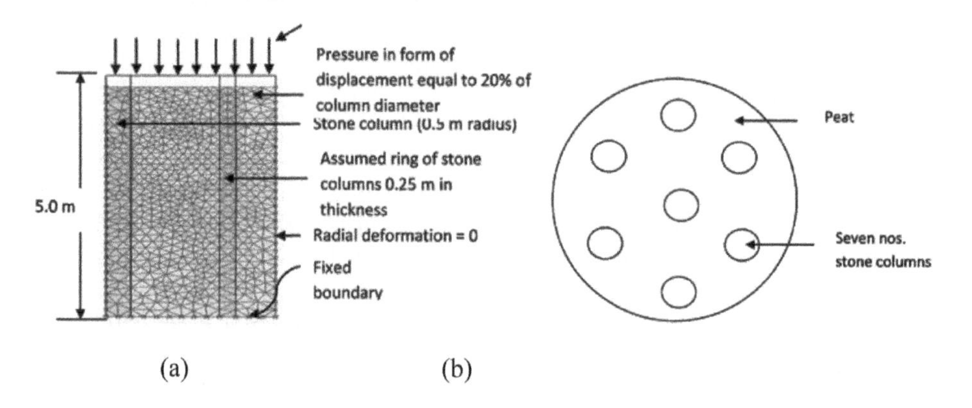

Figure 7.42 (a) Deformed mesh, entire area loaded, group of seven columns (SC), $s/d = 3$, $c = 6$ kPa, diameter = 1.0 m; (b) plan view of the group layout (*after* Prasad *et al.*, 2012).

observed that the failure is by bulging of the column at a depth about 0.5 to 2.0 times the diameter of the column (Figure 7.41(a)). The bulging disappears when the column is encased with geogrid as seen in Figure 7.41(b).

A typical deformed mesh for the group of seven columns is shown in Figure 7.42.

It is observed that the load capacity of the SC is dependent on the cohesive strength of the surrounding clay soil. On the other hand, the influence of the strength of surrounding soil on the load capacity of the GC gradually decreases as the stiffness of the geogrid increases. When the encasement stiffness was increased from 50 to 5,000 kN m^{-1}, the pressure–settlement response of GC was practically independent of the strength of the surrounding clay soil, as shown in Figure 7.43.

As the stiffness of the encasement increases, the lateral bulging of the stone column reduces, thereby reducing the stresses transferred into the surrounding soil. Hence it can be said that the contribution of the surrounding soil to the stability of the encased

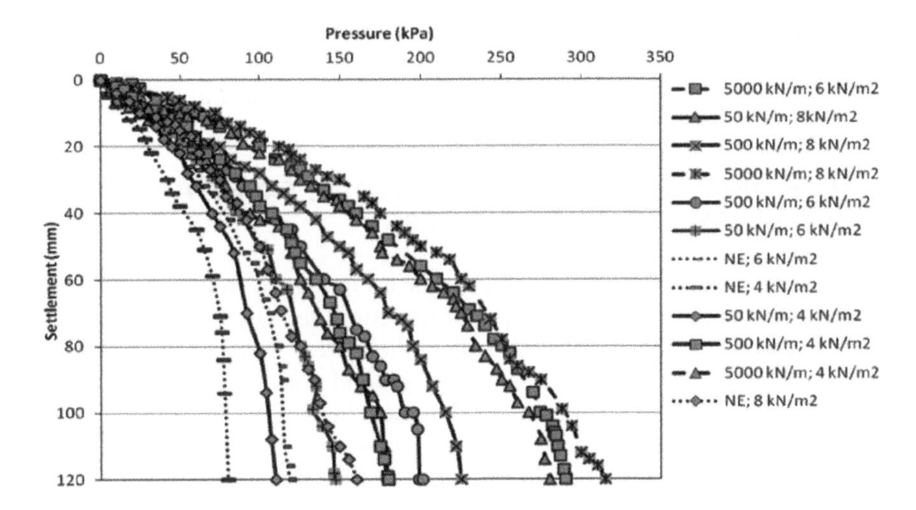

Figure 7.43 Pressure vs. settlement curves; different shear strengths and different geogrid encasement stiffness; column diameter = 0.5 m, $s/d = 3$ (NE = No encasement) (*after* Prasad *et al.*, 2012).

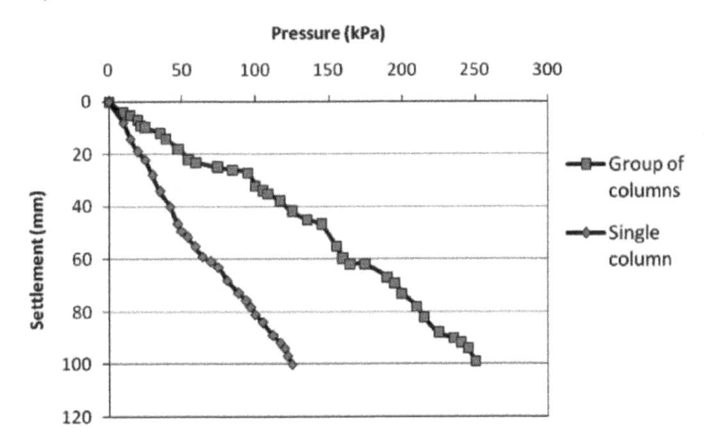

Figure 7.44 Pressure vs. settlement for single column and a group of columns, entire area loaded, $s/d = 3$ (*after* Prasad *et al.*, 2012).

stone column reduces as the stiffness of the encasement increases. This implies that the capacity of encased columns is almost independent of the strength of the surrounding soil for extremely stiff geogrid encasements. Murugesan and Rajagopal (2006, 2010) have also observed that the stiffness of the encasement plays an important role in reducing the bulging of the columns, thus leading to higher bearing capacity of the columns.

Figure 7.44 shows a comparison of axial stress versus settlement behaviour of a group of seven columns and of a single column when the entire area is loaded based on finite-element analysis for $s/d = 3$. It can be seen that the behaviour of a single column and a group of columns is almost comparable.

Prasad *et al.* (2012) observed that the field behaviour of an interior column when a large number of columns are simultaneously loaded can be simulated by the single column behaviour using a unit cell concept. Similar behaviour of a group of columns is also reported by Dhouib and Blondeau (2005) and Maurya *et al.* (2005).

7.11 NEW DEEP MIXING METHODS (DMM) FOR STABILIZATION WITH NEW CHEMICAL BINDERS

DMM, a deep *in situ* soil stabilization technique using cement and/or lime as a stabilizing agent, was developed in Japan and the Nordic countries independently in the 1970s. Numerous research efforts have been made in these areas investigating the properties of treated soil, behaviour of DMM improved ground under static and dynamic conditions, design methods and execution techniques. In the past three to four decades, traditional mechanical mixing has been improved to meet changing needs. New types of technology have also been developed in the last 10 years – e.g. the high-pressure injection mixing method and a method that combines mechanical mixing and high pressure injection mixing technologies (Kitazume and Terashi, 2013).

In this DMM, soil mixing is usually carried out using mixing augers through which a grout is introduced and mixed with the soil, resulting in stabilized soil. The most important factor in such an application is to ensure that the injection and mixing process of the additives with the soil are thorough and effective so that homogeneous well-mixed soil additives in monolithic columns are produced (Al-Tabbaa *et al.*, 1999). Miura *et al.* (1998), Shen (1998) and Shen *et al.* (2003a,b) indicated that there exists an influence zone of property changes in soil surrounding the injection columns, ranging from the edge of the column to a distance of about 2–3 times the columns' radius. Within this influence zone, the moisture content decreased while the pH values and the concentration of cations increased. In this condition, the shear strength of the soil decreased during the installation, which is very helpful for doing injection in the soil, and it was restored after a short curing period. Furthermore, there is a consolidation process involved in the surrounding soil after installation of the injection column. This is due to the dissipation of excess pore-water pressure developed during the installation, which contributes to the strength increase in the surrounding soil (Randolph *et al.*, 1979; Asaoka *et al.*, 1994; Shen, 1998). Four major factors were identified by Shen *et al.* (2003a) that caused property changes in the surrounding soils during and after installation of injection columns: (1) soil fracturing, (2) disturbance, (3) thixotropy and consolidation and (4) cementation effects as a result of the diffusion of the chemical binder.

Since injection columns are constructed by mixing *in situ* soft soils with chemical admixtures using rotating blades, there exist two types of forces acting on the surrounding soil. The first is created by an expanding action caused by the injection pressure of the chemical binder. The second is caused by a shearing action resulting from the blade rotation. These dual effects can generate excess pore-water pressure and disturb the surrounding soil, with the end result being that a plastic zone is formed around the column (Shen *et al.*, 2003a).

There are three binder injection methods: (1) injecting all the binder while penetrating with the mixing blades, (2) injecting all the binder while withdrawing the mixing blades, and (3) injecting part of the binder during penetration and the rest during withdrawal. In the case of peaty ground, to which a large amount of the binder is applied, the viscosity of soil may increase if the entire amount of binder is injected during penetration, thus decreasing its efficiency (Hayashi *et al.*, 2003). Al-Tabbaa *et al.* (1999) and Hayashi and Nishikawa (2003) reported that the number of drilling and mixing cycles in DMM is 400–450 times m^{-1} to achieve high efficiency and adequate mixing. Hayashi and Nishikawa (1999) developed Equation (7.12) to find the number of mixings per metre.

$$T = N \times (R_P/S_P \times W_i/W + R_W/S_W) \tag{7.12}$$

where:
T = number of mixings per metre in depth (times m^{-1}),
N = total number of mixing blades,
R_P & R_W = penetration and withdrawal speeds ($m\,min^{-1}$),
S_P & S_W = rotation speeds of mixing blades during penetration and withdrawal (rpm),
W_i = binder injection amount during penetration, and
W = Total amount of binder ($kg\,m^{-3}$).

Grouting is generally used to fill voids in the ground (fissures and porous structures) with the aim of increasing resistance to deformation; augmenting cohesion, shear-strength and uni-axial compressive strength; and finally (even more frequently) reducing conductivity and interconnected porosity in an aquifer (Moseley and Kirsch, 2004). The vacuum dewatering method (vacuum consolidation method), with or without preloading, is one of the methods for improving soft soils and has been applied in a number of countries (Hayashi *et al.*, 2003; Chai *et al.*, 2006). Kjellmann (1952) stated that when a vacuum is applied to a soil mass, it generates a negative pore-water pressure. If the total stress remains constant (unchanged), the negative pore pressure results in an increase in the effective stress in the soil, which leads to consolidation. In the vacuum dewatering method, a typical on-land system consists of an airtight membrane over the ground surface, anchored along the periphery of the site. Slotted collection pipes are embedded into a sand blanket beneath the membrane. The outlet is a solid pipe connected to a vacuum pump and discharge system. Typically, for an existing soft ground site, vertical wick drains (prefabricated vertical drains – PVDs) are installed beneath the sand blanket up to a depth of about 1 m above the bottom of soft soil deposit (Thevanayagam *et al.*, 1994; Hayashi *et al.*, 2003). This technique is often used in conjunction with vertical drains as well as preloading methods. It enables the construction of a very high embankment on very soft ground to be made over a relatively short period of time by reducing the development of shear strain in the soil. In addition, the amount of surcharge fill may be reduced by several metres if a vacuum pressure of at least 70% of atmospheric pressure is applied and sustained (Rujikiatkamjorn *et al.*, 2008).

Recently, a new technique of applying vacuum pressure to soft clayey subsoil has been developed in which the vacuum pressure is combined with a special prefabricated vertical drain (PVD) consisting of a PVD, a drainage hose and a cap connecting the PVD and the hose, known as cap-drain (CPVD) (Fujii *et al.*, 2002). Chai *et al.* (2008) explained that the method uses a surface or subsurface soil layer as a sealing layer and there is no need to place an airtight sheet on the ground surface, and therefore no worry about air leakage caused by damage to the sheet. The thickness of the surface sealing layer can be determined according to the field conditions, and generally the variation in thickness will not cause additional cost. In this method, the thickness of the surface sealing layer is estimated as suggested by Chai *et al.* (2008).

In comparison with pre-loading with vertical drains, the vacuum consolidation method has more advantages: (1) no/less fill material is required; (2) construction periods are generally shorter; (3) there is no need for heavy machinery; (4) a vacuum pressure up to 600 mm Hg (80 kPa) can be achieved in practice using the vacuum equipment available, which is equivalent to a fill 4.5 m in height; and (5) there is no need to control the rate of vacuum application to prevent bearing capacity failure because applying a vacuum pressure leads to an immediate increase in the effective stress in the soil (Cognon *et al.*, 1994; Jacob *et al.*, 1994; Shang *et al.*, 1998; Chai *et al.*, 2006).

However, this method also has certain disadvantages. A poor skeleton of soft clays and organic soils needs a large number of columns for injection to get high strength, which increases the cost as well as the construction time of the project, and applying a very powerful injection pump is not applicable in all projects because of the large weight of these instruments, requiring especial techniques to access the soft soils in the field to avoid sinking. Further, the poor skeleton of these soils may not be able to tolerate the high pressure of a powerful injection pump. Injection and grouting methods may be less successful when the voids consist of many fine interstices that are not always interconnecting (complete filling being very difficult to achieve). When it is difficult to confine the grouting, it may need either more injection points or more powerful injection instruments, thus increasing the cost.

Vacuum consolidation also has some shortcomings. The applied vacuum is limited by atmospheric pressure and it may cause cracks in the surrounding surface area due to consolidation-induced inward lateral displacement of the ground (Thevanayagam *et al.*, 1994; Tang and Shang, 2000; Chai *et al.*, 2006). Furthermore, due to the complication of air–water separation and badly sealed *in situ* boundary conditions, the efficiency of the system decreases. Theoretically, the maximum vacuum pressure that may be applied is one atmosphere (about 100 kPa), but practically achievable values are normally in the range 60–80 kPa (Tang and Shang, 2000; Qiu *et al.*, 2007). A system with an efficiency of 75% shows results with only 4.5 m of equivalent surcharge, and for the stabilization of very soft and organic soils (which need more negative pressure), pre-loading is necessary. In addition, in very soft soils, with the passage of time and with increasing settlement, the efficiency of the system will reduce. This is because the PVDs will crumple and be unable to discharge water properly.

A new deep mixing method (DMM) for stabilization of peat has been developed by Kazemian (2011), keeping in mind the concepts of injection grouting and vacuum dewatering, as discussed earlier in this chapter. A detailed procedure and its advantages are presented below.

Procedure of the new DMM technique (combination of injection and vacuum technology)

1. The vacuum well is installed and the vacuum is started to remove (any) water from the voids within the soil by using its auger at predefined positions and depths.

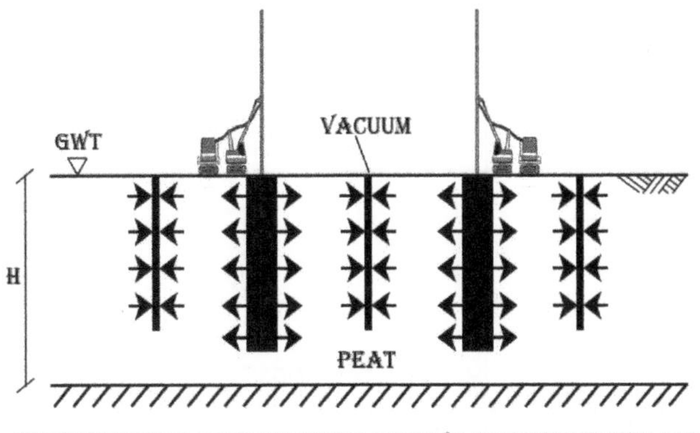

Figure 7.45 Schematic diagram of the new DMM technique (*after* Kazemian, 2011).

2. After arriving at a stable vacuum condition, the injection auger is operated to make injection wells at predefined location and depths (between vacuum slots).
3. The selected new binder is injected in the injection wells by using the injection pump simultaneously with the vacuum.
4. To achieve homogeneity in the migration of chemical binders, and at the same time speedy dispersion from the injection nozzle, a high-quality vacuum pump which is capable of delivering a satisfactory level of stable flow is used (Figure 7.45).

Advantages of the new DMM technique

1. The function of the vacuum technology is mainly to distribute the injected binder in soft and organic soils and also to achieve a lower magnitude of consolidation in the soil. In this method, the vacuum forms a homogenized mixture of soil and binder.
2. To get much higher strength by using a new binder in comparison with conventional ones.
3. No need to lower the water table. (The ground water level rather helps as a distribution tool after achieving a stable vacuum within the soil.)
4. The horizontal permeability of soils is higher than the vertical permeability and this method helps to speed up stabilization and leads to homogenization.
5. Because of the very low strength and loose matrix of soft clays and organic soils, using a high-pressure injection pump for stabilization is impractical. In this method we can use low-pressure injection pumps with high efficiency.
6. By applying this method, higher efficiency is achieved in a short time and the ratio of affected area method is much higher than with ordinary injection. There is no need to use a super jet pump for the soil.

A new DMM apparatus based on the above idea was developed at UPM (Universiti Putra Malaysia) in 2010 by Kazemian (2011). The vacuum part of this method is the same as the cap-drain (CPVD) method (Fujii *et al.*, 2002) and hence there is no need

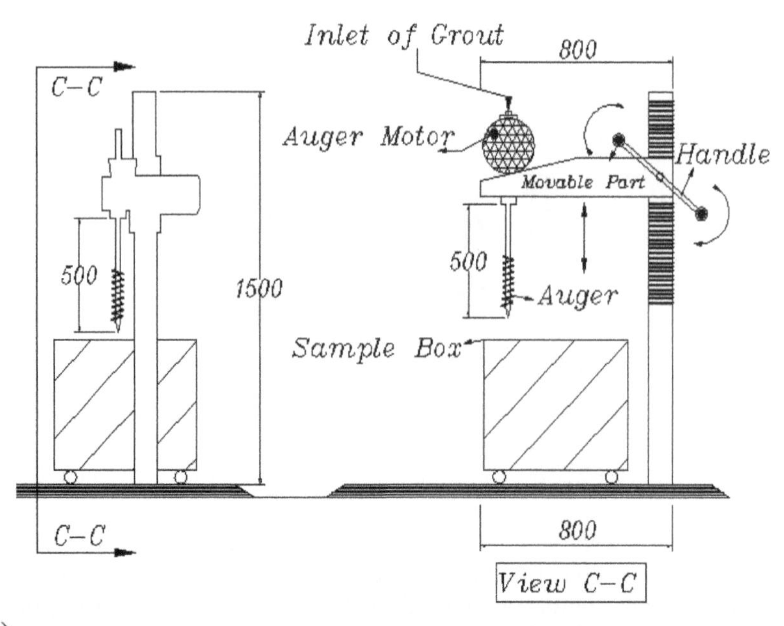

Figure 7.46 (a) Schematic diagram of model DMM equipment; (b) fabricated model DMM apparatus for new DMM technique (*after* Kazemian, 2011).

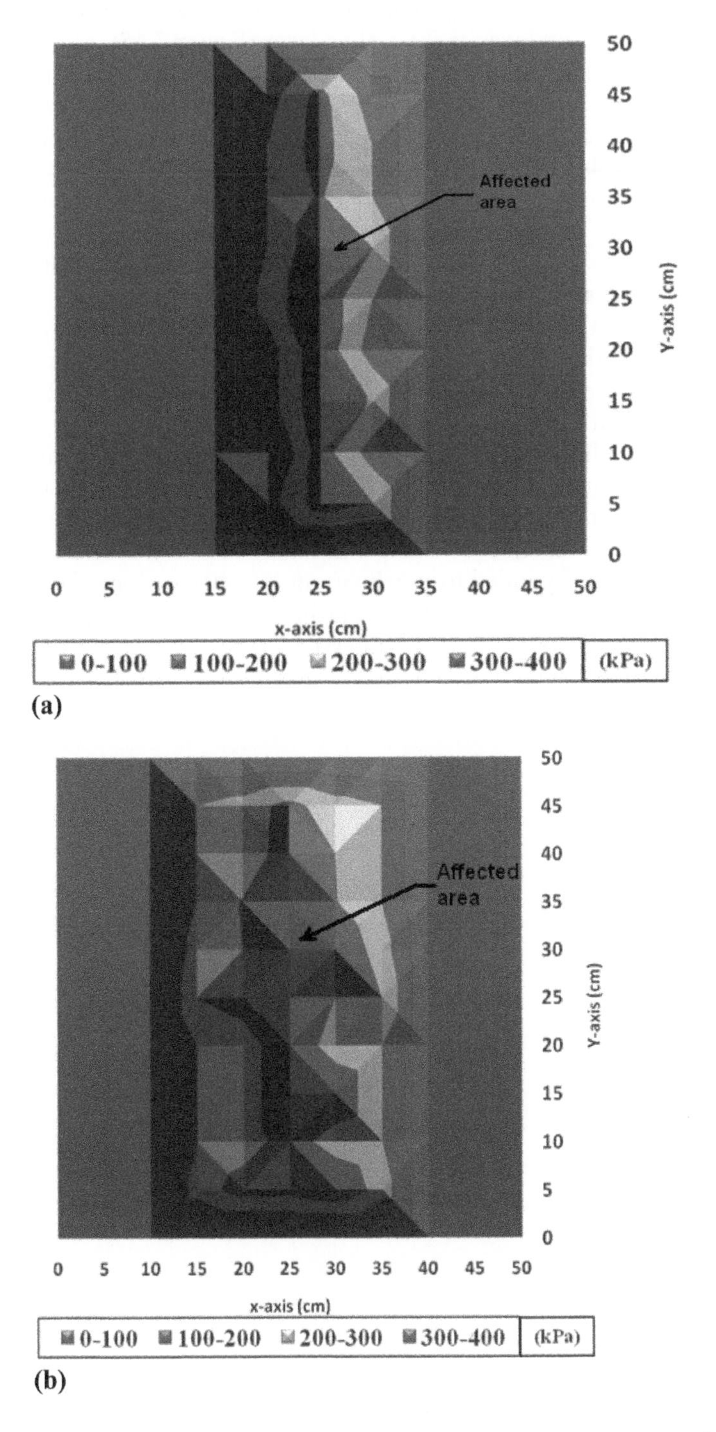

Figure 7.47 Comparison of the affected area: (a) conventional method with injection alone; (b) new DMM technique (*after* Kazemian, 2011).

to place an airtight sheet on the ground surface. The injection can be performed by using the surface soft soil layer as the sealing layer (Figure 7.45).

Kazemian (2011) has investigated the effect of this method on peat by using a new binder (explained in Section 7.9.1) by injected the binder in soft soil after a constant flow in the soil is achieved by the vacuum pipe. The results were compared with conventional methods by pocket penetrometer after curing for 14 days. The unconfined compressive strengths at pre-defined points at a depth of 100 mm were measured by pocket penetrometer and the changes are shown in Figures 7.46(a) and (b). Each tested point was 20 mm apart in a quadrilateral pattern. Figures 7.47(a) and (b) show the affected areas in the samples respectively at a depth of 100 mm.

It was observed that by applying a combination of injection and vacuum technology in the samples, the affected area increased dramatically. The affected areas in the samples were categorized in four groups (1) 0–100 kPa outer layer, (2) 100–200 kPa, (3) 200–300 kPa and (4) 300–400 kPa center core.

Figure 7.47(a) shows that the horizontal extent of the affected area in sample I (stabilized by conventional methods) is nearly 20 cm (15–35 cm along the x-axis), and a small area of the sample had a strength of 300–400 kPa. On the other hand, by using negative pressure by vacuum slots (sample II, Figure 7.47(b)), the horizontal extent of the affected area was expanded to nearly 30 cm (10–40 cm along the x-axis), i.e. the affected area increased from 20 to 30 cm. Furthermore, in this sample, more stabilized areas with strength 300–400 kPa were observed.

Chapter 8

Environmental geotechnics in peat and organic soils

8.1 INTRODUCTION

As described in the earlier chapters, peat lands in recent times have come to be seen as an economic resource to be drained, cleared and then used for building and road construction, agriculture or forestry. With rapid reclamation and development, some may argue to increase attention on conserving what is left for biodiversity, water resources, recreation and other purposes. According to Rieley (1991), peat land (lowland tropical peat land) performs vital ecological and hydrological functions related to hydrology (water storage, flood control and fisheries) and hydrochemistry. Andriesse (1991) listed the ecological assets of pristine peat land, such as playing a role in the delicate hydrological balance existing in the lower region of large river basin, serving as a buffer between salt and fresh water, and absorbing heavy metals in agricultural areas fringing the peat land. The ecological attributes and benefits derived from undisturbed peat land are summarized in Table 8.1.

8.2 PEAT HYDROLOGY

Water plays a fundamental role in the development and maintenance of tropical peat. An adequate supply, determined by the balance of rainfall and evapotranspiration, is critical to its sustainability. A small change in the water level will induce a shift and losses within the peat communities. Hydrology is the pattern of water movement. It is vital to the functioning of the peat ecosystem. The hydrology of peat is extremely important as an environmental determining factor for the maintenance of the peat's structure and function. The hydrological conditions affect many abiotic factors, such as soil aeration, nutrient availability and salinity, which then affect the flora and fauna. Even though the hydrological behaviour of tropical peat is not clearly understood, rainfall and surface topography are believed to regulate the overall hydrological characteristics of peat land.

The peat land is purely rain-fed, thus making it an ombrogenous peat, especially in the middle section (see Chapter 2). The water table is close to or above the peat surface throughout the year and fluctuates with the intensity and frequency of rainfall. The fluctuation is caused by variations in the influx and efflux of water. The effects of precipitation on the mean annual level of the water table in the peat land depend on the peat complex and time span considered. In the long-term, the mean annual water

Table 8.1 Summary of ecological attributes and benefits derived from undisturbed peat swamp (*after* Silvius and Giesen, 1996).

Attributes	*Benefits*
Regulation of hydrology	Prevents or mitigates flooding in adjacent populated or agricultural areas; provides fresh water supply for drinking, washing, irrigation etc.; prevents saline water intrusion to agricultural lands, freshwater aquifers and up-river segments of short coastal peat land rivers
Maintenance of biodiversity	Serves as gene bank of unique, representative biota and haven for some species of other nearby ecosystems (e.g. mangrove and mixed dipterocarp forest) whose home ranges extend into the peat swamp or whose habitats have disappeared or been dramatically altered by human activity; provides wildlife, fish, plant and microbial habitats and niches for many endemic taxa
Source of renewable bioresources	Provides variety of commercially valuable timber, latex, resins, traditional culture foods, dyes, medicinal plants, fungi and microbes (many pharmacological taxa have been used effectively by indigenous cultures yet are unknown to modern medical science; other taxa of potential pharmaceutical value remain undiscovered)
Regulation of climate	Stores carbon in accumulating peat thus mitigating build-up of atmospheric CO_2 contributing to global warning; regulates local climate via forest cover evapotranspiration, heat absorption and shading, oxygen generation, wind break, etc.
Source of natural history	Stores palynological record of ecosystem's natural history via pollen and seed rain, and may have archaeological deposits of interest; based on the peat swamp ecosystem's unique attributes and its continued existence due primarily to its previous inaccessibility as one of the few remaining pristine wildernesses, it is a valuable repository of ecological materials about which little is known and from which we can learn and enjoy via research, educational and wilderness recreational activities

table level depends strongly on the water table level of the previous year. On an annual basis, runoff is the main element of the peat water balance reacting to the variation in influx. Evapotranspiration, which depends mainly on radiation, is quite stable year on year. Over short periods, such as a single rain shower, the effect of precipitation on the water table level depends mainly on the capillary and osmotic moisture content of the unsaturated peat layer and on the water table before the rain.

The runoff from the peat dome depends markedly on the slope of the dome and the hydraulic conductivity of the underlying peat. Water movement in the central part of the peat is slower due to its flatter topography than at the slope of the peat dome (see Chapter 2). Therefore, because of the slower water movement in the central part of the peat dome, the impact of precipitation on the annual water table is greater in the central part than on the slope. The mean annual water table and the effect of yearly precipitation differ over the whole peat complex. Thus, the impact of precipitation on the water level also depends on the peat complex in the short term during rainfall. The water level will start to rise sooner in the part of complex in the central part of the dome, where the distance from the peat surface to the water level is shorter in comparison with the periphery. The amount of water needed to moisten the upper peat layer during rainfall is also related to the pre-rainfall water level. In addition, the nature of the peat itself, in terms of its ability to retain and store water, is important and determines the relative height of the water table.

Drainage is mostly radial (due to the dome's convex surface) and diffuse (i.e. near-surface flow occurs rather than channel flow). Surface runoff concentrates in shallow depressions, from where small rivulets are formed that turn into larger streams and finally into navigable streams.

Peat land can be divided into several catchments areas. The boundaries of these catchments areas change over time as the elevation of the peat surface changes because of the accumulation of organic matter or subsidence caused by oxidation.

Natural peat swamps are important water catchments and control systems. They function as aquifers by absorbing and storing water during wet periods and releasing this water slowly when the rainfall is low. As a result they help reduce flood peaks and provide water in dry periods. Coastal peat lands are a buffer between marine and freshwater systems, maintaining a balance between them and preventing excessive saline intrusion into coastal lands. They also protect offshore fisheries from onshore pollution (Andriesse, 1988). Groundwater can be successfully extracted from peat aquifers and treated to conform to World Health Organization standards for potable water, making peat swamps an important area for water supply.

The physical and chemical properties of peat depend on the species composition of the peat-forming plant communities and their nutritional and hydrological requirements. In its natural state, tropical peat land (for instance) is covered with a variant of tropical evergreen forest known as peat swamp forest. Peat land vegetation is specially and variably adapted to a waterlogged, fluctuating water table, with adaptations such as stilt roots, buttresses and sclerophyllous leaves. However, compared with mixed dipterocarp forest, the number of tree species found in peat land/swamps is rather limited. Notable examples of highly commercial timber trees found in the tropical peat swamp of Malaysia and Indonesia are Ramin, Meranti and Nyatoh. Anderson (1961) divided tropical peat forest into six distinct communities, designated PC1 to PC6, as shown in Table 8.2. PC1 (mixed peat swamp forest) is found at the edges of the peat land or swamp. This is followed by PC2 (Alan forest), PC3 (Alan Bunga forest) and PC4 (Padang Alan forest). At the centre of a highly developed tropical peat swamp, PC5 and the highly stunted vegetation of PC6 (Padang Paya forest) are found. These forest associations, particularly PC3 and PC6, can be easily recognized from aerial photographs.

Water draining from the tropical peat land, as mentioned before, is highly acidic (pH 3–4.5), is low in inorganic ions and oxygen, and has high concentration of humic acids that give it a characteristic 'black water' appearance.

Over-drainage can cause serious ground subsidence problem in peat areas. Tai and Lee (2003) proposed a concept of groundwater monitoring over dry seasons to study historic low groundwater levels to facilitate efforts to minimize risks of ground subsidence due to over-drainage. Standpipe piezometers can be used for this purpose. Figure 8.1 shows an example of ground water monitoring.

From this record, a programme of ground water monitoring works can be devised. This applies to both the construction of new drains and to dewatering works during construction activities. The effect of lowering ground water levels to beyond their limiting values on ground subsidence can be determined by estimating the loss of buoyancy in the dewatered soils. In the Sibu case, for example, the lowering of 1.4

Table 8.2 Anderson's phasic communities (PC) of concentric forest associations within coastal basin peat swamps of Borneo (*after* Anderson, 1961; Lee, 1991).

Forest association	*Area (ha)*	*Descriptive characteristics*
PC1. *Gonsytylus–Dactylocladus-Neoscrotechinia* (Mixed swamp forest)	1,174,000	Found at peat swamp margins, with structure and physiognomy similar to lowland dipterocarp evergreen rain forest on mineral soil; uneven canopy with dominants attaining 40–45 m; 120–150 tree species ha^{-1}; epiphytes and climbers abundant; *Sharea albida* absent
PC2. *Shorea albida–Gonystylu–Stenonurus* (Alan Batu forest)	127,000	Similar to PC1 but dominated by scattered, large (>3.5 m girth) trees of *Shorea albida* (Alan); *Stemonurus umbellatus* indicator species
PC3. *Shorea albida* (Alan Bunga forest)	76,000	Even upper canopy at 50–60 m in height, dominated by *Shorea albida* with 70–100 trees ha^{-1} of this species alone; middle storey commonly absent; moderately dense understorey; *Pandanus andersonii* forms dense thickets in shrub layer; herbs, climbers and epiphytes rare
PC4. *Shorea albida–Litsea–Parastemon* (Padang Alan forest)	41,000	Dense, even canopy at 30–40 m composed of relatively small-sized (<1.8 m DBH) trees which give the forest a pole-like and xeropytic appearance; *Sharea albida* density of >400 trees ha^{-1}; small, prostrate shrubs (*Euthemis minor and Ficus deltoidea var. motleyana) indicator species; herb, fens,* epiphytes and climbers rare or absent
PC5. *Tristania–Palaquium–Parastemon*	NA	Typically found in central domes of peat swamps; narrow transitional forest between PC4 and PC6; dense, even, closed canopy at 15–20 m with high density (850–1,250) stems ha^{-1}) of small-sized tress; herb layer rare or absent
PC6. *Combretacorpus–Dactylocladus* (Padang Paya forest)	37,000	Found in central domes of peat swamps; resembles open savannah woodland with stunted, xeromorphic trees; patchy shrub layer present; pitcher plants *(Nepenthes spp.)* and epiphytic and-plants *(Myrmecodia tuberosa* and *Leconopteris sinuosa)* indicator species

metres of ground water level (GWL) below the limiting value is equivalent, in terms of settlements caused, to the surcharging effect of about 0.75 metres of conventional ground filling.

8.3 PHYSICO-CHEMICAL PROPERTIES OF PEAT

The content of peat may differ from location to location due to factors such as the origin of the fibres, temperature and degree of humification. Decomposition or humification involves the loss of organic matter either in gas or in solution, the disappearance of physical structure and the change in chemical state. As stated earlier, breakdown of

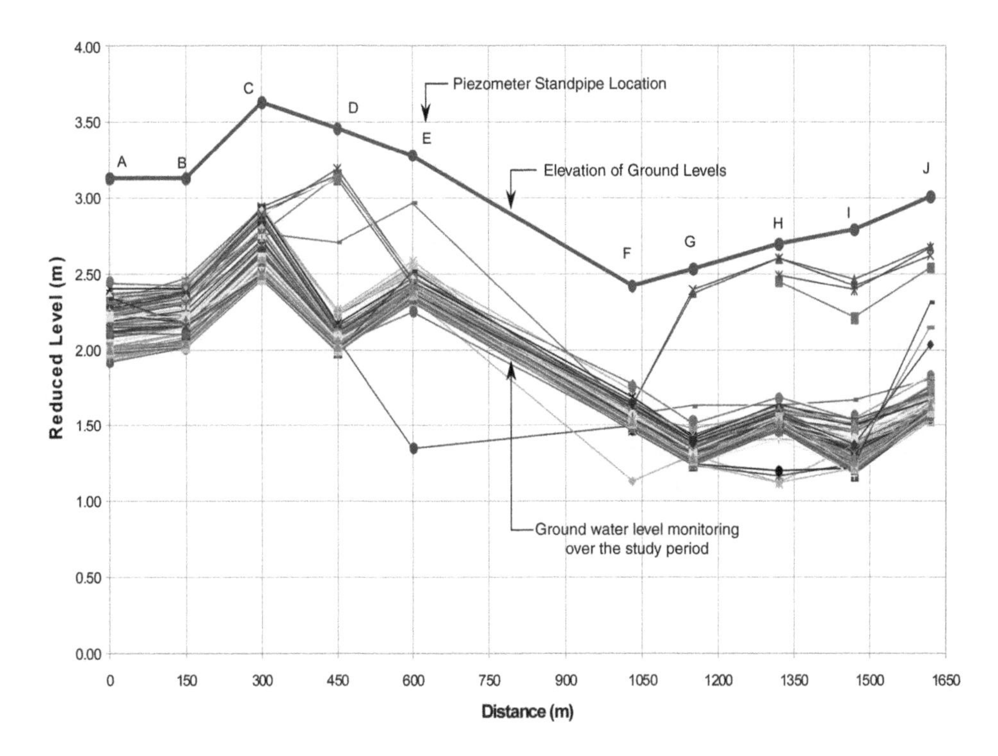

Figure 8.1 Groundwater monitoring of Sibu Town (10 July 1997 to 12 August 1997) (*after* Tai and Lee, 2003).

the plant remains is brought about by soil microflora, bacteria and fungi, which are responsible for aerobic decay. Therefore the end products of humification are carbon dioxide and water, the process being essentially one of biochemical oxidation. Immersion in water reduces the oxygen supply, enormously which in turn reduces aerobic microbial activity and encourages anaerobic decay, which is much less rapid. Consequently, it has been said that peat shows unique geotechnical properties in comparison with those of inorganic soils, such as clay, and sandy soils, which are made up of soil particles only (Hashim and Islam, 2008).

In agricultural use, fens and raised bogs need to be drained to adjust the water and air in the soil to provide suitable conditions for cultivated or pasture plants. The removal of water from the upper peat by drainage and subsequently by oxidation leads to compaction and hence subsidence of the surface. The release of CO_2 and N_2O is increased due to the drainage of peat, but release of CH_4 decreases. The rates of release depend on factors such as the level of groundwater and the temperature of the peat, but gases like CH_4 and N_2O are also released. During the process of peat extraction the greenhouse gas (GHG) sink function of the peat land is lost.

Emissions of these gases also occur while preparing the surface for the removal of vegetation, during ditching, during the extraction, storage and transportation of peat, and also due to combustion and after-treatment of the cutaway area. Combustion

Table 8.3 Physical properties of peat based on location (*after* Huat, 2004).

Soil deposits	*Natural water content ω_o (%)*	*Unit weight, γ (kN m^{-3})*	*Specific gravity G_s*	*Organic content (%)*
Fibrous peat, Quebec	370–450	8.7–10.4	–	–
Fibrous peat Antoniny, Poland	310–450	10.5–11.1	–	65–85
Fibrous peat, Co. Offaly Ireland	865–1400	10.2–11.3	–	98–99
Amorphous peat, Cork, Ireland	450	10.2	–	80
Cranberry bog peat, Massachusetts	759–946	10.1–10.4	–	60–77
Austria	200–800	9.8–13.0	–	–
Japan	334–1320	–	–	20–98
Italy	200–300	10.2–14.3	–	70–80
USA	178–600	–	–	–
Canada	223–1040	–	–	17–80
Hokkaido	115–1150	9.5–11.2	–	20–98
West Malaysia	200–700	8.3–11.5	1.38–1.70	65–97
East Malaysia	200–2207	8.0–12.0	–	76–98
Central Kalimantan	467–1224	8.0–14.0	1.50–1.77	41–99

accounts for more than 90% of the greenhouse gas emissions. Sometimes the plant fibres are visible, but in the advanced stages of decomposition they may not be evident. Peat will ultimately be converted into lignite coal over geologic periods of time provided favourable conditions prevail. Also, the fresher the peat, the more fibrous material it contains, and (as far as engineering is concerned), the more fibrous the peat, the higher are the shear strength, void ratio and water content. In fact, the properties of peat are greatly dependent on the detailed formation of peat deposits. This means that peat at different locations will have different properties (Table 8.3). Commonly, the classification of peat is based on fibre content, organic content and ash content.

The degree of decomposition varies throughout peat, since some plants (or parts of them) are more resistant than others. Also, the degree of decomposition of peat depends on a combination of conditions, such as the chemistry of the water supply, the temperature of the region, aeration and the biochemical stability of the peat-forming plant (Lishtvan *et al.*, 1985).

Much of the information about the physical properties of peat has been included in Chapters 2–5. Tropical peat is generally non-homogeneous compared with temperate (fen) peat. As mentioned earlier, the physical properties of peat are the result of many factors, including wood content, degree of decomposition or humification, bulk density, porosity, water holding properties and hydrology. Tropical peat, for instance, is generally reddish brown to very dark brown in colour depending on the stage of decomposition.

Peat has a high moisture content and water holding capacity. The high water content results in high buoyancy and high pore volume, leading to low bulk density and low bearing capacity of the peat. The bulk density of the peat varies according to its degree of decomposition. Highly decomposed peat generally has a higher bulk density. The degree of decomposition and the bulk density are intrinsically related. The high water content results in high buoyancy and high pore volume, leading to low bulk density and low bearing capacity. Excessive drainage of the peat will cause a transformation of its colloids, resulting in irreversible drying.

In tropical peat there is a vacant zone, which is more common and extensive under the root mat of the Alan forest compared with that found in mixed peat swamp forest. The vacant zone is found within the top 0.5 m of the surface horizon below the root mat. It makes the surrounding of the forest floor dry. In fact, the surface horizon of the peat is actually floating on the inter-winding root mats. The thickness of the vacant zone ranges from 0.25–0.40 m. The surface of tropical peat is often very raw and fibrous. The very fibric and porous nature of the peat results in its low bulk density, thus decreasing the growth media and availability of nutrients.

Peats are very dynamic. This is because they undergo subsidence and oxidation upon drainage. Initially, it involves principally the loss of buoyancy and compaction of the organic column under its own weight. Compaction results in the changes in the hydropedological parameters like the hydraulic conductivity, bulk density, pore volume and moisture content. The subsequent dominant process, which may last for decades, is oxidation and shrinkage. The rate of subsidence varies strongly depending on the peat's profile morphology, composition and depth; the depth of drainage; and land use. Due to the differential rate of subsidence upon drainage, the micro relief of peat surfaces is hummocky (Ambak and Melling, 1999).

Surface topography and peat thickness, however, do not influence the vegetation directly. They operate through changes that are brought about in other characteristics of the peat land, especially hydrology, chemistry and organic matter dynamics (the balance between peat accumulation and peat degradation).

The peat thickness, the nature of the subsoil, the humification level and the peat surface altitudes and gradients influence the chemical composition of the peat. The chemical properties of a typical tropical peat are shown in Table 8.4.

Tropical peat, like temperate bog peat, is very acidic in nature, with a pH of 3 to 4. Aluminium (Al) toxicity, however, is not a problem for deep peat, which has a high CEC. The value is very misleading because it is due to the dissociated carboxyl groups, which determine the high acidity. The amount of exchangeable bases is actually very small. Thus the base saturation is very low and strongly buffered.

The high CEC is not due to the presence of Na, K, Mg or Ca but because of the amount of exchangeable H^+. The N content of peat is rather high, but its availability for plant uptake is rather low. The high C:N ratio, coupled with the low pH, results in low materialization in peat. The ash content can be less than 10%, showing a very high organic matter content. This is indicated by a loss of ignition value exceeding 90% (Ambak *et al.*, 1991; Ambak and Tadano, 1991). Peat is highly deficient in micronutrients such as Cu and B. From the viewpoint of nutrient dynamics, the potential use of reclaimed peat land was rather limited, especially under low input management (Yonebayashi, 2003).

Table 8.4 Chemical properties of (Sarawak) tropical peat (*after* Melling and Hatano, 2003).

Property	*Top soil (0–25 cm)*
pH (H_2O)	3.7
C:N	1:40
Ash content (%)	1.16
Exch. Ca (me/100 g)	6.91
Exch. Mg (me/100 g)	6.17
Exch. K (me/100 g)	0.84
CEC (me/100 g)	154.1
Available P (ppm)	185.73
Available Cu (ppm)	13.27
Available Zn (ppm)	48.36
Available B (ppm)	4.5

Table 8.5 Physico-chemical properties of natural pore water of peats (*after* Asadi, 2010).

Parameter	*H3 peat water*	*H5 peat water*	*H7 peat water*
Water temperature (°C)	31	31	31
pH	5.3	5.8	6.1
Sodium (mg l^{-1})	0.76	0.85	1.2
Calcium (mg l^{-1})	0.98	1.4	1.6
Magnesium (mg l^{-1})	0.3	0.3	0.23
Potassium (mg l^{-1})	0.1	0.3	0.2

8.4 PHYSICO-CHEMICAL PROPERTIES OF PEAT PORE FLUID

Most laboratory tests follow standards and are performed at room temperature with distilled water as the pore fluid; however, in the *in situ* condition, local environmental conditions can influence results significantly (Asadi et al. 2011d, 2011g) . Asadi (2010) characterized peat pore fluids. The fluid can be extracted from disturbed peat using a vacuum pump and filter paper. A motorized extraction machine can be used to extract the peat water from the surface of peat (Figure 8.2). The concentrations of cations can be measured using an atomic absorption spectrometer (Figure 8.3).

The physico-chemical characteristics of the peat pore fluids are given in Table 8.5. The pH ranges between 5.3 and 6.1 (Asadi, 2010).

Viscosity is a measure of the resistance of a fluid which is being deformed by either shear stress or extensional stress. It is a property of a fluid that offers resistance to flow. The viscosity of the peat water is measured with a viscometer (Figure 8.4).

The viscosity of peat water is about 2.2 cP at 24.6°C (Figure 8.5). The results show that the viscosity of peat water is low and there are no pronounced differences compared with water (Asadi, 2010).

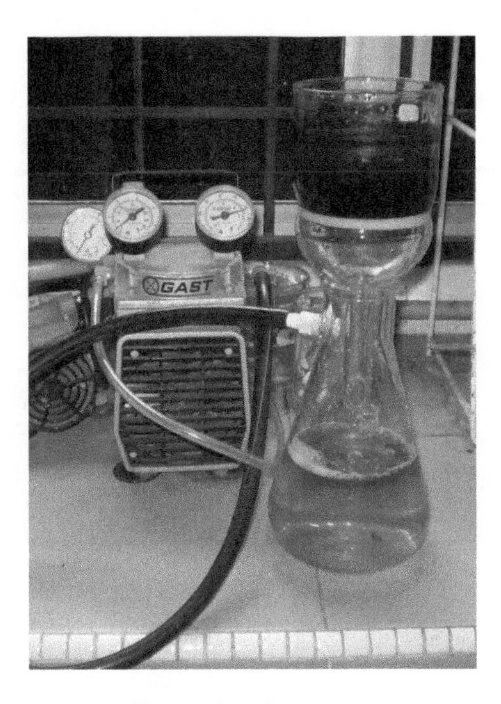

Figure 8.2 Motorized extraction for preparation of peat pore fluid.

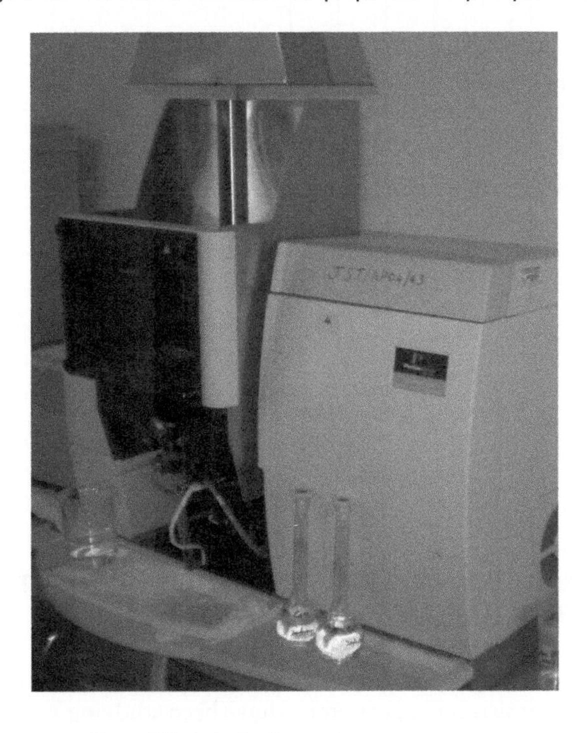

Figure 8.3 Atomic absorption spectrometer.

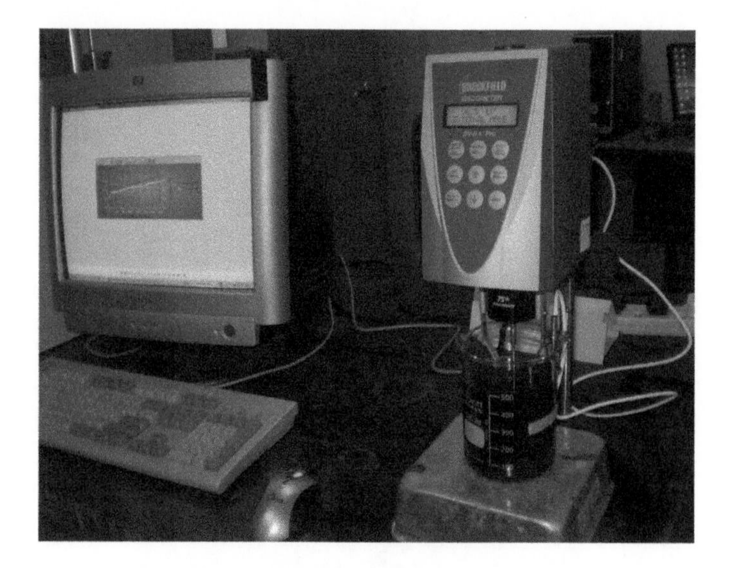

Figure 8.4 Viscometer apparatus.

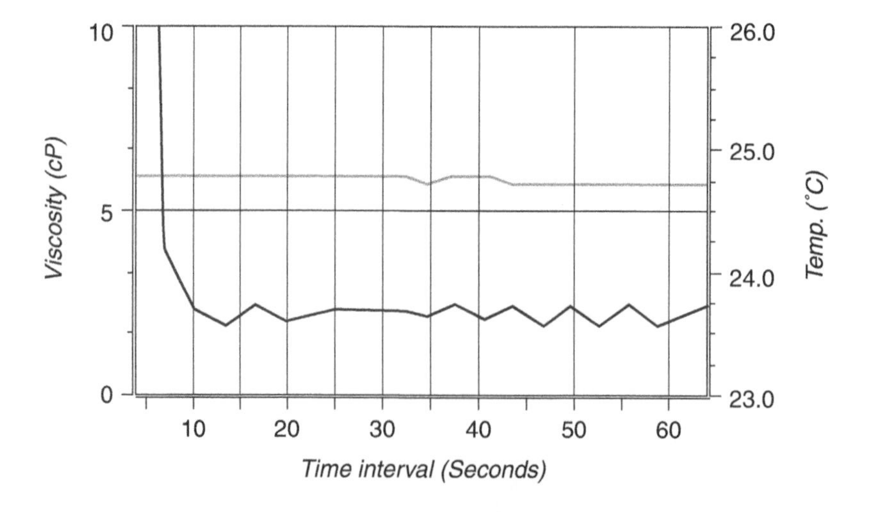

Figure 8.5 Viscosity of peat water.

8.5 COMMON GROUND BETWEEN SOIL SCIENTISTS AND GEOTECHNICAL ENGINEERS

Engineers have been studying the properties of soil and its interaction with structures for as long as soil scientists and agronomists have been studying soil chemistry, biology and the interaction between soil, groundwater, plants and the receiving environment.

There is a lot of common ground with what scientists and engineers need to know about soils and how to use the knowledge.

Peat land is gaining importance as potential land for agriculture due to its occurrence near coastal lowlands, which are usually developed and populated areas. In recent years, extensive utilization of peat swamp areas has started due to the increasing pressure on upland areas. However, from an agriculturalist's point of view, peat soils have generally been recognized as a problem soil, with marginal agricultural capability. Thus with it come a multitude of problems associated with such development.

Drainage is a prerequisite for any agriculture development on peat. Essentially, it removes the excess surface and subsurface water. However, drainage activities are the greatest threat to the ecology of a tropical peat. By lowering the water table and accelerating the rate of decomposition of the peat, drainage results in irreversible changes in the hydrology, morphology and ecology of the peat land. The effect of long-term drainage on peat soil morphology has been described in Chapter 6, whereby drainage resulting in subsidence of 0.8 m over 20 years resulted in the development of a sapric-hemic-fibric profile.

As a result of drainage, the water table in the peat is lowered. The buoyancy of the peat us reduced, which increases the overburden pressure, causing more consolidation and subsidence. The subsidence compression again brings the surface level with the water level.

Drainage and cultivation of peat soil increases soil aeration and reverses the carbon flux into a net carbon dioxide (CO_2) emission into the atmosphere. Cultivated peat soils then become a large source of both carbon dioxide (CO_2) and nitrous oxide (N_2O) emission. Thus agriculture development can contribute significantly to the increase in atmospheric N_2O (Melling and Hatano, 2003).

The original ecosystem will change radically and permanently because uncontrolled drainage bleeds the peat, the very medium that is the basis of its existence. There will be more rapid runoff, higher peak flows and widespread incidence of flooding.

Over-drainage of peat land causes a chain of problems (Melling *et al.*, 1999). These include subsidence, flooding, reduced water holding capacity, increased occurrence of acid-sulphate soils, forest fires, pest infestation and termite attack, nutrient imbalance and moisture stress to the crop (Andriesse, 1991). Excessive drainage of the peat will also cause a transformation of its colloids, resulting in it having the properties of irreversible drying.

The topo-hydrological characteristics of the peat also make water table control in tropical peat development very complex. The complexity is even when the conversion of these lands for agriculture is made in small individual pockets of about 5,000 hectares each within a peat basin and when there are crops with conflicting water table requirements. The heterogeneous nature of tropical peat makes it difficult to do water management. Designing an efficient water management system in peat land is made more complex by the following factors:

- Rainfall is not well distributed throughout the year
- Spatial variation of rainfall
- High variation in hydraulic conductivity or permeability
- High fluctuation of the water table

Land surface subsidence is undoubtedly the next greatest problem (after drainage) when peat is drained for agriculture. As a result of drainage, the continuous and dynamic process of peat subsidence is triggered. Draining the peat initially reduces its buoyancy and leads to compaction of the organic column under its weight. This then increases the pressure exerted by the upper peat layers (whose specific gravity has been increased by drainage) on the peat layers beneath. As a result, all the peat layers subside and it is not only the surface (drained) layers that are affected. Subsidence is greatest in the uppermost peat layers because they are the most aerated and exposed to the effects of land use (shrinkage, materialization and the mechanical pressure of machines).

Peat subsidence is a physical process that depends on factors such as peat depth, peat type, degree of decomposition, mineral matter content, porosity and moisture content, and especially on drainage depth and density. Thus subsidence is a combination of shrinkage, compaction, biochemical oxidation and burning of peat materials. Waterlogged and anaerobic peat becomes aerobic upon drainage. This leads to changes in hydropedological parameters such as hydraulic conductivity, bulk density, pore volume and moisture content. It needs to be realized that a change in ground surface level of a few centimetres, where the natural variation in ground level is less than one metre, can have dramatic hydrological effects.

In the case of tropical peat, subsidence of 0.60 m has been recorded for a drained deep peat (water level of 0.75–1.0 m) (Melling and Hatano, 2003). Subsidence of 1.0–1.5 m can be expected within the first three years of an agricultural development (Melling, 2000). This is especially so in the Alan forest areas because of the widespread existence of a vacant layer (Melling, 2000) and the high porosity of the peat. The heterogeneous nature of peat also causes differential rates of subsidence upon drainage making the peat surface quite hummocky. The micro-relief variation can be as much as 0.5 m. This also complicates the estimation of peat subsidence rate. The height loss of the peat resulting from agricultural development will lead to a complete transformation of the peat land landscape.

Peat subsidence has several serious consequences. Drainage needs to be regularly adapted to new levels and conditions, otherwise inundation and flooding will recur. The rooting systems, particularly of perennial species, become exposed, and top-heavy crops such as oil palms start to lean over and are partly uprooted. Roads and other structures become unstable. Annual subsidence rates of cultivated peat are a function of the degree of aeration and temperature (McAfee, 1989). Subsidence rates are greatest in areas nearest the main drains and decrease with distance from the drain.

Oxidation refers to microbial decomposition of the soil material, ultimately converting it to carbon dioxide and water. The entry of air into the soil causes organic matter to be oxidized at a more or less constant rate, through biochemical processes under the influence of microorganisms. A microorganism that oxidizes the peat and cause the most serious form of subsidence requires oxygen from the soil air. However, the soil profile below the water table contains essentially no air or dissolved oxygen, so the microorganisms are unable to function, and the soil is protected from this type of subsidence.

Peat subsidence due to consolidation and oxidation leads to an increase in peat bulk density. Consolidation does not involve a loss of soil material compared to oxidation. The bulk density is closely associated with the hydrological properties of the peat soils; thus subsidence may further modify soil moisture conditions beyond the effects of lower water table levels due to drainage alone. The total porosity is not sharply reduced with increasing bulk density. However, bulk density is very closely related to peat pore size distribution (which regulates soil water retention), and saturated and unsaturated water transport dynamics.

Greater soil water contents are expected after drainage, which are actually associated with greater peat bulk density. Peat bulk density, water retention and capillary moisture transport from the water table, and saturated hydraulic conductivity are thus related to the degree of peat decomposition.

The woody nature of tropical peat, for instance, is a big hindrance to land preparation. The very low bulk density and bearing capacity presents many problems in the use of heavy machinery. In an undrained condition, the machine tends to sink in the peat due to the low bearing capacity of peat soil. To support mechanization, the water table has to be sufficiently lowered to attain the required bearing capacity. Machinery has to be modified by having lighter weight and wider tracks to reduce ground pressure so that it will not sink into the peat. Construction of farm roads on peat will require fill materials to improve the bearing capacity of the porous peat soil to withstand the load of a vehicle. A 3.5 ton vehicle creates a pressure of 8.75 ton m^{-2} or 88.85 kN m^{-2}. The water table has to be lowered by about 1.8 m at mid-field in order to achieve this bearing capacity on peat (Salmah *et al.*, 1991)

It can be seen that there is a conflict of function between the need to drain for traffic access and the moisture supply required for crop growth. The mechanization problem becomes particularly critical during harvesting. The regular harvesting of oil palm, particularly if heavy machinery is used, will then lead to differences in the height of planting rows and harvesting rows. This will affect the management of the required water table and will also make heavy demands on the surface structure and consistency of the soil. The rate of surface compaction is also accentuated. Mechanization is also more difficult during the high rainfall period because the ground is soggier.

The high acidity and low nutrient levels in tropical peat demand the use of lime and fertilizer. Liming and fertilization will encourage microbial activity and rate of decomposition by increasing the pH and decreasing the C:N ratio. Therefore the rate of subsidence will increase. Newly developed tropical peats, such as those in Sarawak, Malaysia, are still very porous. This, together with the extremely high rainfall and fluctuating water table, gives rise to excessive leaching of applied fertilizers in deep peat. Peat soil, being an organic material, will continuously mineralize and thus cause a dynamic change in its nutrient content, especially the available nitrogen. This will tend to cause an imbalance between nitrogen and potassium.

All of these problems will have an adverse effect on the sustainable utilization of the peat for agricultural development, for which a balance solution has yet to be found. Tie (1990) suggested that agriculture development, where necessary, should be confined to the fringes of the peat dome, using PC1 and PC2. PC3 may be considered for silviculture with proper management, while PC4 and PC5 are best kept under natural conditions.

Figure 8.6 illustrates the sequence of impact to peat land due to logging. Even when drainage is not actually involved, there is still an associated loss in biodiversity.

Stage 1 – Pristine mixed peat swamp forest with many tree species

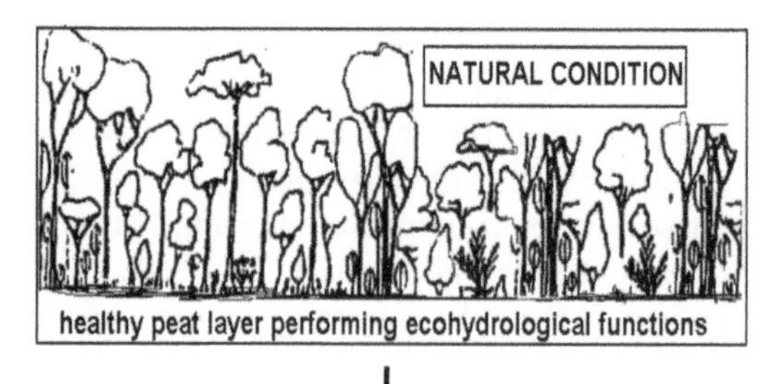

Stage 2 – Merchantable timber harvested from previously undisturbed mixed peat swamp

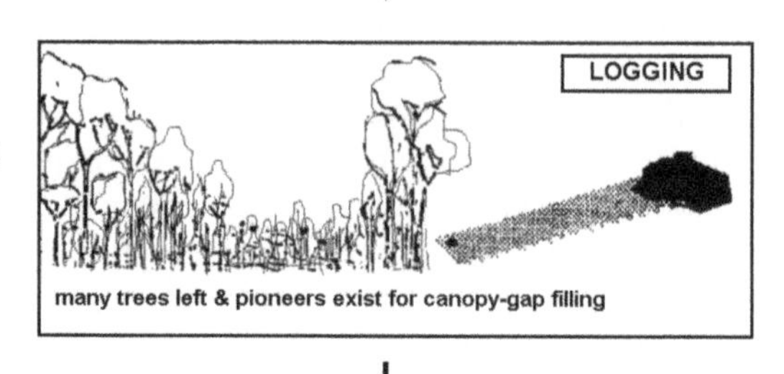

Stage 3 – Controlled burning removes most trees; site is drained and cultivated for planting to agricultural crop; reclamation activities preclude secondary ecological secession in the logged-over forest

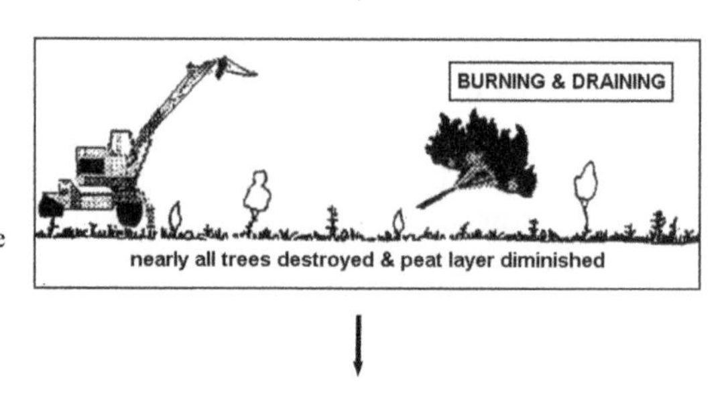

Stage 4 – Short-lived reclamation project abandoned after site conditions deteriorate; crop yields and profits decline

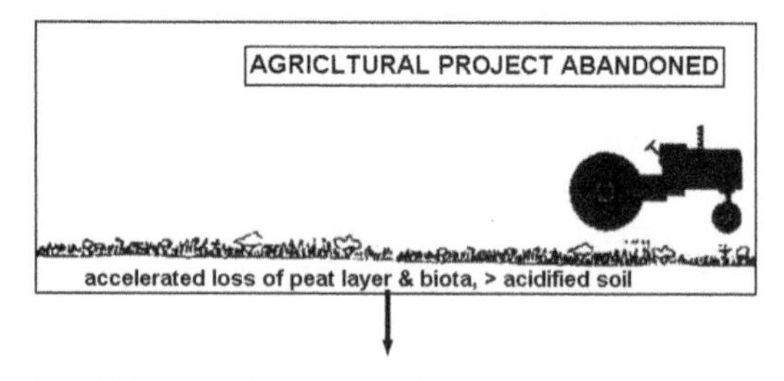

Stage 5 – Very different replacement community occupies permanently altered site; subsidence, change hydrology and acid-sulfate soil prevent original biotic community from re-establishing

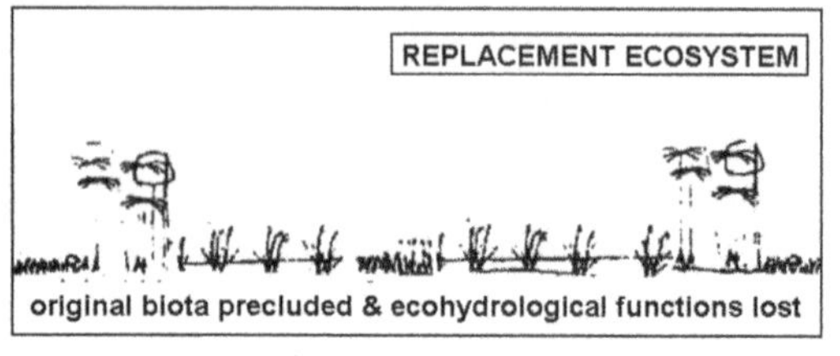

Figure 8.6 Impact degradation sequence of a typical peat swamp (*after* Giesen, 1990; Riely and Ahmad Shah, 1996).

8.6 CHEMICAL AND BIOLOGICAL CHANGES

An important characteristic of peat is its potential chemical and biological changes with time. Further humification of the organic constituents would alter the mechanical properties, such as compressibility, strength and hydraulic conductivity. Lowering of ground water may cause shrinking and oxidation of peat, leading to humification and a consequent increase in permeability and compressibility. Oxidation also leads to gas formation which may contribute to excess pore pressures (Vonk, 1993). The significance of these effects on the long-term performance of structures placed on peat is often not taken rigorously into account. However, in the absence of any significance change in the submergence of these deposits, the long-term chemical and biological degradation of the peats may not be significant.

8.7 EFFECT OF PEAT MEDIA ON STABILIZATION PROCEDURE

Other important characteristics of peats are the presence of carbon dioxide (CO_2) and nitrogen (N), and an acidic environment with a high ground water level. The decomposition of peat is a complicated phenomenon, and in this process the leaf litter (providing the main input of organic matter to the soil) is physically broken down by the larger soil fauna, including earthworms and termites. Microorganisms (bacteria and some fungi) can start their decomposition activities while the leaf is still on the plant. These microbial processes diversify and intensify once the leaf reaches the soil. CO_2 is produced by aerobic and anaerobic decomposition above and below the water table, which causes loss of organic matter and an altered physical structure and chemical state. Glenn *et al.* (1993) have reported the CO_2 production rates of aerobic laboratory incubations to be 0.2–1.4 mg CO_2 $g^{-1}d^{-1}$(0.14–0.98 ppm/min), an average of five times more than the rate under anaerobic conditions. The N content of peat that develops from reeds, sedges and trees is rather high, being an average of three times that developing from Sphagnum mosses and Eriophorum sedges; it is around 0.3–5% for oven dry peat. The range of acidity levels in peat is very wide. The pH of most peats ranges between 2 and 6, but in some conditions, where there is infiltration of brackish water or the peat contains pyritic materials, the pH may be as high as 7.8 or less than 2, respectively (Andriesse, 1988).

Kazemian *et al.* (2011a) described the influence of the characteristics of peat (CO_2, N, and acidic or alkaline media) on cementation and pozzolanic reactions when treating tropical fibrous, hemic and sapric peats with cement and slag.

8.7.1 Effect of CO_2 on treated peat

The effect of CO_2 on cementation and pozzolanic reactions was investigated by producing water containing dissolved CO_2 and curing samples in it for 45, 90 and 180 days. A simple CO_2 injection system was fabricated to produce the water containing dissolved CO_2. It consisted of two parts: a generator to hold a yeast mixture (fine sugar, water and yeast) and a reactor to ensure that the CO_2 was efficiently dissolved in water in the sample container (Figure 8.7). The concentration of CO_2 produced was measured using a CO_2 test kit, and the targeted rate of 0.2 ppm min^{-1} (the same as

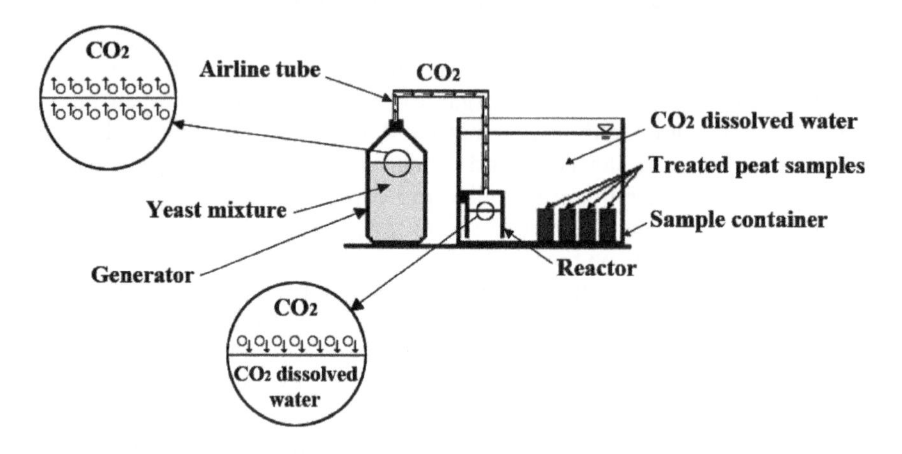

Figure 8.7 Schematic diagram of CO_2 injecting system (*after* Kazemian *et al.*, 2011a).

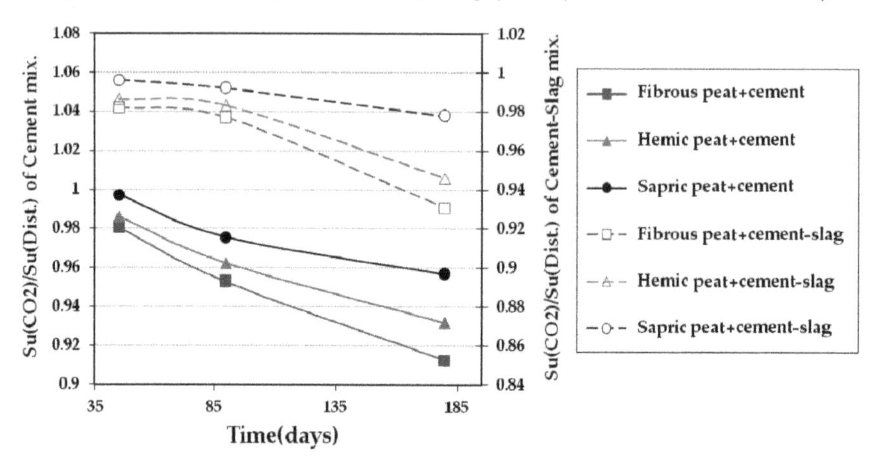

Figure 8.8 Undrained shear strength ratio versus time of peat treated with cement and cement–slag mixtures and cured in water containing dissolved CO_2 (*after* Kazemian *et al.*, 2011a).

that available in natural peat) was accomplished by varying the ratio of components in the generator (yeast, sugar and water).

The depth of carbonation was determined using a solution of 1% phenolphthalein in 70% ethyl alcohol as an indicator. If free calcium hydroxide ($Ca(OH)_2$) is present, it is coloured red or pink, whereas it remains uncoloured along the depth of carbonation.

The effect of CO_2 on cementation and pozzolanic reactions is shown in Figure 8.8. This shows the shear strength ratio of samples (cured in water containing dissolved CO_2 [$Su_{(CO2)}$] to cured in distilled water [$Su_{(Dist.)}$]) versus time, for fibrous, hemic and sapric peats treated with cement or the cement–slag mixture.

A ratio of one indicates that there is no effect on the cementing process; greater than one indicates an improvement and less than one indicates a negative effect. The UCS results indicate that the shear strength of the samples cured in CO_2 dissolved water was less than that for the samples cured in distilled water.

The results showed that the shear strength of sapric peat with cement decreased at 90 and 180 days of curing respectively, compared with the shear strength at 45 days of curing. Similarly, the shear strength of fibrous, hemic and sapric peats with the cement–slag mixture decreased at 90 and 180 days of curing, compared with the shear strength at 45 days of curing.

The reason for such behaviour is that when water containing dissolved CO_2 comes into contact with treated peats, it reacts with $Ca(OH)_2$ (produced from cement hydration) to form calcite ($CaCO_3$). Carbonation has three main effects: (1) a reduction in $Ca(OH)_2$, and a subsequent reduction in pozzolanic reactions between $Ca(OH)_2$ and pozzolanic minerals in the soil, (2) the decomposition and carbonation of calcium silicate hydrate (C–S–H) and (3) a reduction in the pH of pore water in the hardened mixture. The peat in the present study contained few mineral solids, and hence the main reason for the decrease in the shear strength of all samples treated with cement or cement–slag mixtures when they came into contact with CO_2 was the decomposition of the C–S–H gel to form $CaCO_3$ and the reduced pH of the pore water.

It was shown that peats treated with cement–slag mixtures have greater resistance to carbonation with a soft surface of carbonation in comparison with peats treated with cement only (Kazemian *et al.*, 2011a). It was observed that the depth of carbonation increased with increasing fibre content and curing time. Furthermore, the decrease in shear strength was greater in fibrous peat than in hemic and sapric peats (Kazemian *et al.*, 2011a).

8.7.2 Effect of N on treated peat

The effect of N on cementation was observed by adding urea (2.5% and 5% by weight) to fibrous, hemic and sapric peat. The peat was first thoroughly homogenized; urea was then added and mixed again. Finally, the binders were added and mixed for five minutes to ensure homogenization. The effect of N (2.5%) is shown in Figure 8.9.

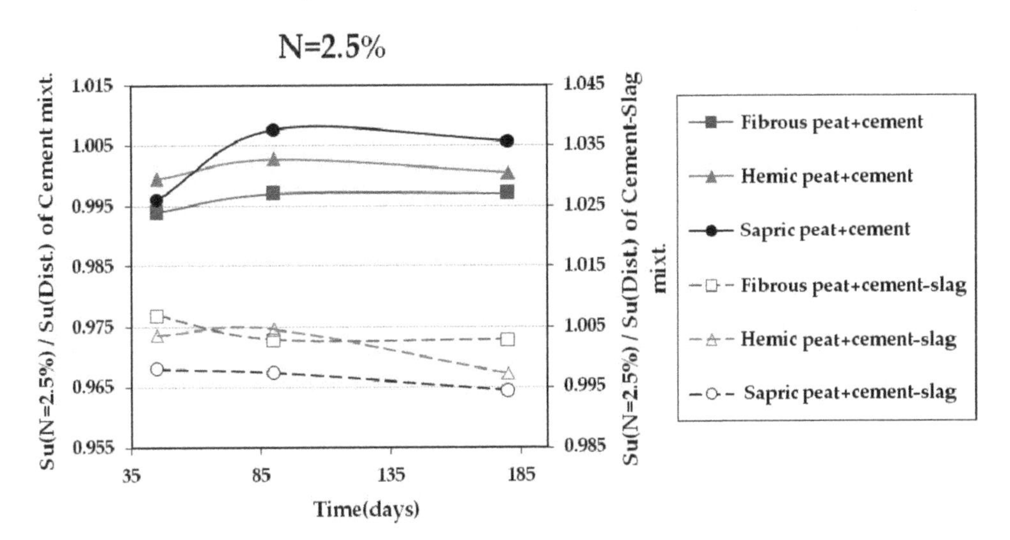

Figure 8.9 Undrained shear strength ratio vs. time of peat with cement and cement–slag mixtures (N = 2.5%).

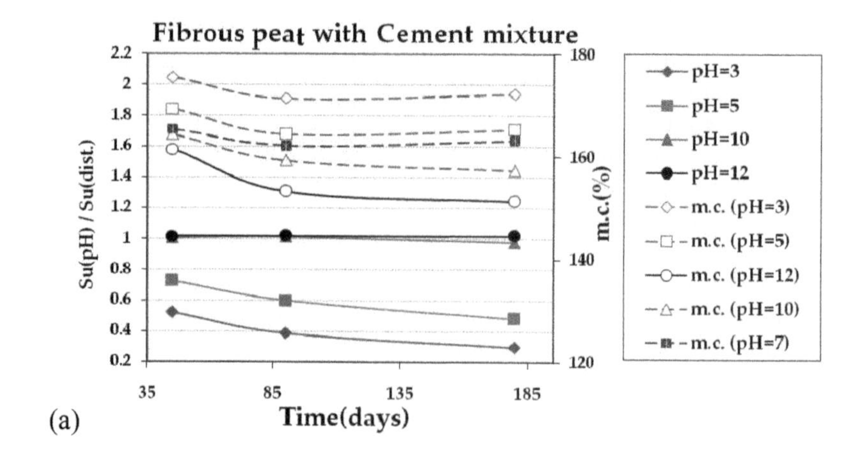

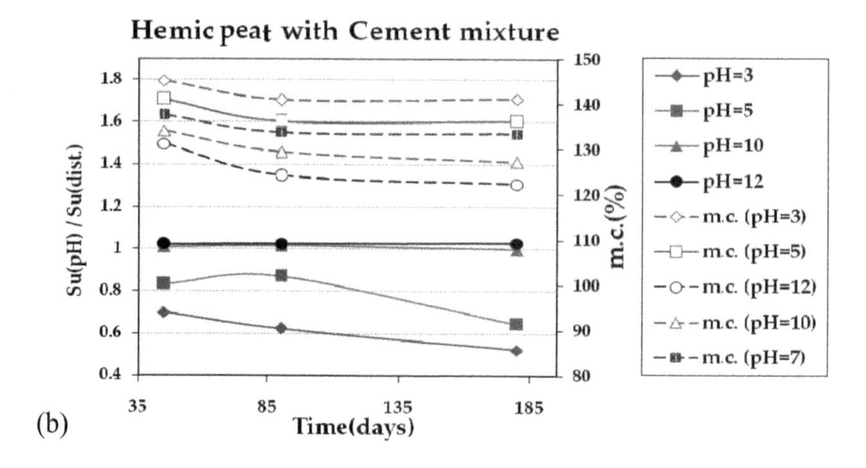

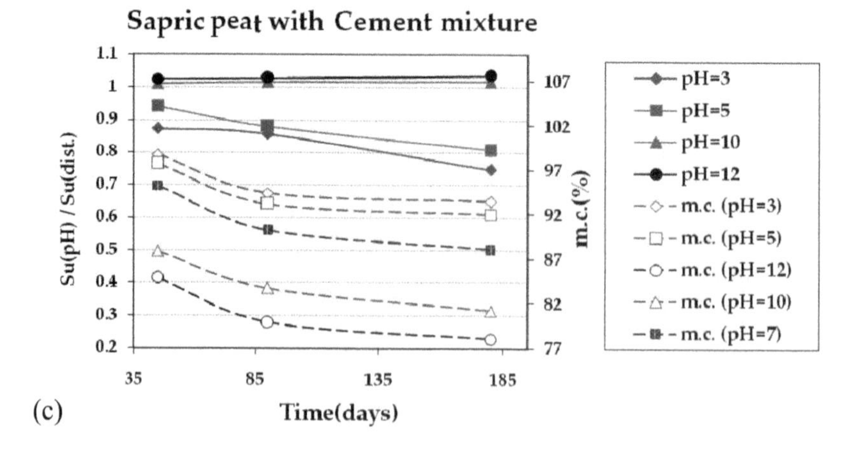

Figure 8.10 Undrained shear strength ratio vs. time and moisture content with different pH of (a) fibrous, (b) hemic and (c) sapric peat treated with cement (*after* Kazemian *et al.*, 2011a).

The ratios of shear strength for all peats were close to one. This indicates that there was practically no change in the shear strength of peat with or without N. This implies that N has little effect on the cementation and pozzolanic reactions. Furthermore, the dissolution and consumption of calcium, sulphate and potassium ions in N mixtures were similar to the control mixture.

8.7.3 Effect of pH on treated peat

The impact of pH on cementation and pozzolanic reactions was investigated by varying the pH of the media. This was carried out by preparing media with different pH values (3, 5, 7, 10 and 12) by mixing the appropriate amount of hydrogen chloride (HCl) and sodium hydroxide (NaOH) in distilled water, and measuring the pH with an electronic pH meter.

It was observed that at low pH (3 and 5), the shear strength ratio of treated peats (fibrous, hemic and sapric) decreased substantially with curing time, and to a greater extent in cement than in cement–slag mixtures (Kazemian *et al.*, 2011s). The decrease in shear strength ratio occurred for up to 90 days of curing for fibrous peat, but showed a gradual decrease for hemic and sapric peats. In contrast, at higher pH (10 and 12), the shear strength ratio showed a nominal increase with curing time for up to 180 days and the increase was greater for peat treated with cement (Figure 8.10).

The shear strength of the treated peats increased in alkaline media (pH 10 and 12) because the OH^- ions liberated during the hydration process were easily utilized when producing C–S–H gel during the curing process. The negative effect of low pH (3, 5) on fibrous peat was tangible and was more obvious than on hemic and sapric peats with both cement and cement–slag mixtures.

8.8 CONTINUING RESEARCH IN PEAT LAND DEVELOPMENT

From the perspective of agriculture, one of the major concerns is that plantation activities are carried out on relatively unknown terrain. The peat type in the development areas will not have been properly characterized and mapped. Proper economic evaluation cannot therefore be carried out when data or information on the peat characteristics, hydrology and agriculture potential are lacking or insufficient. It also results in a lack of direction in the planning and management of the peat development, which becomes disorganized. This points to a need to have proper planning. There is also a need to increase knowledge about the ecology of peat land among government planners, policy makers and developers when tackling the complex problem of peat land development. In particular, they need to know about the inter-relationships between peats, the vegetation it supports, hydrology, nutrient input and cycling, and decomposition processes.

From the engineering perspective, new construction methods, which are cheaper and more efficient, and more refined analytical techniques and theories still need to be found. Continued effort in research and development is vital for the sustainable development of peat land.

References

Abdullah, A., John, A. N. and Arulrajah, A. (2003) Augeo pile system used as piled embankment foundation in soft soil environment. In: *Proc. 2nd International Conference on Advances in Soft Soil Engineering and Technology* (eds. B. Huat *et al.*), Putrajaya, Malaysia. pp. 703–14.

Aboshi, H., Ichimoto, E., Harada, K., and Emoki, M. (1979) The composer-A method to improve the characteristics of soft clays by inclusion of large diameter sand columns. In: *Proc. Int. Conf. on Soil Reinforcement*, E.N.P.C., 1, Paris, pp. 211–16.

Ab. Rahim, M. A. (2003) Settlement monitoring and prediction with Asaoka method. In: *Proc. 2nd International Conference on Advances in Soft Soil Engineering and Technology* (eds. B. Huat *et al.*), Putrajaya, Malaysia. pp. 431–8.

Abuel-Naga, H. and Bouazza, A. (2009) Equivalent diameter of a prefabricated vertical drain. *Geotextiles and Geomembranes*, **27**, 227–31.

Acar, Y. B. and Alshawabkeh, A. N. (1993) Principles of electrokinetic remediation. *Environmental Science & Technology*, **27**(13), 2638–47.

Acar, Y. B., Gale, R. J., Alshawabkeh, A. N., Marks, R. E., Puppala, S., Bricka, M. and Parker, R. (1995) Electrokinetic remediation: basics and technology status. *Journal of Hazardous Materials*, **40**, 117–37.

Acar, Y. B., Gale, R. J. Putnam, G., Harned, J. and Wong. R. (1990) Electrochemical processing of soils: theory of pH gradients development by diffusion and linear convection. *Journal of Environmental Science and Health, Part (a)*, **25**(6), 687–712.

Achal, V. and Pan, X. (2011) Characterization of urease and carbonic anhydrase producing bacteria and their role in calcite precipitation. *Current Microbiology*, **62**(3), 894–902.

Achal, V., Mukherjee, A. and Reddy, M. S. (2010) *Journal of Industrial Microbiology and Biotechnology*, **20**(11), 571–6.

Adams, J. I. (1965) The engineering behavior of a Canadian Muskeg. In: *Proc. 6th International Conference on Soil Mechanics and Foundation Engineering*, Montreal, Canada, 1, pp. 3–7.

Agarwal, S. K. (2006) Pozzolanic activity of various siliceous materials. *Cement and Concrete Research*, **36**(9), 1735–9.

Åhnberg, H. (2006) Strength of stabilized soils: A laboratory study on clays and organic soils stabilized with different types of binder. *Report 16, Swedish Deep Stabilization Research Centre*, Linköping, Sweden.

Åhnberg, H., Johansson, S. E., Retelius, A., Ljungkrantz, C., Holmqvist, L. and Holm, G. (1995a). *Cement och kalkk for djupstabilisering av jord. en kemisk-fysikalisk studie av stabiliserings effekter*. Swedish Geotechnical Institute. Report No. 48 (in Swedish with English summary).

Åhnberg, H., Ljungkratnz, C. and Holmqvist, L. (1995b). Deep stabilization of different types of soft soils. In: *Proc. 11th European Conference on Soil Mechanics and Foundation Engineering*, Copenhagen, pp. 167–72.

Aiban, S.A. (1994) A study of sand stabilization in Eastern Saudi Arabia. *Engineering Geology*, 38, 65–97.

Ajlouni, M. A. (2000) Geotechnical properties of peat and related engineering problems. *PhD Dissertation*, University of Illinois at Urbana-Champaign.

Alexiew, D., Brokemper, D. and Lothspeich, S. (2005) Geotextile encased columns (gec): load capacity, geotextile selection and pre-design graphs. In: *Proc. Geo- Frontiers 2005*, Austin, Texas, pp. 497–510.

Al-Homud, A. and Degen, W. (2006) Marine stone columns to prevent earthquake induced soil liquefaction. *Geotechnical and Geological Engineering*, **24**(3), 775–90.

Ali, F. H. and Huat, B. B. K. (1992) Performance of composite and monolithic prefabricated vertical drains. *PERTANIKA, Journal of Biological, Physical and Social Sciences*, **15**(3), 255–64.

Alkan, M., Demirbas, O. and Dogan, M. (2005) Electrokinetic properties of kaolinite in mono- and multivalent electrolyte solutions. *Microporous and Mesoporous Materials*, **83**, 51–9.

Al-Raziqi, A. A., Huat, B. B. K. and Munzir, H. A. (2003) Potential usage of hyperbolic method for prediction of organic soil settlement. *Proc. 2nd International Conference on Advances in Soft Soil Engineering and Technology* (eds. B. Huat *et al.*), Putrajaya, Malaysia, pp. 439–45.

Al Refeai, T. O. (1992) Strengthening of soft soil by fibre reinforced sand column. In: *Proc. International Symposium of Earth Reinforcement Practise*, Fukuoka.

Alshawabkeh, A. N. and Acar, Y. B. (1992) Removal of contaminants from soils by electrokinetics: a theoretical treatise. *Journal of Environmental Science and Health (A)*. **27**(7), 1835–61.

Al-Tabbaa, A., Al-Tabbaa, M. B. and Ayotamuno, J. M. (1999) Laboratory-scale dry soil mixing of sand. In *Dry Mixing Methods for Deep Soil Stabilization*. Balkema, Netherlands.

Al Wahab, R. M. and El-Kedrah, M. M. (1995) Using fibres to reduce tension cracks and shrink/swell in compacted clays. In: *Proc. Conference on Geoenvironment 2000*. Geotechnical Special Publication No. 46, ASCE, New York, pp. 791–805.

Alwi, A. (2008) Ground improvement on Malaysian peat soils using stabilized peat column techniques. *PhD Dissertation*, University of Malaya, Kuala Lumpur.

Amaryan, L. S., Sorokina, G. V. and Ostoumova, L. V. (1973) Consolidation laws and mechanical-structural properties of peat soils. *Proc. 8th International Conference on Soil Mechanics and Foundation Engineering*, Moscow, 2, pp. 1–6.

Ambak, K. and Melling, L. (1999) Management practices for sustainable cultivation of crop plants on tropical peatland. *Proc. International Symposium on Tropical Peatlands*, Bogor, Indonesia, 22–23 November, pp. 119–34.

Ambak, K. and Tadano, T. (1991) Effect of micronutrient application on the growth and occurrence of sterility in barley and rice in a Malaysian deep peat soil. *Soil Science and Plant Nutrition*, 37, 715–24.

Ambak, K., Zahari, A. B. and Tadano, T. (1991) Effect of liming and micronutrient application on the growth and occurrence of sterility in maize and tomato in a Malaysian deep peat soil. *Soil Science and Plant Nutrition*, 37, 689–98.

Ambily, A. P. and Gandhi, S. R. (2007) Behavior of stone columns based on experimental and FEM analysis. *Journal of Geotechnical and Geoenvironmental Engineering*, **133**(4), 405–15.

Andersen, A. (1981) Exploration, sampling and *in situ* testing of soft clay. In: *Soft Clay Engineering* (eds E. W. Brand and R.P. Brenner). *Proc. International Symposium on Soft Clay*, Bangkok, 1977. SOA Reports, Amsterdam, Elsevier, pp. 239–308.

Anderson, J. A. R. (1961) The ecology and forest types of the peat swamp forest of Sarawak and Brunei in relation to their silviculture. *PhD Dissertation*. University of Edinburgh.

Anderson, J. A. R. (1983) Tropical peat swamp of Western Malaysia. *Ecosystem of the World 4B. Mires, Swamp, Bog, Fen and Moor. Regional Studies*. Elsevier, pp. 181–99.

Andrejko, M. J., Fiene, F. and Cohen, A. D. (1983) Comparison of ashing techniques for determination of the inorganic content of peats. *Testing of Peats and Organic Soils*, ASTM STP 820, pp. 5–20.

Andriesse, J. P. (1988) Permanent constraint in peat reclamation. In: *Nature and Management of Tropical Peat Soils*, FAO Soil Bulletin 59, pp. 81–91.

Andriesse, J. P. (1991) Constraints and opportunities for alternative use options of tropical peat land. *Proc. International Symposium on Tropical Peat Land*, Kuching, Sarawak, Malaysia (ed. B. Y. Aminuddin), pp. 1–6.

Anggraini, V. (2006) Shear strength improvement of peat soil due to consolidation. *Master Dissertation*, Universiti Teknologi Malaysia, Malaysia.

Arman, A. (1969) A definition of organic soils (an engineering identification). *Engineering Research Bulletin No. 101*, Louisina State University, Division of Engineering Research, for Louisiana Department of Highways.

Arman, A. (1971) Discussion to "Ignition loss and other properties of peats and clays from Avonmouth King's Lynn and Cranberry Moss". *Géotechnique*, **21**, 418–21.

Asadi, A. (2010) Electro-osmotic properties and effects of pH on geotechnical behaviour of peat. *PhD Dissertation*, University Putra, Malaysia.

Asadi, A. and Huat, B. B. K. (2009a) Electrokinetic phenomena in tropical peat. *Proc. International Symposium on Ground Improvement Technologies and Case Histories, ISGI09*, Singapore.

Asadi, A. and Huat, B. B. K. (2009b) Study of electroosmotic mechanisms on peat. *Regional Conference on Environmental and Earth Resources, RCER09*, University Malaysia Pahang, Malaysia.

Asadi A. and Huat B. B. K. (2009c) Electrical resistivity of tropical peat. *Electronic Journal of Geotechnical Engineering*, **14**, 1–9.

Asadi, A. and Huat, B. B. K. (2010) Ground improvement technologies and case histories. Electrokinetic phenomena in tropical peat. *Research Publishing GEOSS*, pp. 525–31.

Asadi, A., Huat, B. B. K. and Mohamed, T. A. (2007) Electrokinetic and its applications in geotechnical and environmental engineering. civil and environmental engineering. *World Engineering Congress 2007*. Institution of Engineers, Penang, Malaysia, pp. 348–54.

Asadi A., Huat, B. B. K., Hanafi, M. M., Mohamed, T. A. and Shariatmadari, N. (2009d) Role of organic matter on electroosmotic properties and ionic modification of organic soils. *Geosciences*, **13**(2), 175–81.

Asadi, A., Huat, B. B. K. and Shariatmadari, N. (2009e) Keeping electrokinetic phenomena in tropical peat into perspective. *European Journal of Scientific Research*, **29**(2), 281–8.

Asadi, A., Huat, B. B. K., Hassim, M. M., Mohamed, A. T., Hanafi, M. M. and Shariatmadari, N. (2009f) Electroosmotic phenomena in organic soils. *American Journal of Environmental Sciences*, **5**(3), 310–14.

Asadi, A., Huat, B. B. K., Hanafi, M. M., Mohamed, T. A. and Shariatmadari, N. (2010) Physicochemical sensitivities of tropical peat to electrokinetic environment. *Geosciences*, **14**(1), 65–75.

Asadi, A., Huat, B. B. K., Hanafi, M. M., Mohamed, T. A. and Shariatmadari, N. (2011a). Chemico-geomechanical effects of pore fluid pH on tropical peat related to controlling electrokinetic phenomena. *Journal of the Chinese Institute of Engineers*, **34**(4), 481–7.

Asadi, A., Moayedi, H., Huat, B. B. K., Zamani, B. F., Parsaie, A. and Sojoudi, S. (2011b) Prediction of zeta potential for tropical peat in the presence of different cations using artificial neural networks. *International Journal of Electrochemical Science*, **6**(4), 1146–58.

Asadi, A., Huat, B. B. K., Moayedi, H., Shariatmadari, N. and Parsaie, A. (2011c) Electro-osmotic permeability coefficient of peat with different degree of humification. *International Journal of Electrochemical Science*, **6**(10), 4481–92.

Asadi, A., Huat, B. B. K., Moayedi, H., Shariatmadari, N. and Parsaie, A. (2011d) Changes of hydraulic conductivity of silty clayey sand soil under effects of MSW leachate. *International Journal of the Physical Sciences*, **6**(12), 2869–76.

Asadi, A., Moayedi, H. and Huat, B. B. K. and Parsaie, A. (2011e) Probable electrokinetic phenomena in peat: a review. *International Journal of the Physical Sciences*, **6**(9), 2184–8.

Asadi, A., Moayedi, H., Huat, B. B. K., Parsaie, A. and Taha, M. R. (2011f) Artificial neural networks approach for electrochemical resistivity of highly organic soil. *International Journal of Electrochemical Science*, **6**(4), 1135–45.

Asadi, A., Shariatmadari, N., Moayedi, H. and Huat, B. B. K. (2011g) Effect of MSW leachate on soil consistency under influence of electrochemical forces induced by soil particles. *International Journal of Electrochemical Science*, **6**(7), 2344–51.

Asadi, A., Huat, B. B. K., Fong, T. M. and Arumugam, T. (2013) Assessing the contribution of vetiver grass roots and biochar to slope stability, *Proc. 18th Southeast Asian Geotechnical and Inaugural AGSSEA*, Singapore.

Asaoka, A. (1978) Observational procedure of settlement prediction. *Soils and Foundation*, **18**(4), 87–101.

Asaoka, A., Kodaka, T. and Nozu, M. (1994) Undrained shear strength of clay improved with sand compaction piles. *Soils and Foundations*, **34**(4), pp. 23–32.

ASTM Standard (1990) *Annual Book of ASTM Standards*. ASTM International, West Conshohocken, PA, 2004 www.astm.org.

ASTM Standard D 4764-04 (2004) *Standard Test Method for Consolidated Undrained Triaxial Compression Test for Cohesive Soil*," ASTM International, West Conshohocken, PA, 2004. www.astm.org.

Axelsson, K., Johansson, S. E. and Anderson, R. (2002) *Stabilization of Organic Soils by Cement and Puzzolanic Reactions – Feasibility Study*. Swedish Deep Stabilization Research Centre, Linkoping, Sweden, Report 3, pp. 1–51 (English translation: Goran Holm).

Ayadat, T. and Hanna, A. M. (2005) Encapsulated stone columns as a soil improvement technique for collapsible soil. *Ground Improvement*, **9**(4), 137–47.

Azzouz, A. S., Krizek, R. J. and Corotis, R. B. (1976) Regression analysis of soil compressibility. *Soils and Foundation*, **16**(2), 19–29.

Balaam, N. P. and Booker, J. R. (1985) Effect of stone columns yield on settlement of rigid foundations in stabilized clay. *International Journal for Numerical and Analytical Methods in Geomechanics*, **9**(4), 351–81.

Balvac Ltd. (2008) Retrieved from: http://www.balvac.co.uk/

Barker JE, Rojers CDF, Boardman DI, and Peterson J (2004). Electrokinetic stabilization: an overview and case study. *Ground Improvement*, **8**, 47–58.

Barron, R. A. (1948) Consolidation of soil using vertical drain wells. *Géotechnique*, **31**, 718–42.

Basha, E. A., Hashim, R., Mahmud, H. B. and Muntobar, A. S. (2005) Stabilization of residual soil with rice husk ash and cement. *Construction and Building Materials*, **19**(6), 448–53.

Bauer, G. E. and Al-Joulani, N. (2005) Laboratory and analytical investigation of sleeve reinforced stone columns. In: *Geosynthetics: Application, Design and Construction*, pp. 463–6.

Beddiar, K., Chong, T. F., Dupas, A., Berthaud, Y. and Dangla, P. (2005) Role of pH in electroosmosis: experimental study on NaCl-water saturated kaolinite. *Transport in Porous Media*, **61**(1), 93–107.

Bergado, D. T. (1996) Soil compaction and soil stabilisation by admixtures. *Proc. Seminar on Ground Improvement Application to Indonesian Soft Soils*, Jakarta, Indonesia, pp. 23–6.

Bergado, D. T., Chai, J. C., Mirua, N. and Balasubrananiam, A. S. (1998) PVD improvement of soft Bangkok clay with combined vacuum and reduced sand embankment preloading. *Geotechnical Engineering*, **29**(1), 95–121.

Berry, P. L. (1983) Application of consolidation theory for peat to the design of a reclamation scheme by preloading. *Quarterly Journal of Engineering Geology*, **16**(9), 103–12.

Berry, P. L. and Poskitt, T. J. (1972) The consolidation of peat. *Géotechnique*, **22**(1), 27–52.

Berry, P. L. and Vickers, B. (1975) Consolidation of fibrous peat, *Journal of Geotechnical Engineering*, ASCE, **101**, 741–53.

Biringen, E. and Edil, T. B. (2003) Improvement of soft soil by radial preloading. *Proc. 2nd International Conference on Advances in Soft Soil Engineering and Technology* (eds. B. Huat *et al.*), Putrajaya, Malaysia, 741–52.

Black, A. J., Sivakumar, V., Madhav, M. R. and Hamil, A. G. (2007) Reinforced stone column in weak deposit: laboratory model study. *Journal of Geotechnical and Geoenvironmental Engineering*, ASCE, 9, 1154–61.

Bornside, G. H. and Kallio, R. E. (1956) Urea-hydrolyzing bacilli I. *Journal of Bacteriology*, **71**(6), 627–34.

Bosscher, P. J. and Edil, T. B. (1988) Performance of lightweight waste-impoundment dikes. *Proc. 2nd International Conference on Case Histories in Geotechnical Engineering*, St Louis.

Bouassida, M., de Buhan, P. and Dormieux, L. (1995) Bearing capacity of a foundation resting on a soil reinforced by a group of columns. *Géotechnique*, **45**(1), 25–34.

Bowen, W., Jacob, P. and Dupas, A. (1986) Electro-osmosis and the determination of zeta potential: the effect of particle concentration. *Journal of Colloid and Interface Science*, **111**(1), 223–9.

Bowles, J. E. (1978) *Engineering Properties of Soil and Their Measurements*. McGraw-Hill, USA.

Brady, N. C. and Weil, R. R. (2007) *The Nature and Properties of Soils*. Prentice Hall, New Jersey.

Brinkgreve, R. B. and Vermeer, P. A. (1998) *Plaxis-Finite Element Code for Soil and Rocks Analysis*, Version 8. Brookfield, Rotterdam.

British Standards Institution (2006) *Geotechnical Design, Part 2*. BS EN 1997-2: 2006, Eurocode 7.

Brown, R., Shukla, A. and Natarajan, K. R. (2002) Fiber reinforcement of concrete structures, *URITC Project*, University of Rhode Island.

BS1377: 1990. Soils for Civil Engineering Purposes. BSI.

BS SP 36 (Part 1): 1987. *Compendium of Indian Standards on Soil Engineering*. Bureau of Indian Standards.

Bunke, D. (1988) ODOT's experiences with silica fume (microsilica) concrete. *67th Annual Meeting of the Transportation Research Board*, paper no. 870340 (January).

Carlsten, P. (1988) Peat geotechnical properties and up to date methods of design and construction. *Proc., 2nd Baltic CSMFE*. Swedish Geotechnical Institute.

Carlsten, P. (2000) Geotechnical properties of some Swedish peats. *Proc. NGM-2000, XIII Nordisk Geotechnikermötet*, Helsinki, **1**, 51–60.

Carlsten, P. and Ekstrom, J. (1995) Kalk-och kalkcementpelare. vägledning for projektering, utförandeoch konroll, *SGF Rapport 4(95)*, *Svenska Geotekniska Foreningen*, Sweden.

Cartier, G., Allaeys, A., Londez, M. and Ropers, F. (1989) Secondary settlement of peat during a load test. *Proc., 12th International Conference on Soil Mechanics and Foundation Engineering*, Rio de Janeiro, 3, pp. 1721–2.

Casagrande, L. (1949) Electro-osmosis in soils. *Géotechnique*, **1**, 159–177.

Casagrande, A. and Fadum, R. E. (1940) *Notes on Soil Testing for Engineering Purposes*. Publication 268, Graduate School of Engineering, Harvard University, Cambridge, MA.

Cashman, P. M. and Preene, M. (2002) *Groundwater Lowering in Construction. A Practical Guide*. Taylor & Francis, London and New York.

Chai, J. C., Hayashi, S. and Carter, J. P. (2006) *Ground Modification and Seismic Mitigation*. American Society of Civil Engineering (ASCE).

Chai, J., Miurab, N. and Bergado, D. T. (2008) Preloading clayey deposit by vacuum pressure with cap-drain: analyses versus performance. *Journal of Geotextiles and Geomembranes*, **26**, 220–30.

Chapman, H.D. (1965) Cation exchange capacity. In: *Methods of Soil Analysis* (eds. C. A. Black *et al.*). *Agronomy*, **9**, 891–901. American Society of Agronomy, Inc., Madison, WI.

Chauhan, M. S., Mittal, S. and Mohanty, B. (2008) Performance evaluation of silty sand subgrade reinforced with fly ash and fibre. *Geotextiles and Geomembranes*, **26**(5), 429–35.

Chen, C. S. and Tan, S. M. (2003) Some engineering properties of soft clay from Klang area. *Proc. 2nd International Conference on Advances in Soft Soil Engineering and Technology* (eds. B. Huat *et al.*), Putrajaya, Malaysia, pp. 79–87.

Chen, H. and Wang, Q. (2006) The behaviour of organic matter in the process of soft soil stabilization using cement. *Bulletin of Engineering Geology and the Environment*, **65**, 445–8.

Chen, S. P., Lam, S. K. and Tan, Y. K. (1989) Geology of urban planning and development in Sarawak. *Seminar on Urban Geology for Planners and Decision Makers Developing the Urban Environment*. Geology Survey of Malaysia.

Chen, Y. M., Cao, W. P. and Chen, R. P. (2008) An experimental investigation of soil arching within basal reinforced and unreinforced piled embankments. *Geotextiles and Geomembranes*, **26**(2), 164–74.

Cherepy, N. and Wildenschild, D. (2003) *Environmental Science & Technology*, **37**, 3024–30.

Chin, F. K. (1975) The seepage theory of primary and secondary consolidation. *Proc. 4th Southeast Asian Conference on Soil Engineering*, Kuala Lumpur, pp. 21–8.

Christoulas, S., Giannaros, C. and Tsiambaos, G. (1997) Stabilization of embankment foundations by using stone columns. *Geotechnical and Geological Engineering*, **15**(3), pp. 247–58.

Chu, J. and Yan, S. W. (2005) Estimation of degree of consolidation for vacuum preloading projects, *International Journal of Geomechanics*, **5**(2), 158.

Chu, J., Yan, S. W. and Yang, H. (2000) Soil improvement by the vacuum preloading method for an oil storage station. *Géotechnique*, **50**, 625–32.

Clarke, W. (1984) Performance characteristics of microfine cement. *Proc. ASCE Geotechnical Conference*, Atlanta, May, pp. 14–18.

Coduto, D. P. (1998) *Geotechnical Engineering – Principles and Practice*. Prentice Hall, New Jersey.

Cognon, J. M., Juran, I. and Thevanayagam, S. (1994) Vacuum consolidation technology: principles and field experience. *Proc. Vertical and Horizontal Deformations of Embankments (Settlement '94)*, p. 1237. Texas: ASCE Special Publication.

Colley, B. E. (1950) Construction of highways over peat and muck areas. *American Highway*, **29**(1), 3–7.

Conklin, A. R. (2005) *Introduction to Soil Chemistry, Analysis and Instrumentation*. John Wiley & Sons, Inc., Canada.

Consoli, N. C., Prietto, P. D. M. and Ulbrich, L. A. (1998) Influence of fibre and cement addition on behavior of sandy soil. *Journal of Geotechnical and Geoenvironmental Engineering*, **124**(12), 1211–14.

Consoli, N. C., Montardo, J. P., Prietto, P. D. M. and Pasa, G. S. (2002) Engineering behavior of a sand reinforced with plastic waste. *Journal of Geotechnical and Geoenvironmental Engineering*, **128**(6), 462–72.

Consoli, N. C., Vendruscolo, M. A. and Prietto, P. D. M. (2003) Behavior of plate load tests on soil layers improved with cement and fibre. *Journal of Geotechnical and Geoenvironmental Engineering*, **129**(1), 96–101.

Consoli, N. C., Vendruscolo, M. A., Fonini, A. and Rosa, F. D. (2009) Fibre reinforcement effects on sand considering a wide cementation range. *Geotextiles and Geomembranes*, **27**, 196–203.

Costas, A. A. and Chatziangelou, M. (2008) Compressive strength of cement stabilized soils – a new statistical model. *Electronic Journal of Geotechnical Engineering*, **13B**.

Craig, R. F. (1992) *Soil Mechanics*, 5th Edn. Chapman and Hall, London.

Croney, D. and Croney, P. (1998) *Design and Performance of Road Pavement.* MacGraw-Hill, New York.

Culloch, F. M. (2006) Guidelines for the risk management of peat slips on the construction of low volume/low cost road over peat. *Forestry Civil Engineering Forestry Commission*, Scotland, pp. 1–46.

Dai, J. Y. (2004) Comparison study on extraction of soil humic acid with acidic dmso and with mixture of NaOH and $Na_4P_2O_7$. *Acta Pedologica Sinica*, **2**, 310–13.

Das, B. M. (2008) *Advanced Soil Mechanics.* Taylor & Francis, New York.

Deboucha, S., Hashim, R. and Alwi, A. (2008) Engineering properties of stabilized tropical peat soils. *Electronic Journal of Geotechnical Engineering*, **13E.**

De Flaun, M. F. and Condee, C. W. (1997) *Journal of Hazardous Materials*, **55**, 263–77.

De Jong, J., Fritzges, M. and Nüsslein, K. (2006) Microbially induced cementation to control sand response to undrained shear.*Journal of Geotechnical and Geoenvironmental Engineering*,**132**(11), 1381–92.

de Mello, L. G., Mondolfo, M., Montez, F., Tsukahara, C. N. and Bilfinger, W. (2008) First use of geosynthetic encased sand columns in South America. *Proc. 1st Pan-American Geosynthetics Conference*, Cancun, pp. 1332–41.

de Muynck, W., Cox, K. and De Belie, N. (2007) *Sustainable Construction Materials and Technologies.* Taylor & Francis, London, pp. 411–16.

den Haan, E. J. (1997) An overview of the mechanical behaviour of peats and organic soils and some appropriate construction techniques. In: *Proc. Conference on Recent Advances in Soft Soil Engineering* (eds. B. B. K. Huat and H. M. Bahia), Kuching, Sarawak. pp. 17–45.

den Haan, E. J. and El Amir, L. S. F. (1994) A simple formula for final settlement of surface loads on peat. In: *Proc. Conference on Advances in Understanding and Modelling the Mechanical Behaviour of Peat* (eds. E. J. den Haan *et al.*). Balkema, pp. 35–48.

den Haan, E. J. and Kruse, G. A. M. (2006) Characterisation and engineering properties of Dutch peats. In: *Proc. 2nd International Workshop on Characterisation and Engineering Properties of Natural Soils*, Singapore, **3**, 2101–33.

Detwiler, J. R. and Metha, P. K. (1989) Chemical and physical effects of silica fume on the mechanical behavior of concrete. *Material Journal ACI*, **86**(6), 609–14.

Dhouib, A. and Blondeau, F. (2005) *Colonnes Ballastées.* Edition Presses de L'école Nationale des Ponts et Chaussées, Paris.

Dhowian, A. W. and Edil, T. B. (1980) Consolidation behavior of peats. *ASTM Geotechnical Testing Journal*, **3**(3), 105–14.

Dick, J., De Windt, W., De Graef, B., Saveyn, H., Van der Meeren, P., De Belie, N. and Verstraete, W. (2006) *Biodegradation*, **17**, 357–67.

Djajaputra, A. A. and Shouman, M. (2003) Performance of the trial embankment on pile mat – JHS system founded in peaty soils. *Proc. 2nd International Conference on Advances in Soft Soil Engineering and Technology* (eds. B. Huat *et al.*), Putrajaya, Malaysia, pp. 553–60.

Douglas, R. P. (2004) Properties of self-consolidating concrete containing type F fly ash: with a verification of the minimum paste volume method. *MTech. Dissertation*, Northwestern University, Illinois.

Duraisamy, Y., Huat, B. B. K. and Aziz, A. A. (2007a) Methods of utilizing tropical peat land for housing scheme. *American Journal of Environmental Sciences*, **3**(4), 258–63.

Duraisamy, Y., Huat, B. B. K. and Aziz, A. A. (2007b) Engineering properties and compressibility behavior of tropical peat soil. *American Journal of Applied Sciences*, **4**(10), 768–73.

Duraisamy, Y., Huat, B. B. K. and Muniandy, R. (2009) Compressibility behavior of fibrous peat reinforced with cement columns. *Geotechnology and Geological Engineering*, **27**(5), 619–29.

Edil, T. B. (1994) Immediate issues in engineering practise. *Proc. Conference on Advances in Understanding and Modelling the Mechanical Behaviour of Peat.* (eds. E. J. den Haan *et al*). Balkema, pp. 403–44.

Edil, T. B. (1997) Construction over peats and organic soils. *Proc. Recent Advances in Soft Soil Engineering* (eds. B. B. K. Huat and H. M. Bahia), Kuching, Sarawak, pp. 85–108.

Edil, T. B. (1999) Construction over soft organic ground: issues and recent directions. *Proc. 47th Minnesota Conference on Soils Mechanics and Foundation Engineering*, University of Minnesota, St. Paul, Minnesota, pp. 99–126.

Edil, T. B. (2001) Site characterization in peat and organic soils. *Proc. International Conference on* in situ *Measurement of Soil Properties and Case Histories*, Bali, Indonesia, pp. 49–59.

Edil, T. B. (2002) Mechanical properties and mass behavior of shredded tire-soil mixtures. *Proc. International Workshop on Lightweight Geo-Materials (IW-LGM2002)*, Tokyo, Japan, pp. 17–32.

Edil, T. B. (2003) Recent advances in geotechnical characterization and construction over peat and organic soils. *Proc. 2nd International Conference on Advances in Soft Soil Engineering and Technology* (eds. B. Huat *et al.*), Malaysia, pp. 3–25.

Edil, T. N. and den Haan, E. J. (1994) Settlement of peats and organic soils, *Proc. Settlement '94*, ASCE, New York, 2, pp. 1543–72.

Edil, T. B. and Dhowian, A. W. (1979) Analysis of long-term compression of peats. *Geotechnical Engineering*, **10**, 159–78.

Edil, T. B. and Dhowian, A. W. (1981) At-rest lateral pressure of peat soils. *Journal of the Geotechnical Engineering Division*, **107**(GT2), 201–17.

Edil, T. B. and Mochtar, N. E. (1984) Prediction of peat settlement. *Proc. Sediment Consolidation Models Symposium Prediction Validation*, ASCE, San Francisco, CA, pp. 411–24.

Edil, T. B. and Simon-Gilles, D. A. (1986) Settlement of embankment on peat. Two case histories. *Proc. Conference on Advances in Peat Lands Engineering*, Ottawa, pp. 147–54.

Edil, T. B. and Wang, X. (2000) Shear strength and K_0 of peats and organic soils. *Geotechnics of High Water Content Materials, ASTM STP 1374* (eds. T. B. Edil and P. J. Fox), American Society for Testing and Materials, West Conshohocken, PA, pp. 209–25.

Edil, T. B., Fox, P. J. and Lan, L. T. (1991) End-of-primary consolidation of peat. *Proc. 10th ECSMFE*, Florence, 1, pp. 65–8.

Edil, T. B., Fox, P. J. and Lan, L. T. (1994a) An assessment of one-dimensional peat compression. *Proc. 13th International Conference on Soil Mechanics and Foundation Engineering*, New Delhi, India, 1, pp. 229–32.

Edil, T. B., Fox, P. J. and Lan, L. T. (1994b) Stress-induced o]ne-dimensional creep of peat. *Proc. International Workshop on Advances in Understanding and Modelling the Mechanical Behaviour of Peat* (eds. E. J. den Haan, R. Termaat and T. B. Edil), Balkema, Rotterdam, pp. 3–18.

Elshazly, H. A., Elkasabgy, P. and Elleboudy, A. (2007) Effect of inter-column spacing on soil stresses due to vibro-installed stone columns: interesting findings. *Geotechnical and Geological Engineering*, **26**(2), 225–36.

Elshazly, H. A., Hafez, D. H. and Mossaad, M. E. (2008) Reliability of conventional settlement evaluation for circular foundations on stone columns. *Geotechnical and Geological Engineering*, **26**, 323–34.

Esterle, J. S., Calvert, D., Durig, D., Tie, Y. L. and Supardi, M. N. N. (1991) Characterization and classification of tropical woody peats from Baram river, Sarawak and Jambi Sumatra. *Proc. International Symposium on Tropical Peat Land*, Kuching, Sarawak, Malaysia, pp. 33–48.

EuroSoilStab (2002) Development of design and construction methods to stabilize soft organic soils: design guide soft soil stabilization, CT97-0351. Project No. BE 96-3177, *Industrial and Materials Technologies Programme (Brite-EuRam III), European Commission*, pp. 15–60.

Eykholt, G. R. and Daniel, D. E. (1994) Impact of system chemistry on electroosmosis in contaminated soil. *Journal of Geotechnical Engineering*, ASCE, **120**(5), 797–815.

Fang, H. Y. (1997) *Introduction to Environmental Geotechnology*, CRC Press, Boca Raton, FL.

Fang, H. Y. and Daniels, J. L. (2006) *Introductory Geotechnical Engineering. An Environmental Perspective.* Taylor & Francis, London and New York.

Farrell, E. R. (1997) Some experience in the design and performances of roads and road embankment on organic soils and peats. *Proc. Conference on Recent Advances in Soft Soil Engineering* (eds. B. B. K. Huat and H. M. Bahia), Kuching, Sarawak, pp 66–84.

Farrell, E. R. (1998) The determination of geotechnical parameters of organic soils. *Problematic Soils* (eds. E. Yanagisawa, N. Moroto and T. Mitachi), Balkema, Rotterdam, pp. 33–6.

Farrell, E. R., O'Neill, C. and Morris, A. (1994) Changes in the mechanical properties of soils with variation in organic content. *Proc. Conference on Advances in Understanding and Modelling the Mechanical Behaviour of Peat* (eds. E. J. den Haan *et al.*), Balkema, pp. 19–25.

Fleri, A. M. and Whetstone, G. T. (2007) *In situ* stabilization/solidification: project life cycle. *Journal of Hazardous Materials*, **14**, 441–56.

Food and Agriculture Organisations of the United Nations (1988) Soil of the world revised legend, world soil resources, Report 60, F.A.O., Rome.

Forrest, J. B. and MacFarlane, I. C. (1969) Field studies of response of peat to plate loading. *Journal of the Soil Mechanics and Foundation Engineering Division, ASCE*, **95**(SM4), 949–67.

Forsberg, S. and Alden, L. (1988) Dewatering of peat: characterization of colloidal and subcolloidal particles in peat. *Colloids and Surfaces*, **34**, 335–43.

Fox, P. J. and Edil, T. B. (1994) Temperature induced one dimensional creep of peat. *Proc. Conference on Advances in Understanding and Modelling the Mechanical Behaviour of Peat.* (eds. E. J. den Haan *et al*), Balkema, pp. 27–34.

Friedrich, B. and Magasanik, B. (1977) *Journal of Bacteriology*, **8**, 313–22.

Frydenlund, T. E. and Aaboe, R. (1997) Expanded polystyrene – the light solution. *Proc. Conference on Recent Advances in Soft Soil Engineering* (eds. B. B. K. Huat and H. M. Bahia), Kuching. Sarawak, pp. 309–24.

Fuchsman, C. (1986a) The peat-water problem: reflections, perspectives, recommendations. In: *Peat and Water: Aspects of Water Retention and Dewatering in Peat.* Elsevier Applied Science, New York, pp. 331–60.

Fuchsman, C. H. (1986b) *Peat and Water: Aspects of Water Retention and Dewatering in Peat.* Elsevier Applied Science, New York, pp. 95–118.

Fujii, A., Tanaka, H., Tsuruya, H. and Shinsha, H. (2002) Field test on vacuum consolidation method by expecting upper clay layer as sealing-up material. *Proc. Symposium on Recent Development about Clayey Deposit – From Microstructure to Soft Ground Improvement.* Japanese Geotechnical Society, pp. 269–74 (in Japanese).

Gabr, M. A., Wang, J. and Bowders, J. J. (1996a) Model for efficiency of soil flushing using PVD-enhanced system. *Journal of Geotechnical Engineering*, **122**(11), 914–19.

Gabr, M. A., Wang, J. and Bowders, J. J. (1996b) *In situ* soil flushing using prefabricated vertical drain. *Journal of Geotechnical Engineering*, **33**, 97–105.

Gan, C. H. and Tan, S. M. (2003) Some construction experiences on soft soil using light weight materials. *Proc. 2nd International Conference on Advances in Soft Soil Engineering and Technology.* (eds. B. Huat *et al.*), Putrajaya, Malaysia, pp. 609–16.

Garga, V. K. and Medeiros, L. V. (1995) Field performance of the Port of Sepetiba test fills. *Canadian Geotechnical Journal*, **33**, 106–22.

Gautschi, M. A. (1965) *Peat as a Foundation Soil.* Research summary report, NGI, Oslo.

Gillman, G. P. and Sumpter, E. A. (1986) Modification to compulsive exchange method for measuring exchange characteristics of soils. *Australian Journal of Soil Research*, **24**, 61–6.

Glenn, S., Heyes, A. and Moore, T. (1993) Carbon dioxide and methane fluxes from drained peat soils, southern Quebec. *Global Biogeochemical Cycles*, **7**, 247–57.

Gniel, J. and Bouazza, A. (2009) Improvement of soft soils using geogrid encased stone columns. *Geotextiles and Geomembranes*, **27**(3), 167–75.

Goughnour, R. R. (1983) Settlement of vertically loaded stone columns in soft ground. *Proc. 8th European Conference on Soil Mechanics and Foundations Engineering*, Helsinki, Finland, 1, pp. 235–40.

Goughnour, R. R. and Bayuk, A. A. (1979) A field study of long-term settlements of loads supported by stone columns in soft ground. *Proc. International Conference on Soil Reinforcement*, Paris, pp. 279–85.

Gray, D. H. and Al-Refeai, T. (1986) Behavior of fabric versus fibre-reinforced sand. *Journal of Geotechnical Engineering*, **112**(8), 804–20.

Gray, D. H. and Mitchell, J. K. (1967) Fundamental aspects of electro-osmosis in soils. *Journal of Soil Mechanics and Foundation Division*, **93**(SM6), 209–36.

Gray, D. H. and Ohashi, H. (1983) Mechanics of fiber reinforcement in sand. *Journal of Geotechnical Engineering*, **109**(3), 335–53.

Greenacre, M. (1996) Banking on soft options. *Ground Engineering*, October, 34–6.

Guetif, Z., Bouassida, M. and Debats, J. M. (2003a) Parametric study of the improvement due to vibrocompacted columns installation in soft soils. *Proc. 13th African Regional Conference of Soil Mechanics and Geotechnical Engineering*, Marrakech (Morocco), 8–11 December, pp. 463–6.

Guetif, Z., Bouassida, M. and Debats, J. M. (2003a) Soft soil improvement due to vibrocompacted columns installation. *Proc. Workshop Geotechnics of Soft Soils. Theory and Practice* (eds. P. A. Vermeer *et al.*), The Netherlands, 17–19 September, pp. 551–7.

Guetif, Z., Bouassida, M. and Debats, J.M. (2007) Improved soft clay characteristics due to stone column installation. *Computers and Geotechnics*, **34**, 104–11.

Gunther, J. (1983) The present state of peat production and use of peat in the federal republic of Germany, p. 11–12, in G.W. Luttig (ed.), Recent technologies in the use of peat: E. Schweizerbart sche Verlagsbuchhandlung. Stuttgart, 223 p.

Hamed, J. T. and Bhadra, A. (1997) Influence of current density and pH on electrokinetics. *Journal of Hazardous Materials*, **55**, 279–94.

Hamed, J, Acar, Y. B. and Gale, R. J. (1991) Pb(II) removal from kaolinite by electrokinetics. *Journal of Geotechnical Engineering*, ASCE, **117**(2), 241–71.

Hammes, F. and Verstraete, W. (2002) *Environmental Science & Bio/Technology*, **1**, 3–7.

Han, J. and Gabr, M. A. (2002) Numerical analysis of geosynthetic-reinforced and pile-supported earth platforms over soft soil. *Journal of Geotechnical and Geoenvironmental Engineering*, **128**(1), 44–53.

Hanrahan, E. T. (1954) An investigation of some physical properties of peat. *Géotechnique*, 4(2), 108–23.

Hanrahan, E. T. and Rogers, M. G. _(1981)_ Road on peat: observations and design. *Journal of Geotechnical Engineering Division*, **107**(10), 1403–15.

Hansbo, S. (1991) Full-scale investigations of the effect of vertical drains on the consolidation of a peat deposit overlying clay. *de Mello Volume, Editoria Edgard Bldcher LTDA, Caixa*, Sao Paolo, Brasil, pp. 1–15.

Hanzawa, H., Kishida, T., Fukusawa, T. and Asada, H. (1994) A case study of the application of direct shear and cone penetration tests to soil investigation, design and quality control for peaty soils. *Soils and Foundations*, **4**, 13–22.

Hartlen, J. and Wolski, J. (eds.) (1996) *Embankments on Organic Soils*. Elsevier, Philadelphia.

Hashim, R. and Islam, M. S. (2008) Properties of stabilized peat by soil-cement column method. *Electronic Journal of Geotechnical Engineering*, **13J**, 1–9.

Hausmann, R. M. (1990) *Engineering Principles of Ground Modification*. McGraw-Hill, USA.

Hayashi, H. and Nishikawa, J. (2003) Mixing efficiency of dry jet mixing methods applied to peaty soft ground. In: *Dry Mixing Methods for Deep Soil Stabilization*. Balkema, The Netherlands.

Hayashi, H., Nishikawa, J. and Sawai, K. (2003) Improvement effect of vacuum consolidation and prefabricated vertical drain in peat ground. *Proc. 2nd International Conference on Advances in Soft Soil Engineering and Technology* (eds. B. Huat *et al.*), Putrajaya, Malaysia. pp. 625–33.

Hebib, S. (2001) Experimental investigation of the stabilisation of Irish peat. *PhD Dissertation*, Trinity College Dublin, Ireland. Hebib, S. and Farrell, R. E. (2003) Some experiences on the stabilization of Irish peats. *Canadian Geotechnical Journal*, **40**, 107–20.

Helmholtz, H. (1879) Studien über elektrische Grenschichten, *Annalen der Physik*, 7, 337–82.

Hobbs, N. B. (1986) Mire morphology and the properties and behaviour of some British and Foreign Peats. *Quarterly Journal of Engineering Geology*, **19**, 7–80.

Holtz, R. D. and Kovacs, W. D. (1981) *An Introduction to Geotechnical Engineering*. Prentice-Hall, New Jersey.

Horowitz, A. J. (1991) *A Primer on Sediment-Trace Element Chemistry*. Lewis Publishers, Chelsea, Michigan.

Horvath, J. S. (1995) *Geofoam Geosynthetic*. Horvath Engineering, Scarsdale, NY.

Hooton, R. D. (1993) Influence of silica fume replacement of cement on physical properties and resistance to sulphate attack, freezing and thawing and alkali reactivity,*ACI Materials Journal*, **90**, 143–51.

Houlsby, A. C. (1990) *Construction and Design of Cement Grouting*. John Wiley &, New York, p. 65.

Huang, J. T. and Airey, D. W. (1998) Properties of artificially cemented carbonate sand. *Journal of Geotechnical and Geoenvironmental Engineering*, **124**(6), 492–9.

Huang, J. and Han, J. (2009) 3D coupled mechanical and hydraulic modeling of a geosynthetic-reinforced deep mixed column-supported embankment. *Geotextiles and Geomembranes*, **27**, 272–80.

Huat, B. B. K. (1990) Model studies of piled embankment – centrifuge and finite elements. *British Geotechnical Society Young Geotechnical Engineers Symposium*, University of Sheffield, UK, 6–8 March.

Huat, B. B. K. (1994) Behaviour of soft clay foundation beneath an embankment. *Pertanika Journal of Science and Technology*, **2**(2), 215–35.

Huat, B. B. K. (1995) Stability of embankment on soft ground: lesson from failures. *Pertanika Journal of Science and Technology*, **3**(1), 123–39.

Huat, B. B. K. (1996) Observational method of predicting the settlement. *Proc. Twelfth Southeast Asian Geotechnical Conference*, Kuala Lumpur, pp. 191–6.

Huat, B. B. K. (2002a) Hyperbolic method for predicting embankment settlement. *Proc. 2nd World Engineering Congress*, Kuching, Sarawak, 1, pp. 228–32.

Huat, B. B. K. (2002b) Some mechanical properties of tropical peat and organic soils.*Proc. 2nd World Engineering Congress*, Kuching, Sarawak, pp. 82–7.

Huat, B.B.K. (2004) *Organic and Peat Soils Engineering*. 1st edn. University Putra Malaysia Press, Serdang, Malaysia.

Huat, B. B. K. and Ali, F. H. (1992a) Embankment on soft ground reinforced with piles. *Institution of Engineers, Malaysia*, **51**, 57–69.

Huat, B. B. K. and Ali, F. H. (1992b) Piled embankment on soft ground. *Proc. Second International Conference on Deep Foundation Practise*, Singapore, 4–6 November, 27–32.

Huat, B. B. K. and Ali, F. H. (1993a) Piled embankment on soft clay. Comparison between model and field performance. *Proc. Third International Conference on Case Histories in Geotechnical Engineering*, St Louis, 1, 433–6.

Huat, B. B. K. and Ali, F. H. (1993b) A contribution to the design of a piled embankment. *Pertanika Journal of Science and Technology*, **1**(1), 79–92.

Huat, B. B. K. Ali, F. H. and Low, T. H. (2006) Water infiltration characteristics of unsaturated soil slope and its effect on suction and stability. *Geotechnical and Geological Engineering*, **24**, 1293–1306.

Huat, B. B. K. and Craig, W. H. (1994) Simulation of a sand column trial embankment. In: *Centrifuge 94* (eds. C. F. Leung, F. H. Lee and T. S. Tan), Balkema, Rotterdam, pp. 561–6.

Huat, B. B. K., Craig, W. H. and Merrifield, C. M. (1991) Simulation of a field trial structure in Malaysia. In: *Centrifuge '91* (eds. H.-Y. Ko and F. G. McLean), Balkema, Rotterdam, pp. 51–7.

Huat, B. B. K., Craig, W. H. and Ali, F. H. (1993a) Behaviour of sand columns improved foundation. *Institution of Engineers, Malaysia*, **53**(1), 65–74.

Huat, B. B. K., Craig, W. H. and Ali, F. H. (1993b) Reinforcing of soft soils with granular columns. *Proc. Eleventh South East Asian Geotechnical Conference*, Singapore, pp. 357–61.

Huat, B. B. K., Craig, W. H. and Ali, F. H. (1994) The mechanics of piled embankment. *FHWA International Conference of Design and Construction of Deep Foundation*, Orlando, 2, pp. 1069–82.

Huat, B. B. K, Othman, K. and Jaffar, A. A. (1995) Geotechnical properties of some marine clays. *Institution of Engineers, Malaysia*, **56**(1), 23–36.

Huat, B. B. K., Asadi, A. and Kazemian, S. (2010) Electro-osmotic properties of tropical peat. *Proc. 3rd International Conference on Problematic Soils* (eds. D. Cameron and W. Kaggwa), 7–9 April, Adelaide, Australia, pp. 171–6.

Huat, B. B. K., Kazemian, S. and Kuang, W. L. (2011) Effect of cement-sodium silicate grout and kaolinite on undrained shear strength of reinforced peat. *Electronic Journal of Geotechnical Engineering*, **16K**.

Hunter, R. J. (1981) *Zeta Potential in Colloid Science*. Academic Press, London.

Hunter, R. J. (1993) *Introduction to Modern Colloid Science*. Oxford University Press, New York.

Hunter, R. N. (2000) *Asphalt in Road Construction*. Thomas Telford, London.

Hussein, A. N. and Mustapha, A. H. (2003) Investigation of embankment failures at Jalan Ranca-Ranca to Bebuluh, Wilayah Persekutuan Labuan. *Proc. 2nd International Conference on Advances in Soft Soil Engineering and Technology* (eds. B. Huat *et al.*), Putrajaya, Malaysia, pp. 423–9.

Huttunen, E., Kujala, K. and Vesa, H. (1996) Assessment of the quality of stabilized peat and clay. *Symposium Grouting and Deep Mixing*. Balkema, Rotterdam, pp. 607–12.

Hvorslev, M. J. (1949) *Subsurface Exploration and Sampling of Soils for Civil Engineering Purposes*. U.S. Army Waterways Experimental Station, Vicksburg, MS.

Hwan Lee, B. K. and Lee, S. (2002) Mechanical properties of weakly bonded cement stabilized kaolin. *KSCE Journal of Civil Engineering*, **6**(4), 389–98.

Ismail, M. A., Joer, H. A., Sim, W. H. and Randolph, M. (2002) Effect of cement type on shear behavior of cemented calcareous soil. *Geotechnical and Geoenvironmental Engineering*, **128**(6), 520–9.

ISSMFE Subcommittee on Soil Sampling (1981) *International Manual for the Sampling of Soft Cohesive Soils*. Tokyo University Press. Ivanov, V. and Chu, J. (2008) *Reviews in Environmental Science & Biotechnology*, 7(2), 139–53.

Jacob, A., Thevanayagam, S. and Kavazanjian, E. (1994) Vacuum assisted consolidation of a hydraulic landfill. *Geotechnical Special Publication*, **40**(2), 1249–61.

Janz, M. and Johansson, S. E. (2002) The function of different binding agents in deep stabilization. *Swedish Deep Stabilization Research Center*, SGI, Report No. 9, Linkoping.

Jarrett, P. M. (1995) *Geoguide 6. Site Investigation for Organic Soils and Peat*. JKR Document 20709-0341-95. Institut Kerja Raya, Malaysia.

Jones, D. B., Beasley, D. H., and Pollock, D. J. (1986) Ground treatment by surcharging on deposits of soft clays and peat, *Proc. Conference on Building on Marginal and Derelict Land*, LCE, Glasgow, pp. 679–95.

Juran, I. and Guermazi, A. (1988) Settlement response of soft soils reinforced by compacted sand columns. *Journal of Geotechnical and Geoenvironmental Engineering*, **114**(8), 903–43.

Kabai, J. and Farkas, K. (1988) Strength and deformation tests of Hungarian peats. *Proc. 2nd Baltic CMSFE Conference*, Tallin, pp. 48–54.

Kalantari, B. and Huat, B. B. K. (2008) Stabilization of peat soil using ordinary Portland cement, polypropylene fibres, and air curing technique. *Electronic Journal of Geotechnical Engineering*, **13J**.

Kalantari, B. and Huat, B. B. K. (2009a) Effect of fly ash on the strength values of air cured stabilized tropical peat with cement, *Electronic Journal of Geotechnical Engineering*, **14N**, 1–14.

Kalantari, B. and Huat, B. B. K. (2009b) Precast stabilized peat columns to reinforce peat soil deposits. *Electronic Journal of Geotechnical Engineering*, **14B**.

Kalantari, B. and Huat, B. B. K. (2010) UCS evaluation tests for cement treated peat soil with polypropylene fibres. *Proc. International Symposium on Ground Improvement Technologies and Case Histories, ISGI'09*, Geotechnical Society of Singapore, pp. 587–94.

Kalantari, B., Prasad, A, and Huat, B. B. K. (2010) Peat stabilization using cement, polypropylene and steel fibres. Geomechanics and Engineering, **2**(4), 321–35.

Kalantari, B., Prasad, A. and Huat, B. B. K. (2011) Stabilising peat soil with cement and silica fume. *Proceedings of the Institution of Civil Engineers: Geotechnical Engineering*, **164**(1), 33–9

Kalantari, B., Prasad, A. and Huat, B. B. K. (2012) Use of cement, polypropylene fibers and optimum moisture content values to strengthen peat. *International Journal of Physical Sciences*,7(8), 1276–85.

Kallstentiusatenius, T. (1963) Studies on clay samples taken with standard piston sampler. *Swedish Geotechnical Institute Proceedings*, No. 21. Kaltwasser, H., Kramer, J. and Conger, W. R. (1972) *Archives of Microbiology*, **81**, 178–96.

Kaniraj, S. R. and Gayathri, V. (2003) Geotechnical behaviour of fly ash mixed with randomly oriented fiber inclusions. *Geotextiles and Geomembranes*, **21**, 123–49.

Kaniraj, S. R. and Huong, H. L. (2008) Electroosmotic consolidation experiments on a north Sarawak peat. *IGC Conference*, Bangalore, India, pp. 70–3.

Kaniraj, S. R. and Yee, J. H. S. (2011) Electro-osmotic consolidation experiments on an organic soil. *Geotechnical and Geological Engineering*, **29**(4), 505–18.

Kaniraj, S. R., Huong, H. L. and Yee, J. H. S. (2011) Electro-osmotic consolidation studies on peat and clayey silt using electric vertical drain. *Geotechnical and Geological Engineering*, **29**(3), 277–95.

Kanmuri, H., Kato, M., Suzuki, O. and Hirose, E. (1998) Shear strength of K_o consolidated undrained peat. *Problematic Soil* (eds. E. Yanagisawa, N. Moroto and T. Mitachi), Balkema, Rotterdam, pp. 25–9.

Karlsson, R. and Hansbo, S. (1981) *Soil Classification and Identification*. Swedish Council for Building Research, St Göransg, 66, S-112 33, Stockholm, Sweden.

Karol, R. H. (2003) *Chemical Grouting and Soil Stabilization*, 3rd edn. Marcel Dekker, New Jersey.

Karunawardena, W. A. and Kulatilaka, S. A. S. (2003) Field monitoring of a fill on peaty clay and its modeling. *Proc. 12th Asian Regional Conference on Soil Mechanics and Geotechnical Engineering*, Singapore, pp. 159–62.

Kazemian, S. (2011) Stabilization of peat by cement-sodium silicate grout using injection-vacuum technology. *PhD Dissertation*, University Putra Malaysia, Malaysia.

Kazemian, S. and Huat, B. B. K. (2009a) Compressibility characteristics of fibrous tropical peat reinforced with cement column. *Electronic Journal of Geotechnical Engineering*, **14C**.

Kazemian, S. and Huat, B. B. K. (2009b) Assessment and comparison of grouting and injection methods in geotechnical engineering. *European Journal of Scientific Research*, **27**(2), 234–47.

Kazemian, S., Asadi, A. and Huat, B. B. K. (2009) Laboratory study on geotechnical properties of tropical peat soils. *International Journal of Geotechnics and Environment*, **1**, 69–79.

Kazemian, S., Prasad, A., Ghiasi, V. and Huat, B. B. K. (2010a) Experimental study of engineering behavior of fibrous peat soil reinforced by cement column. *Proc. International Symposium on Ground Improvement Technologies and Case Histories, ISGI'09*, Singapore, Geotechnical Society of Singapore, pp. 533–9.

Kazemian, S., Huat, B. B. K., Prasad, A. and Barghchi, M. (2010b) A review of stabilization of soft soils by injection of chemical grouting. *Australian Journal of Basic and Applied Sciences*, **4**(12), 5862–8.

Kazemian, S., Prasad, A., Huat, B. B. K. and Barghchi, M. (2011a) A state of an art review of peat from general perspective. *International Journal of the Physical Sciences*, **6**(8), 1988–96.

Kazemian, S., Prasad, A., Huat, B. B. K., Bolouri, B. J., Farah, N. A. A. and Thamer, A. M. (2011b) Influence of cement-sodium silicate grout admixed with calcium chloride and kaolinite on sapric peat. *Journal of Civil Engineering and* Management, **17**(3), 309–18.

Kazemian, S., Huat, B. B. K., Farah, N. A. A., Thamer, A. M., Moayedi, H. and Barghchi, M. (2011c) Influence of the peat characteristics on cementation and pozzolanic reactions in dry mixing method. *Arabian Journal for Science and Engineering*, **6**(36), 919–1169.

Kazemian, S., Prasad, A., Huat, B. B. K., Ghiasi, V. and Ghareh, S. (2012a) Effect of cement-sodium silicate grout on organic soils. *Arabian Journal for Science and Engineering*, **37**(8), 2137–48.

Kazemian, S., Huat, B. B. K. and Moayedi, H. (2012b) Undrained shear characteristics of tropical peat reinforced with cement stabilized soil column, *Geotechnical and Geological Engineering Journal*, DOI:10.1007/s10706–012-9492-7.

Keene, P. and Zawodniak, C. D. (1968) Embankment construction on peat utilizing hydraulic fill. *Proc. 3rd International Peat Conference*, National Research Council of Canada, Ottawa, pp. 45–50.

Keller Holding GmbH (2005) Retrieved from http://www.kellergrundbau.com/.

Kenneth, B. and Andromalos, W. E. (2003) The application of various deep mixing methods for excavation support systems. *Proc. Third International Conference*, New Orleans, Geo-Institute of the ASCE and Deep Foundation Institute, pp. 512–26.

Keykha, H. A., Huat, B. B. K., Asadi, A. and Kawasaki, S. (2012) Electro-biogrouting and its challenges. *International Journal of Electrochemical Science*, 7(2), 1196–204.

Khayat, K. H. and Sonebi, M. (2001) Effect of mixture composition on washout resistance of highly flowable underwater concrete. *ACI Material Journal*, 289–95.

Khayat, K. H., Sonebi, M., Yahia, A. and Skaggs, C. B. (1996) Statistical models to predict flow ability, washout resistance and strength of underwater concrete. *RILEM Conference on Production Methods and Workability of Concrete*, Glasgow, USA, pp. 463–81.

Kilian, A. P. and Ferry, C. D. (1993) Long term performance of wood fiber fills. *Transportation Research Board Record 1422*.

Kirov, B. (1994) Experience of peat preloading in the Varna West Harbor. *Proc. Conference on Advances in Understanding and Modelling the Mechanical Behaviour of Peat* (eds. E. J. den Haan *et al.*), Balkema, Rotterdam, pp. 341–5.

Kirov, B. (2003) Deformation properties of soft soil: oedometer testing. *Proc. 2nd International Conference on Advances in Soft Soil Engineering and Technology* (eds. B. Huat *et al.*) Putrajaya, Malaysia, pp. 49–51.

Kitazume, M. and Terashi, M. (2013) *The Deep Mixing Methods.*, CRC Press, Routledge, New York.

Kjellmann, W. (1952) Consolidation of clay soil by means of atmospheric pressure.*Proc. Soil Stabilization Conference*, Boston, p. 258.

Koda, E. (1997) The influence of vertical drains on consolidation process in organic subsoil. *Proc. Conference on Recent Advances in Soft Soil Engineering* (eds. B. B. K. Huat and H. M. Bahia), Kuching, Sarawak, pp. 251–62.

Koda, E. and Wolski, W. (1994) The influence of strip drains on the consolidation performance of organic soils. *Proc. Conference on Advances in Understanding and Modelling the Mechanical Behaviour of Peat* (eds. E. J. den Haan *et al.*), Balkema, Rotterdam, pp. 347–59.

Kolias, S., Kasselouri-Rigopoulou, V. and Karahalios, A. (2005) Stabilisation of clayey soils with high calcium fly ash and cement. *Cement and Concrete Composites*, **27**(2), 301–13.

Krishnaswamy, N. R. and Isaac, N. T. (1994) Liquefaction potential of reinforced sand. *Geotextiles and Geomembranes*, **13**(1), 23–41.

Kumar, P. and Singh, S. P. (2008) Fiber-reinforced fly ash subbases in rural roads. *Journal of Transportation Engineering*, **134**(4), 171–80.

Kumar, S. and Tabor, E. (2003) Strength characteristics of silty clay reinforced with randomly oriented nylon fibres. *Electronic Journal of Geotechnical Engineering*, **8B**.

Kurihara, N., Isoda, T., Ohta, H. and Sekiguchi, H. (1994) Settlement performance of the central Hokkaido expressway built on peat. *Proc. Conference on Advances in Understanding and Modelling the Mechanical Behaviour of Peat.* (eds. E. J. den Haan *et al.*), Balkema, Rotterdam, pp. 361–7.

Ladd, R. S. (1978) Preparing test specimen using under compaction, *Geotechnical Testing Journal*, **1**, pp. 16–23.

Lam, S. K. (1989) Quaternary geology of Sibu town area, Sarawak. Report 23 (part II). *Geological Survey Malaysia.*

Landva, A. O. And La Rochelle, P. (1983) Compressibility and shear characteristics of Radforth Peats. In P. M. Jarett (ed.), *Testing of peats and organic soils, ASTM STP 820*, pp. 157–191.

Landva, A. O. and Pheeney, P. E. (1980) Peat fabric and structure. *Canadian Geotechnical Journal*, **3**, pp. 416–35.

Landva, A. O., Pheeney, P. E. and Mersereau, D. E. (1983) Undisturbed sampling of peat. *Testing of Peats and Organic Soils, ASTM STP 820* (ed. P. M. Jarrett), pp. 141–56.

Larson, A. D. and Kallio, R. E. (1954) Purification and properties of bacterial urease, *Journal of Bacteriology*, **68**, 67–73.

Lauritzsen, R. and Lee, L. T. (2002) EPS – foundation for marshland Sarawak. *Proc. 2nd World Engineering Congress* (eds. B. Huat *et al.*), Kuching, Sarawak, pp. 203–7.

Lea, F. M. (1956) *The Chemistry of Cement and Concrete* (rev. edn of Lea and Desch), Edward Arnold, London, p. 637.

Lea, N. D. (1958) Notes on mechanical properties of peat. *Proc. 4th Muskeg Research Conference, Ottawa*, ACSSM, National Research Council of Canada, Technical Memo 54, pp. 53–7.

Lea, N. and Brawner, C. O. (1963) Highway design and construction over peat deposits in the lower British Colombia, *Highway Research Record*, 7, 1–32.

Lee, H. S. (1991) Utilization and conservation of peat swamp forests in Sarawak. *Proc. International Symposium on Tropical Peat Land* (ed. B. Y. Aminuddin), Kuching, Sarawak, Malaysia, pp. 286–92.

Lee, I. K., White, W. and Ingles, O. G. (1983) *Geotechnical Engineering, Soil Mechanics.* Pitman, Marshfield, MA.

Leelavathamma, B., Mini, K. M. and Pandian, N. S. (2005) California bearing ratio behavior of soil-stabilized class F fly ash. *Journal of Testing and Evaluation*, **33**(6), 406–10.

Leete, R. (2006) *Malaysia's Peat Swamp Forests, Conservation and Sustainable Use.* United Nations Development, Kuala Lumpur, Malaysia.

Lefebvre, G., Langlois, P., Lupien, C. and Lavallee, J. G. (1984) Laboratory testing and *in situ* behavior of peat as embankment foundation. *Canadian Geotechnology Journal*, **21**, 322–37.

Leonards, G. A. and Girault, P. (1961) A study of the one-dimensional consolidation test. *Proc., 9th CSMFE*, Paris, 1, pp. 116–30.

Leong, F. C., Soemitro, R. A. and Rahardjo, H. (2000) Soil improvement by surcharge loading and vacuum preloading. *Géotechnique*, **50**(5), 601.

Levesqe, M., Jacquin. F. and Polo, A. (1980) Comparative biodegradability of sphagnum and sedge peat from France. *Proc. 6th International Peat Congress*, Duluth, Minnesota, pp. 584–90.

Lewis, W. A. (1956) The settlement of the approach embankments to a new road bridge at Lockford, West Suffolk. *Géotechnique*, **6**(3), 106–114.

Li, A. L. and Rowe, R. K. (2008) Effects of viscous behaviour of geosynthetic reinforcement and foundation soils on embankment performance.*Geotextiles and Geomembranes*,**26**(4), 317–34.

Li, S. Z. and Xu, R. K. (2008) Electrical double layers' interaction between oppositely charged particles as related to surface charge density and ionic strength. *Colloids and Surfaces A: Physicochemical and Engineering Aspects*, **326**, 157–61.

Lien, B., Fox, N. S. and Kwong, H. K. (2002) Geopier floating foundation, a solution for roadway embankment over soft soils in Asia. *Proc. 2nd World Engineering Congress* (eds. B. Huat *et al.*), Kuching, Sarawak, pp. 182–7.

Lioliou, M. G., Paraskeva, C. A., Koutsoukos, P. G. and Payatakes, A. C. (2007) *Journal of Colloid and Interface Science*, **308**(2), 421–8.

Lishtvan, I. I., Bazin, E. T. and Kosov, B. I. (1985) *Physical Properties of Peat and Peat Deposits* (in Russian). Nauka I Technika Press, Miñsk, pp. 134–45.

Long, R. and Covo, A. (1994) Equivalent diameter of vertical drains with oblong cross section. *Journal of Geotechnical Engineering*, **120**(9), 1625–30.

Lorenz, P. B. (1969) Surface conductance and electrokinetic properties of kaolinite beds. *Clays and Clay Minerals*, **17**, 223–31.

Lu, R. K. (2000) *Chemical Analyzing Method of Soil Agriculture*. China Agriculture Scientific and Technical Publishing House, Beijing, pp. 121–2.

Madaeni, S. S., Naghdi, S. and Nobili, M. D. (2006) Ultrafiltration of humic substances in the presence of protein and metal ions. *Transport in Porous Media*, **65**, 469–84.

Maddison, J. D., Jones, D. B., Bell, A. L. and Jenner, C. G. (1996) Design and performance of an embankment supported using low strength geogrids and vibro compacted columns. In: *Geosynthetics: Application, Design and Construction*, Balkema, Rotterdam, pp. 325–32.

Magdoff, F. and Weil, R. R. (2005) *Soil Organic Matter in Sustainable Agriculture*. Taylor & Francis, New York.

Magill, D. and Berry, R. (2006) Comparison of chemical grout properties: which grout can be used where and why? *Avanti International and Rembco Geotechnical Contractors*. Retrieved from http://www.pilemedic.com/pdfs/comparison-of-chemical-grout-properties.pdf

Magnan, J. P. (1980) Classification geotechnique des sols: 1 – A propos de la classification LPC. *Bulletin de Liaison des Laboratoires des Ponts et Chaussees*, Paris, pp. 19–24.

Magnan, J. P. (1994) Construction on peat. State of the art in France. *Advances in Understanding and Modelling the Mechanical Behaviour of Peat* (eds. E. J. den Haan *et al.*), Balkema, Rotterdam, pp. 369–80.

Maher, M. H. and Ho, Y. C. (1993) Behavior of fibre-reinforced cement sand under static and cyclic loads. *Geotechnical Testing Journal*, **16**(3), 330–8.

Majzik, A. and Tombacz, E. (2007) Interaction between humic acid and montmorillonite in the presence of calcium ions II. Colloidal interactions: charge state, dispersing and/or aggregation of particles in suspension. *Organic Geochemistry*, **38**, 1130–340.

Maurya, R. R., Sharma, B. V. R. and Naresh, D. N. (2005) Footing load tests on single and group of stone columns. *Proc. 16th International Conference on Soil Mechanics and Geotechnical Engineering*, Osaka, Japan, 3, pp. 1385–8.

Mayne, P. W. and Kulhawy, H. F. (1982) K_0-OCR relationship in soils. *ASCE Journal of Geotechnical Engineering*, **108**(GT6), 851–72.

McAfee, M. (1989) Aeration studies on peat soil 1. The effects of water-table depth and crop cover. *International Peat Journal*, **3**, 157–74.

McManus, K., Hassan, R. and Sukkar, F. (1997) Founding embankments on peat and organic soils. *Proc. Conference on Recent Advances in Soft Soil Engineering* (eds. B. B. K. Huat and H. M. Bahia), pp. 351–67.

Melling, L. (2000) Dalat and Mukah sago plantation peat soil study. *Land Custody and Development Authority (LCDA) Report*, Sarawak, Malaysia.

Melling, L., and Hatano, R. (2003) Characteristics of tropical peat land and its implication to agriculture development. *Proc. 2nd International Conference on Advances in Soft Soil Engineering and Technology* (eds. B. Huat *et al.*), Putrajaya, Malaysia.

Melling, L., Ambak, K., Osman, J. and Ahnad H. (1999) Water management for the sustainable utilization of peat soils for agriculture. *Proc. International Conference and Workshop on Tropical Peat Swamps*, Penang, Malaysia.

Mesri, G. (1973) Coefficient of secondary compression. *Journal of Soil Mechanics and Foundation Engineering Division*, **99**(1), 123–37.

Mesri, G. and Ajlouni, M. (2007) Engineering properties of fibrous peats. *Journal of Geotechnical and Geoenvironmental Engineering*, **133**(7), 850–66.

Mesri, G. and Godlewski, P. M. (1977) Time- and stress-compressibility interrelationship. *Journal of Geotechnical Engineering*, **103**(GT 5), 417–30.

Mesri, G., Statark, T. D., Ajlouni, M. A. and Chen, C. S. (1997) Secondary compression of peat with or without surcharging. *Journal of Geotechnical and Geoenvironmental Engineering*, **123**(5), 411–21.

Mitachi, T., Yamazoe, N. and Fukuda, F. (2003) FE analysis of deep peaty soft ground during filling followed by vacuum preloading. *Proc. 2nd International Conference on Advances in Soft Soil Engineering and Technology* (eds. B. Huat *et al.*), Putrajaya, Malaysia, pp. 267–76.

Mitchell, J. K. (1993) *Fundamentals of Soil Behavior*. John Wiley & Sons, New York.

Mitchell, J. K. and Huber, T. R. (1985) Performance of a stone column foundation. *Journal of Geotechnical Engineering*, **111**(2), 205–23.

Mitchell, J. M. and Jardine, F. M. (2002) A guide to ground treatment, *CIRIA C573*, ISBN 0 86017 573 1.

Mitchell, J. K. and Santamarina, J. C. (2005) Biological considerations in geotechnical engineering. *Journal of Geotechnical and Geoenvironmental Engineering*, **131**(10), 1222–33.

Mitchell, J. K. and Soga, K. (2005) *Fundamentals of Soil Behavior*. John Wiley & Sons, New Jersey.

Miura, N., Shen, S. L., Koga, K. and Nakamura, R. (1998) Strength change of clay in the vicinity of soil cement column. *Journal of Geotechnical Engineering*, **III-43**(596), 209–21.

Miyakawa, J. (1960) Soils engineering research on peats alluvia. Reports 1–3. *Hokkaido Development Bureau Bulletin No. 20*. Civil Engineering Research Institute.

Moayedi, H., Kazemian, S., Parasad, A. and Huat, B. B. K. (2009) Effect of geogrid reinforcement location in paved road improvement. *Electronic Journal of Geotechnical Engineering*, **14P**, 1–11.

Moayedi, H., Asadi, A,. Huat, B. B. K. and Moayedi, F. (2011) Optimizing stabilizers enhanced electrokinetic environment to improve physicochemical properties of highly organic soil. *International Journal of Electrochemical Science*, **6**(5), 1277–93.

Mohamed, A. M. O. and Anita, H. E. (1998) *Developments in Geotechnical and Geoenvironmental Engineering*. Elsevier, Netherlands.

Mohamedelhassan, E. and Shang, J. Q. (2002) Vacuum and surcharge combined one dimensional consolidation of clay soils. *Canadian Geotechnical Journal*, **39**, 1126–38.

Molenkamp, F. (1994) Investigation of requirements for plane strain elements tests on peat. In: *Advances in Understanding and Modeling the Mechanical Behaviour of Peat* (eds. E. J. den Haan *et al.*), Balkema, Rotterdam.

Moseley, M. P. and Kirsch, K. (2004) *Ground Improvement*, 2nd edn. Taylor & Francis, Basingstoke, pp. 26–165.

Müller-Vonmoos, M. (1983) Die bedeutung der ton minerale für das boden mechanische verhalten. *Mitteilungen der Schweizerischen Gesellschaftfür Boden-und Felsmechanik, Studientag*, 4, Fribourg, Switzerland.

Mullik, K. A., Walia, P. and Sharma, N. S. (2006) Application of polypropylene fiber reinforced concrete (PFRC) with vacuum processing. *Advances in Bridge Engineering*, Walia International Machines Corp, New Delhi.

Munro, R. (2004) *Dealing with Bearing Capacity Problems on Low Volume Roads Constructed on Peat*. The Highland Council, Transport, Environmental & Community Service, Scotland, pp. 1–136.

Murugesan, S. and Rajagopal, K. (2006) Geosynthetic-encased stone columns: numerical evaluation. *Geotextiles and Geomembranes*, **24**, 349–58.

Murugesan, S. and Rajagopal, K. (2009) Shear load tests on stone columns with and without geosynthetic encasement. *Geotechnical Testing Journal*, **32**(1), 35–44.

Murugesan, S. and Rajagopal, K. (2010) Studies on the behavior of single and group of geosynthetic encased stone columns. *Journal of Geotechnical and Geoenvironmental* Engineering, **136**(1), 129–39.

Muskeg Engineering Handbook (1969) (ed. I. C. MacFarlane), University of Toronto Press.

Mutalib, A. A., Lim, J. S., Wong, M. H. and Koonvai, L. (1991) Characterisation, distribution and utilisation of peat in Malaysia. *Proc. International Symposium on Tropical Peat Land*, Kuching, Malaysia, pp. 7–16.

Myślińska, E. (2003) Development of mucks from the weathering of peats: it's importance as an isolation barrier, *Bulletin of Engineering Geology and the Environment*, **62**, 389–92.

Nakayama, M., Yamaguchi, H. and Kougra, K. (1990) Change in pore size distribution of fibrous peat under various one-dimensional consolidation conditions. *Memoirs of the Defense Acad*emy, **30**(1), 1–27.

Narasimha Rao, S., Madhiyan, M. and Prasad, Y. V. S. N. (1992) Influence of bearing area on the behavior of stone columns. *Proc. Indian Geotechnical Conference*, Calcutta, India, pp. 235–7.

Nataraj, M. S. and McManis, K. L. (1997) Strength and deformation properties of soils reinforced with fibrillated fibres. *Geosynthetics International*, **4**(1), 65–79.

Nemati, M. and Voordouw, G. (2003) *Enzyme Microbiology Technology*, **33**, 635–42.

Neville, A. M. (1999) *Properties of Concrete*. Longman, Malaysia.

Ng, S. Y. and Eischens, G. R. (1983) Repeated short-term consolidation of peats. *Testing of Peat and Organic Soils, ASTM STP 820*, 192–206.

Nonveiller, E. (1989) *Grouting: Theory and Practice*. Elsevier, New York.

Norris, V., Grant, S., Freestone, P., Canvin, J., Sheikh, F. N., Toth, I., Trinei, M., Modha, K. and Norman, N. (1996) *Journal of Bacteriology*, **178**(13), 3677–3682.

Noto, S. (1990) *Revised Formula for Prediction of Settlement on Soft Peaty Deposits*. Monthly Report of Civil Engineering Research Institute, No. 446, p. 209 (in Japanese).

Noto, S. (1991) *Peat Engineering Handbook*. Civil Engineering Research Institute, Hokkaido Development Agency, Prime Minister's Office, Japan.

Ochiai, H., Hayashi, S., Umezaki, T. and Otani, J. (1991) Model test on sheet-pile countermeasures for clay foundation under embankment. *Proc. Developments in Geotechnical Aspects of Embankment, Excavations and Buried Structures*, Netherlands, pp. 277–91.

Oikawa, H., and Igarashi, M. (1997) A method for predicting e-log p curve and log c_v-log p curve of a peat from its natural water content. *Proc.*, Recent Advances in Soft Soil Engineering. (eds) Huat and Bahia, Kuching. Sarawak, Malaysia, pp. 201–209.

O'Loughlin, C.D., and Lehane, B.M. (2003) A study of the link between composition and compressibility of peat and organic soils. *Proc., 2nd International Conference on Advances in Soft Soil Engineering and Technology.* (eds). Huat *et al.* Putrajaya, Malaysia, pp. 135 – 152.

Olson, R. E. and Mesri, G. (1970) Mechanisms controlling compressibility of clays. *Journal of Soil Mechanics and Foundation Engineering Division*, **96**(SM6), 1863–78.

O'Mahony, M. J., Ueberschaer, A., Owende, P. M. O. and Ward, S. M. (2000) Bearing capacity of forest access roads built on peat soils. *Journal of Terramechanics*, **37**(3), pp. 127–38.

Ong, B. Y. and Yogeswaran, M. (1991) Peatland as a resource water supply in Sarawak. *Proc. International Symposium on Tropical Peat Land*, Kuching, Sarawak, Malaysia, pp. 255–68.

Orr, T. L. L. and McEnaney, T. (1994) Friction piles for a walkway on a peat bog. *Advances in Understanding and Modelling the Mechanical Behaviour of Peat* (eds. E. J. den Haan *et al.*), Balkema, Rotterdam, pp. 381–7.

Othman, K. H., Huat, B. B. K., Chin, C. and Ali, F. H. (1994) Observed performance of some vertical drain embankments in Malaysia. *Proc. Conference on Geotechnical Engineering, Geotropika 94*, Malacca, Malaysia.

Paikowsky, S., Elsayed, A. and Kurup, P. U. (2003) Engineering properties of Cranberry bog peat. *Proc. 2nd International Conference on Advances in Soft Soil Engineering and Technology* (eds. B. Huat *et al.*), Putrajaya, Malaysia, pp. 153–71.

Parish, F., Sirin, A., Charman, D., Joosten, H., Minayeva, T. and Silvius, M. (2008) *Assessment on Peatland, Biodiversity and Climate Change: Main Report.* Global Environment Centre, Kuala Lumpur and Wetlands International, Wageningen.

Park, T., and Tan, S.A. (2005) Enhanced performance of reinforced soil walls by the inclusion of short fibre. *Geotextiles and Geomembranes*, **23**, 348–61.

Peat Testing Manual (1979) (eds. Day, Rennie, Stanek and Raymond), Technical Memorandum No. 125, National Research Council Canada, Ottawa.

Poungchompu, P. (2009) Development of a timber raft & pile foundation for embankments on soft ground. *PhD Dissertation*, Saga University, Japan.

Poungchompu, P., Hayashi, S., Du, Y. J. and Nagao, S. (2005) Laboratory model test on raft & pile system in soft Ariake clay. *Proc. JSCE Annual Meeting*, 60(Disk 1), pp. ROMBUNNO.3–363.

Poulos, H. (2005) Pile behavior – consequences of geological and construction imperfections 1. *Journal of Geotechnical and Geoenvironmental Engineering*, **131**, 538–63.

Prabakar, J. and Sridhar, R. S. (2002) Effect of random inclusion of sisal fibre on strength behaviour of soil. *Construction and Building Materials*, **16**(2), 123–31.

Prasad, A., Kazemian, S., Kalantari, B. and Huat, B. B. K. (2012) A behavior of reinforced. vibro-compacted stone columnin peat.*Pertanika Journal of Science & Technology*, **20**(2), 221–41.

Priebe, H. J. (1995) The design of vibro replacement. *Ground Engineering*, **28**(12), 31–7.

Probstein, R. F. and Hicks, R. E. (1993) Removal of contaminants from soils by electric fields. *Science*, **260**, 498–504.

Puppala, A. J., Pokala, S. P., Intharasombat, N. and Williammee, R. (2007) Effects of organic matter on physical, strength, and volume change properties of compost amended expansive clay. *Journal of Geotechnical and Geoenvironmental Engineering*, **133**(11), 1449–61.

Qiu, Q. C., Mo, H. H. and Dong, Z. L. (2007) Vacuum pressure distribution and pore pressure variation in ground improved by vacuum preloading. *Canadian Geotechnical Journal*, **44**, 1433–45.

Rahadian, H., Taufik, R. and Moelyanni, D. A. (2001) Field and laboratory data interpretation of peats at Berengbengkel trial site. *Proc. International Conference on in-situ Measurement of Soil Properties and Case Histories*, Bali, Indonesia, pp. 145–51.

Rahadian, H., Satriyo, B. and Peryoga, T. (2003) Settlement behaviour of a peat deposit subjected to embankment loading. *Proc. 2nd International Conference on Soft Soil Engineering and Technology* (eds. B. Huat *et al.*), Putrajaya, Malaysia, 387–97.

Rahim, A. M. and George, K. P. (2004) Subgrade soil index properties to estimate resilient modulus. *Proc. 83th Annual Meeting of the Transportation Research Board*, CDROM.

Raithel, M. and Kempfert, H. G. (2000) Calculation models for dam foundations with geotextile coated sand columns. *Proc. International Conference on Geotechnical and Geological Engineering*, GeoEngg, Melbourne.

Raithel, M., Kempfert, H. G. and Kirchner, A. (2002) Geotextile-encased columns (GEC) for foundation of a dike on very soft soils. *Proc. Seventh International Conference on Geosynthetics*, Nice, France, pp. 1025–8.

Raito Kogyo Co. Ltd (2006) Retrieved from: http://www.raito.co.jp/english/.

Raj, J. K. (1993) Clay minerals in the weathering profile of a quartz-muscovite schist in the Seremban area, Negeri Sembilan, Pertanika.*Journal of Tropical Agricultural Science*, **16**(2), 129–36.

Raju, V. R. (2009) Ground improvement principles and applications in Asia. In: *Ground Improvement Technologies and Case Histories* (eds. C. F. Leung, J. Chu and R. F. Shen), pp. 43–66. Geotechnical Society of Singapore (GeoSS).

Raju, V. R., Abdullah, A. and Arulrajah, A. (2003) Ground treatment using dry deep soil mixing for a railway embankment in Malaysia. *Proc. 2nd International Conference on Advances in Soft Soil Engineering and Technology* (eds. B. Huat *et al.*), Putrajaya, Malaysia, pp. 589–600.

Ramachandran, S. K., Ramakrishnan, V. and Bang, S. S. (2001) *ACI Materials Journal*, **98**, 3–9.

Ramesh, H. N., Manoj Krishna, K. V. and Mamatha, H. V. (2010) Compaction and strength behavior of lime-coir fiber treated Black Cotton soil. *Geomechanics & Engineering*, 2(1).

Randolph, M. F., Carter, J. P. and Wroth, C. P. (1979) Driven piles in clay – the effect of installation and subsequent consolidation, *Géotechnique*, **29**(4), 361–93.

Ranjan, G., Vasan, R. M. and Charan, H. D. (1996) Probabilistic analysis of randomly distributed fibre-reinforced soil. *Journal of Geotechnical Engineering*, **122**(6), 419–26.

Rawlings, C. G., Hellawell, E. E. and Kilkenny, W. M. (2000) *Grouting for Ground Engineering*. CIRIA, UK.

Reuss, F. F. (1809) Sur un nouvel effet de le électricité glavanique. *Mémoires de la Societé Impériale des Naturalistes de Moscou*, **2**, 327–337.

Rieley, J. O. (1991) The ecology of tropical peat swamp forest – a South East Asian perspective. *Proc. International Symposium on Tropical Peat Land* (ed. B.Y. Aminuddin, Kuching, Sarawak, Malaysia, pp. 244–54.

Rieley, J. O. and Ahmad-Shah, A.A. (1996) The vegetation of tropical peat swamp forest. *Proc. Workshop on Integrated Planning and Management of Tropical Lowland Peat Lands* (eds. E. Maltby *et al.*, Cisarua, Indonesia, pp. 55–73.

Roslan, H. and Shahidul, I. (2008) Properties of stabilized peat by soil-cement column method. *Electronic Journal of Geotechnical Engineering*, **13J**.

Rowe, K. H. and Mylleville, B. L. J. (1986) A geogrid reinforced embankment on peat over organic silts – a case history. *Canadian Geotechnical Journal*, **33**, 106–22.

Rujikiatkamjorn, C., Indraratna, B. and Chu, J. (2008) 2D and 3D numerical modeling of combined surcharge and vacuum preloading with vertical drains. *International Journal of Geomechanics*, **8**(2), 144–56.

Russell, E. J. (1952) *Soil Conditions and Plant Growth*, 8th edn, Longman, Green and Co., London.

Salmah, Z., Spoor, Z. G., Zahari, A. B. and Welch, D. N. (1991) Importance of water management in peat soil at farm level. *Proc. International Symposium on Tropical Peat Land* (ed. B.Y. Aminuddin), Kuching, Sarawak, Malaysia.

Samson, L. (1985) Post construction settlement of an expressway built on peat by precompression. *Canadian Geoetchnical Journal*, **22**, 308–12.
Samson, L. and La Rochelle, P. (1972) Design and performance of an expressway constructed over peat by preloading, *Canadian Geoetchnical Journal*, **9**(4), 447–66.
Samsuri, A. (1997) The application of palm oil fly ash in improving a petroleum well cement characteristics, *Proc. Regional Symposium on Chemical Engineering*, Johar Baru, Malaysia, 13–14 October.
Santagata, M., Bobet, A., Johnsto, C. T. and Hwang, J. (2008) One-dimensional compression behavior of a soil with high organic matter content, *Journal of Geotechnical and Geoenvironmental Engineering*, **134**(1), 1–13.
Santoni, R. L. and Webster, S. L. (2001) Airfields and road construction using fiber stabilization of sands. *Journal of Transportation Engineering*, **127**(2), 96–104.
Schmidt, B. (1966). Discussion of 'Earth pressures at rest related to stress history' by Brooker & Ireland (1965), *Canadian Geotechnical Journal*, 3(4), 239–42.
Schmidt, B. (1967) Lateral stresses in uniaxial strain. *Danish Geotechnical Institute Bulletin*, No. 23, pp. 5–12.
Schnitzer, M., and Khan, S.U. (1989) *Soil Organic Matter.* Elsevier, New York.
Scholl, M. A., Mills, A. L., Herman, J. S. and Hornberger, G. M. (1990) *Journal of Contaminant Hydrology*, **6**(4), 321–36.
Schulz, K. F. and Herak, M. J. (1957) The determination of the charge of some inorganic thorium complexes with the ion exchange method. *Croatica Chemica Acta*, **29**, 49–52.
Soilmec Drilling and Foundation Equipment Company (SDFEC) (2007) Retrieved September 2007 from: http://www.soil mec.it/.
Seaby, D. (2001) Designs for one-man, two-stage, samplers for obtaining undisturbed cores of peat over 1 m long. *Forestry*, **74** (1), 79–83.
Shang, J. Q., Tang, M. and Miao, Z. (1998) Vacuum preloading consolidation of reclaimed land: a case study. *Canadian Geotechnical Journal*, **35**, 740–9.
Shen, S. L. (1998) Behavior of deep mixing columns in composite clay ground. *PhD Dissertation*, Saga University, Japan.
Shen, S. L., Miura, N., Han, J. and Koga, H. (2003a) Evaluation of property changes in surrounding clays due to installation of deep mixing column. In: *Grouting and Ground Treatment* (eds. L. F. Johnsen, D. A. Bruce and M. J. Boyle), pp. 634–45.
Geotechnical Special Publication, ASCE Press. Shen, S., Miura, N. and Koga, H. (2003b). Interaction mechanism between deep mixing column and surrounding clay during installation. *Canadian Geotechnical Journal*, **40**, 293–307.
Shogaki, T. and Kaneko, M. (1994) Effects of sample disturbance on strength and consolidation parameters of soft clay. *Soils and Foundations*, **34**(3), 1–10.
Shroff, A.V. and Shah, D. L. (1999) *Grouting Technology in Tunneling and Dam Construction.* Balkema, Rotterdam.
Sika Fibres (2005) *Polypropylene Fibres, Technical Data Sheet*, 3rd edn. Version No. 0010, Sika Fibres, Malaysia.
Silvius, M. J. and Giesen, W. (1996) Towards integrated management of swamp forests: a case study from Sumatra. *Proc. Workshop on Integrated Planning and Management of Tropical Lowland Peat Lands* (eds. E. Maltby *et al.*), Cisarua, Indonesia, pp. 247–67.
Sin, P. T. (2003) Economical solution for roadway embankment construction on soft compressible soil at Putrajaya, Selangor. *Proc. 2nd International Conference on Advances in Soft Soil Engineering and Technology* (eds. B. Huat *et al.*), Putrajaya, Malaysia, pp. 649–57.
Singh, H. and Huat, B. B. K. (2003) Tropical peat and its geotechnics. *Proc. 2nd International Conference on Advances in Soft Soil Engineering and Technology* (eds. B. Huat *et al.*), Putrajaya, Malaysia, pp. 203–19.

Singh, H., Bahia, H. M. and Huat, B. B. K. (1997) Varying perspectives on peat, its occurrence in Sarawak and some geotechnical properties. *Proc. Conference on Recent Advances in Soft Soil Engineering* (eds. B. Huat and H. Bahia), Kuching, Sarawak, pp. 135–49.

Sivakumar Babu, G. L., Vasudevan, A. K. and Haldar, S. (2008) Numerical simulation of fiber-reinforced sand behavior. *Geotextiles and Geomembranes*, **26**, 181–8.

Skempton, A. W. and Petley, D. J. (1970) Ignition loss and other properties of peats and clays from Avonmouth, King's Lynn and Cranberry Moss. *Géotechnique*, **20**(4), 343–56.

Smoluchowski, M. von (1921) *Handbuch der Elektrizität und des Magnetismus*. Barth, Leipzig, pp. 366–28.

Sposito, G. (2008) *The Chemistry of Soils*. Oxford University Press, New York.

Stevenson, F. J. (1994) *Humus Chemistry: Genesis, Composition, Reactions*. John Wiley & Sons, New York.

Stocks-Fischer, S., Galinat, J. K. and Bang, S. S. (1999) *Soil Biology and Biochemistry*, **31**, 1563–71.

Sukumar, B., Nagamani, K. and Srinivasa Raghavan, R. (2008) Evaluation of strength at early ages of self-compacting concrete with high volume fly ash. *Construction and Building Materials*, **22**(7), 1394–401.

Sulaeman, A. (2003) An alternative foundation system on very soft soil. *Proc. 2nd International Conference on Advances in Soft Soil Engineering and Technology* (eds. B. Huat *et al.*), Putrajaya, Malaysia, pp. 617–23.

Tai, L.Y. and Lee, K. W. (2003) Some interesting apects of peat deformations in Sibu Town and related engineering issues. *Proc. 2nd International Conference on Advances in Soft Soil Engineering and Technology* (eds. B. Huat *et al.*), Putrajaya, Malaysia, pp. 725–40.

Tajudin, S. A., Jefferson, I., Rogers, C. D. F. and Boardman, D. I. (2008) *Proc. International Conference on Recent Advances in Engineering Geology*, Kuala Lumpur, Malaysia, pp. 143–50.

Tan, S. B. (1971) Empirical method for estimating secondary and total settlement. *Proc. 4th Asian Regional Conference on Soil Mechanics and Foundation Engineering*, Bangkok, 2, pp. 147–51.

Tan, K. H. (2008) *Soils in the Humid Tropics and Monsoon Region of Indonesia*. Taylor & Francis, London.

Tan, S. A. and Oo, K. K. (2005) Finite element modeling of stone columns – a case history. *Proc. 16th International Conference on Soil Mechanics and Foundation Engineering*, Osaka, 3, pp. 1425–8.

Tanaka, H., Oka, F. and Yashima, A. (1996) Sampling of soft soil in Japan. *Marine Georesources and Geotechnology*, **14**, 283–95.

Tang, M. and Shang, J. Q. (2000) Vacuum preloading consolidation of Yaogiang airport runway, *Géotechnique*, **50**(6), 613–23.

Tang, C., Shi, B., Gao, W., Chen, F. and Cai, Y. (2007) Strength and mechanical behavior of short polypropylene fibre reinforced and cement stabilized clayey soil. *Geotextiles and Geomembranes*, **25**, 194–202.

Taylor, D.W. (1942) Research on consolidation clays. Department of Civil and Sanitation Engineering, Massachusetts Institute of Technology, *Report 82*.

Taylor, H. F. W. (1997) *Cement Chemistry*. Thomas Telford, London.

Termatt, R. and Topolnicki, M. (1994) Biaxial tests with natural and artificial peat. *Proc. International Workshop on Advances in Understanding and Modeling the Mechanical Behavior of Peat*, Delft, Netherlands, pp. 241–51.

Thevanayagam, S., Kavazanjian, E., Jacob, A. and Juran, I. (1994) Prospect of vacuum-assisted consolidation for ground improvement of coastal and offshore fills. In: *In-Situ Deep Soil Improvement* (ed. K. M. Rollins). ASCE.

Tie, Y. L. (1990) Studies of peat swamp forest in Sarawak with particular reference to soil-forest relationship and development of dome shaped structures. *PhD Dissertation*, Polytechnic of North London, UK.

Timuran Engineering. (2007) *Steel Fibres Reinforcements*, Data Sheet, No. 503626-k, Selangor, Malaysia.

Toh, C. T., Chee, S. K., Lee, C. H. and Wee, S. H. (1994) Geotextile-bamboo fascine mattresses for filling over very soft soils in Malaysia. *Geotextile and Geomembranes*. **13**, 357–69.

Torkzaban, S., Tazehkand, S. S., Walker, S. L. and Bradford, S. A. (2008) *Water Resources Research*, **44**(4), 12.

Toutanji, H. A., Liu, L. and El-Korchi, T. (1999) The role of silica fume in the direct tensile strength of cement-based materials. *Materials and Structures*, **32**, 203–9.

Tremblay, H., Duchesne, J., Locat, J. and Leroueil, S. (2002) Influence of the nature of organic compounds on fine soil stabilization with cement. *Canadian Geotechnical Journal*, **39**, 535–46.

US Army Corps of Engineers (USACE). (1995) *Chemical Grouting*. US Army Corps, Washington DC.

Usui, S. (1984) Electrical phenomena at interfaces fundamentals, measurements, and applications: electrical double layer. In: *Surfactant Science Series* (eds. A. Kitahara and A. Watanabe), 15, pp. 15–46, Marcel Dekker, New York.

Vane, M. L. and Zang, G. M. (1997) Effect of aqueous phase properties on clay particle zeta potential and electroosmotic permeability: implications for electrokinetic remediation processes, Electrochemical Decontamination of Soil and Water. *Journal of Hazardous Material (special issue)*, **55**(1–3), pp. 1–22.

Van Paassen, L. A., Dazaa, C. M., Staal, M., Sorokina, D. Y., van der Zon, W. and van Loosdrechta, M. C. M. (2010) *Ecological Engineering*, **36**, 168–75.

Van Paassen, L. A., Harkes, M. P., Van Zwieten, G. A., Van der Zon, W. H., Van der Star, W. R. L. and Van Loosdrecht, M. C. M. (2009) *Proc., 17th International Conference on Soil Mechanics & Geotechnical Engineering (ICSMGE)*, M. Hamza, M. Shahien, and Y. E. Mossallamy (eds.), pp. 2328–2333.

Vinson, T. (1970) *The Application of Polyurethane Formed Plastics in Soil Grouting*. University of California, Berkeley.

Volarovich, M. P. and Churaev, N. V. (1968) Application of the methods of physics and physical chemistry to the study of peat. *Proc. 2nd International Peat Congress*, Leningrad, pp. 819–31.

Vonk, B. F. (1993) Some aspects of the engineering practice regarding peat in small polder dikes. *Advances in Understanding and Modeling the Mechanical Behaviour of Peat* (eds. E. J. den Haan *et al.*), pp. 389–402.

von Post, L. (1922) Sveriges geologiska undersoknings torvinventering och nagre av dess hittills vunna resultat, Sr. Mosskulturfor. *Tidskrift*, **1**, 1–27.

Welker, A.L., Gilbert, R.B., and Bowders, J.J. (2000) Using a reduced equivalent diameter for a prefabricated vertical drain to account for smear, *Geosynthetics International, 7* (1), pp. 47–57.

West, L. J. and Stewart, D. I. (1995) Effect of zeta potential on soil electrokinetics. Characterization, containment, remediation, and performance in environmental geotechnics. *Geotechnical Special Publication No. 46*, ASCE, New York, 2, 1535–49.

West, L. J., Stewart, D. I., Binley, A. M. and Shaw, B. (1999) Resistivity imaging of soil during electrokinetic transport. *Engineering Geology*, **53**, 205–15.

Westers, L., Westers, H. and Quax, W. J. (2004) *Biochimica et Biophysica Acta*, **1694**, 299–310.

Whiffin, V. S. (2004) *PhD Dissertation*, School of Biological Sciences and Biotechnology, Murdoch University, Perth, Australia.

Whiffin, V. S., Van Paassen, L. A. and Harkes, M. P. (2007) *Geomicrobiology Journal*, **24**(5), 417–23.

Wiley, W. R. and Stokes, J. L. (1962) Requirement of an alkaline pH and ammonia for substrate oxidation by *Bacillus pasteurii. Bacteriology*,**84**(4), 730–4.

Wiley, W. R. and Stokes, J. L. (1963) Effect of pH and ammonium ions on the permeability of *Bacillus pasteurii. Bacteriology*, **86**(6), 1152–6.

Williamson, C. (1988) Remedial grout injection of building. *Construction and Building Materials*, Vol. 2, No. 3.

Wong, K. M. (2003) Earth-filling experiences on peat soils at Sri Aman, Sibu and Bintulu. *Proc. 2nd International Conference on Advances in Soft Soil Engineering and Technology* (eds. B. Huat *et al.*), Putrajaya, Malaysia, pp. 669–79.

Wong, L. S., Hashim, R. and Ali, F. H. (2008) Strength and permeability of stabilised peat soil, *Journal of Applied Science*, **8**(17), 1–5.

Woods, K. B., Berry, D. S. and Goetz, W. H. (1960) *Highway Engineering Handbook*. McGraw-Hill, New York.

Yamaguchi, H., Ohira, Y., Kogure, K. and Mori, S. (1985) Deformation and strength properties of peat. *Proc. 11th International Conference on Soil Mechanics and Foundation Engineering*, San Francisco, 4, pp. 2461–4.

Yamaguchi, H. (1990) Physicochemical and mechanical properties of peats and peaty ground. *Proc. 6th International Congress of the International Association for Engineering Geology*, Balkema, Rotterdam, pp. 521–6.

Yamaguchi, H., Ohira, Y., Kogure, K. and Mori, S. (1985) Undrained shear characteristics of normally consolidated peat under triaxial compression and extension conditions. *Soils and Foundations*, **25**(3), 1–18.

Yamaguchi, H., Yamauchi, K. and Kawano, K. (1987) Simple shear properties of peat. *Proc. 6th International Symposium on Geotechnical Engineering in Soft Soils*, Mexico, pp. 163–70.

Yamamoto, K. (1998) Failure mechanism of reinforced foundation ground and its bearing capacity analysis.*PhD Dissertation*, Kumamoto University, Japan.

Yasuhara, K., Oikawa, H., and Noto, S. (1994) Large strain cyclic behavior of peat. *Advances in Understanding and Modeling the Mechanical Behaviour of Peat* (eds. E. J. den Haan *et al.*), pp. 301–11.

Yasuhara, K., Horiuchi, S. and Murakami, S. (2003) Weight reducing geo-techniques for construction of coastal structures. *Proc. 2nd International Conference on Advances in Soft Soil Engineering and Technology* (eds. B. Huat *et al.*), Putrajaya, Malaysia, pp. 523–64.

Yetimoglu, T. and Salbas, O. (2003) A study on shear strength of sands reinforced with randomly distributed discrete fibres. *Geotextiles and Geomembranes*, **21**, 103–10.

Yetimoglu, T., Inanir, M. and Inanir, O. E. (2005) A study on bearing capacity of randomly distributed fiber-reinforced sand fills overlying soft clay. *Geotextiles and Geomembranes*, **23**(2), 174–83.

Yeung, A. T. and Datla, S. (1995) Fundamental formulation of electrokinetic extraction of contaminants from soil. *Canadian Geotechnical Journal*, **32**(4), 569–83.

Yogeswaran, M. (1995) Geological considerations in the development of Kuching area dialogue session. *Geological and Geotechnical Considerations in Civil Works*, Geological Survey of Malaysia, Kuala Lumpur.

Yonebayashi, K. (2003) Morphological characteristics and chemical properties of tropical woody peat in Southeast Asia. *Proc. 2nd International Conference on Advances in Soft Soil Engineering and Technology* (eds. B. Huat *et al.*), Putrajaya, Malaysia, pp. 27–38.

Yoo, C., and Kim, S. B. (2009) Numerical modeling of geosynthetic-encased stone column-reinforced ground. *Geosynthetics International*, **16**(3), 116–26.

Younger, J. S., Barry, A. J., Harianti, S. and Hardy, R. P. (1997) Construction of road over soft and peaty ground. *Proc. Recent Advances in Soft Soil Engineering* (eds. B. Huat and H. Bahia), Kuching, Sarawak, Malaysia, pp. 109–34.

YTL Cement (2008) Retrieved from http://www.ytlcement.com.

Yu, T. R. (1997) *Chemistry of Variable Charge Soils*. Oxford University Press, New York.
Yukselen, Y. and Erzin, Y. (2008) Artificial neural networks approach for zeta potential of Montmorillonite in the presence of different cations. *Environmental Geology*, 54, 1059–66.
Yulindasari, S. (2006) Compressibility characteristics of fibrous peat soil. *Masters Dissertation*, Universiti Teknologi Malaysia.
Zainorabidin, A. and Bakar, I. (2003) Engineering properties of in situ and modified hemic peat soil in western Johor. *Proc. 2nd International Conference on Advances in Soft Soil Engineering and Technology* (eds. B. Huat *et al.*), Putrajaya, Malaysia, pp. 173–9.

Subject index